Walter — with
much love,
Date: Feb. 1994

Groups and Geometry

Groups and Geometry

PETER M. NEUMANN

The Queen's College, Oxford

GABRIELLE A. STOY

Lady Margaret Hall, Oxford

and the late

EDWARD C. THOMPSON

Jesus College, Oxford

Oxford   New York   Tokyo

OXFORD UNIVERSITY PRESS

1994

Oxford University Press, Walton Street, Oxford OX2 6DP

Oxford New York Toronto
Delhi Bombay Calcutta Madras Karachi
Kuala Lumpur Singapore Hong Kong Tokyo
Nairobi Dar es Salaam Cape Town
Melbourne Auckland Madrid
and associated companies in
Berlin Ibadan

Oxford is a trade mark of Oxford University Press

Published in the United States
by Oxford University Press Inc., New York

A catalogue record for this book is available from the British Library

Library of Congress Cataloging in Publication Data
Neumann, P. M.
Groups and Geometry/Peter M. Neumann, Gabrielle A. Stoy,
and Edward C. Thompson.
Includes index.
1. Group theory; geometry. I. Stoy, Gabrielle A. II. Thompson,
E. C. III. Title.
QA183.N48 1993 512'.2–dc 93-270
ISBN 0 19 853452 3 (Hbk)
ISBN 0 19 853451 5 (Pbk)

Typeset by Cotswold Typesetting Ltd, Gloucester

Printed in Great Britain on acid-free paper by
Bookcraft (Bath) Ltd
Midsomer Norton, Avon

Preface

Few of us can have a hand in the writing of our own memorial. We (ΠMN and GAS) dedicate this book to the memory of our much loved and greatly respected late colleague and collaborator Edward Thompson, so that he may become one of those select few. He deserves and has earned it. He was an enthusiast who taught us much delightful mathematics. But he was much more than a mathematician, he was a wise man, and we both learned from him much more than mathematics.

This book is about the measurement of symmetry. That is what groups are for. Symmetry is visible in all parts of mathematics—and in many other arts too—and geometrical symmetry is the most visible of all symmetries. That is why groups and geometry are such close neighbours.

In its first edition (1980, reprinted with corrections 1982) the book was a two-volume set of cyclostyled notes produced and distributed by the Oxford Mathematical Institute for second- and third-year undergraduates and first-year postgraduate students. The main changes to this OUP edition are a complete re-write of Chapter 16 (completed by ECT before his demise), and, on advice from a benevolently critical referee, the addition of specimen solutions to some more-or-less randomly selected exercises in the first half of the book. Although we were guided in our choice of material by the Oxford syllabus of that earlier time, we allowed ourselves to present much other material. It was included for various reasons. Some was introductory or preparatory, some was intended to round out topics that had been, for excellent pedagogical reasons, excluded or curtailed in the syllabus, some was intended to establish links with other subjects, and, although we did not acknowledge in the preface of that earlier work what many readers will have recognised, some was included for the best of all possible reasons, namely, that we liked it. The result may be an idiosyncratic work. We make no apology for that: there is no point in publishing a book that is just like many others.

One of its idiosyncrasies is that it divides clearly into two halves: what you have purchased *amice lector* is really two books for the price of one. Nevertheless, those two books enjoy a symbiotic relationship which more than justifies the union. For, although Edward Thompson's geometrical part, Chapters 10–18, is written in a different style from Chapters 1–9 and 19, it is based on our treatment of group actions, which in turn was adapted to geometrical requirements. He and we worked closely together. There are small discrepancies of notation in the two halves, but we estimate that these should not be confusing: therefore we have chosen to maintain our own usages and not to interfere in any way with Edward's preferences and style. We hope that his enthusiasm for geometry and its history, about which he was writing a treatise when he died aged 72 in May 1991, will shine through.

The learning of theories and theorems is only a small part of the process of learning mathematics, the larger part of which involves acquiring the art of solving problems, hence of thinking creatively in mathematics. For this reason we have included a number of exercises. Some are intended to be easy, most require some thought, and a few are intended to be hard (nevertheless, we have tried to avoid open problems). One or two of the exercises develop parts of the theory for which we (all three of us) felt there was no space in the text, and on occasion we have taken the liberty of citing these where we needed the results in our exposition. The exercises therefore form an important part of the book, arguably the most important, and we hope that the reader will enjoy and benefit from them. The specimen answers in Chapters 1–9 are intended to be of the nature of an extension of the text and are placed between the main body of each chapter and the exercises. The wise and conscientious reader will, however, try the exercises unaided before attempting to evaluate what we have offered.

We record our warm thanks to innumerable people who have helped with advice and helpful criticism. Since the OUP referees were anonymous it is only fair that the many pupils and colleagues whose advice we have accepted and rejected (as may be) should also not be named. We carry forward from the earlier edition our gratitude to Mrs Sheila Robinson, Administrator of the Mathematical Institute, Oxford, and the members of staff there who worked so hard to produce it. That gratitude is extended for this present edition also to Mrs Anne Hodgson who undertook the rather difficult task of typing the revisions of the geometry from Edward's last manuscripts, and to Miss Pat Lloyd for her excellent help with the proof-reading and indexing. And we add our warm thanks to the staff of OUP without whom this work would not have been placed before a wider public.

Queen's College, Oxford   Π.M.N.
Lady Margaret Hall, Oxford   G.A.S.
July 1992 and July 1993

Edward Crossley Thompson

Our late colleague Edward Thompson was born on 5 May 1919 and died on 7 May 1991. Apart from some time during the years of the Second World War, his adult life and mathematical career were centred on Oxford. He went up to New College in 1937 as an undergraduate and, on achieving first class honours in his Final examinations in 1940, he migrated briefly to Merton College as a Harmsworth Senior Scholar. After one term there he left to make his wartime contribution alongside many other distinguished mathematicians in the operational intelligence establishment at Bletchley Park. In 1945 he was elected to a Fellowship at Jesus College, where he stayed until his retirement.

He made important contributions both to College and to University life in Oxford. As a wise and far-sighted Estates Bursar for sixteen years he skilfully guided his College through a major new building programme. In the Faculty of Mathematics he was generous in the help he gave new members, and held leading offices as Chairman of the Faculty Board and as its representative in the higher ranks of University administration. Led by his energetic chairmanship of what came to be known as the Thompson Committee, the Oxford faculty introduced major reforms to its teaching arrangements in the early 1970s.

Mathematics was always a great joy to him. He is remembered by generations of his former pupils at Jesus College for his wise and thoughtful guidance, and by his colleagues for his leadership, his gentleness, and his old-fashioned courtesy.

Contents

Chapter 1

A survey of some group theory

In this chapter we survey the beginnings of the theory of groups, and we establish the language and notation that will be used throughout the book. The reader is assumed not to be entirely new to the subject and therefore some proofs will be omitted, others will merely be sketched.

Groups. A group is a set with an associative binary operation, an identity element, and inverses.

Since mathematics is now built on a set-theoretic foundation, at least for expository purposes, it is of some importance to see that the definition can be formulated using only the notions of set theory. These begin with the concept of set and the membership relation $\in$, but include also derivative concepts such as those of ordered pair (a, b), cartesian product $A \times B$, function (or mapping: we use these words synonymously) $\alpha: A \to B$, etc; usually we shall use notation like $f: a \mapsto b$ or $af = b$ to indicate that the value of the function f on the element a is b. In this language a group may be formally defined as a quadruple $(G, \alpha, \beta, 1)$ of sets, in which α is a function $G \times G \to G$, β is a function $G \to G$, and $1 \in G$, and for which the following conditions are satisfied:

$$((a, b)\alpha, c)\alpha = (a, (b, c)\alpha)\alpha \quad \text{for all } a, b, c \in G; \tag{1.1}$$

$$(1, a)\alpha = a = (a, 1)\alpha \quad \text{for all } a \in G; \tag{1.2}$$

$$(a, a\beta)\alpha = 1 = (a\beta, a)\alpha \quad \text{for all } \alpha \in G. \tag{1.3}$$

A definition such as this is fine for saying in precise terms just what a group is supposed to be, but the notation involved is far too cumbersome for ordinary use. Therefore we write

$$ab, \text{ or sometimes } a.b, \text{ for } (a, b)\alpha, \quad \text{and } a^{-1} \text{ for } a\beta,$$

and we return to the more formal language only on those very rare occasions when these simplifications threaten to produce ambiguity. The axioms now take their familiar forms:

$(ab)c = a(bc)$ for all $a, b, c \in G$ [associative law];

$1a = a = a1$ for all $a \in G$ [1 is a two-sided identity];

$aa^{-1} = 1 = a^{-1}a$ for all $a \in G$ [a^{-1} is a two-sided inverse of a].

Notice that we do not have an explicit 'closure' axiom. The condition that for every $a, b \in G$ the product ab lies in G is implicit in the assertion that $\alpha: G \times G \rightarrow G$, that is, that α is a binary operation on G.

Although we are adopting multiplicative notation when we write ab for $(a, b)\alpha$ and a^{-1} for $a\beta$, we never write these as $a \times b$ or $1/a$ because $\times$ and $/$ are reserved for other purposes. Another simplification is to use the same name G for the group as for its underlying set; the relevant operations will almost always be clear from the context. Also, we usually write 1 for the identity element in any group, in spite of the fact that different groups generally have different identity elements.

If our group satisfies the extra condition

$$ab = ba \qquad \text{for all } a, b \in G \quad \text{[commutative law]},$$

it is said to be commutative (or abelian, in honour of a Norwegian mathematician N. H. Abel, whose work in 1827 and 1828 on certain special types of equations was later interpreted in terms of commutative groups). In this case we often use additive notation and write $a + b$ for ab, $-b$ for b^{-1}, $a - b$ for $a + (-b)$, and 0 instead of 1.

Subgroups.   A subset H of G is said to be a subgroup, and we write $H \leqslant G$, if

$$1 \in H, \quad a \in H \Rightarrow a^{-1} \in H, \quad \text{and} \quad a, b \in H \Rightarrow ab \in H,$$

that is to say, H is *closed* under the operations of the group G. (It is here, rather than in the definition of the notion of group, that closure appears explicitly.) The following is a very easy but useful consequence of the definition.

LEMMA 1.1: *The subset H of the group G is a subgroup if and only if $H \neq \varnothing$ (empty set) and $ab^{-1} \in H$ for every pair a, b of elements of H.*

If H is a subgroup of a group G then $(H, \alpha \restriction H, \beta \restriction H, 1)$, where $\restriction H$ indicates the appropriately restricted functions, is a group in its own right. Notice that $\{1\} \leqslant G$ and that $G \leqslant G$. We talk of non-trivial, proper subgroups (respectively) when we wish to exclude these extremes.

If H_1, H_2 are subgroups then so is their intersection $H_1 \cap H_2$. More generally, the intersection $\bigcap_{i \in I} H_i$ of any family $(H_i)_{i \in I}$ (where I is a non-empty set) of subgroups of G will be a subgroup. In particular, for any subset S of G, we can take $(H_i)_{i \in I}$ to be the family $\{H \mid S \subseteq H \leqslant G\}$ of subgroups of G that contain S. Defining

$$\langle S \rangle := \bigcap_{S \subseteq H \leqslant G} H,$$

we have that $\langle S \rangle$ is a subgroup of G. It is the unique smallest subgroup of G that contains S, and is known as the subgroup generated by S. If S happens to be a subgroup then $\langle S \rangle = S$; in general, as is not very hard to see,

$$\langle S \rangle = \{s_1^{\varepsilon_1} s_2^{\varepsilon_2} \ldots s_k^{\varepsilon_k} \mid k \in \mathbb{N}, \, s_i \in S, \, \varepsilon_i = \pm 1\}.$$

That is, $\langle S \rangle$ consists of all elements of G that can be obtained by multiplying finitely many elements of S and their inverses together. By convention the empty product ($k = 0$) denotes the identity element 1 of G.

Cosets and Lagrange's Theorem.   Suppose that $H \leqslant G$. For $a \in G$ the right coset Ha is defined by the formula

$$Ha := \{ha \mid h \in H\}.$$

Notice that since $1 \in H$ we have $a \in Ha$. An important point about cosets is that $Ha = Hc$ for any $c \in Ha$. For, if $c \in Ha$ then $c = h_1 a$ for some $h_1 \in H$, so that, for any $h \in H$, $hc = h(h_1 a) = (hh_1)a \in Ha$: therefore $Hc \subseteq Ha$. On the other hand, for any $h \in H$ we have $ha = h(h_1^{-1}c) = (hh_1^{-1})c \in Hc$, so that also $Ha \subseteq Hc$: therefore in fact $Ha = Hc$. It follows that if $Ha \cap Hb \neq \varnothing$ then $Ha = Hb$; equivalently, distinct right cosets are disjoint.

A family $\{x_i\}_{i \in I}$ (where I is a suitable index set) of elements of G, one from each coset of H, is known as a complete set of coset representatives, or as a transversal, for H in G. Then

$$G = \bigcup_{i \in I} Hx_i \qquad \text{and} \qquad Hx_i \cap Hx_j = \varnothing \quad \text{when} \quad i \neq j.$$

Furthermore, as the map $h \mapsto hx_i$ gives a one–one correspondence between H and Hx_i, we have $|Hx_i| = |H|$ for all i (we use $|S|$ for the size—i.e. the cardinality—of the set S). A diagram such as Fig. 1.1 can help one to visualize a coset decomposition. It illustrates how each of the cosets is a sort of 'translate' in G of the subgroup H, and how the cosets all have the same size and neatly cover G without overlapping. But such diagrams have just as many drawbacks as Venn diagrams do, and can never offer a substitute for thought.

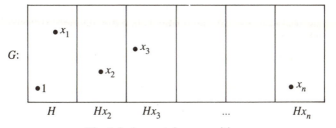

Fig. 1.1. A coset decomposition.

The number of cosets is known as the index of H in G, written $|G{:}H|$. If G is finite then of course $|H|$ and $|G{:}H|$ are also finite and the coset decomposition gives us the important

THEOREM 1.2: *If G is a finite group then* $|G| = |G{:}H| \cdot |H|$.

In group theory, if G is a group it is usual to call $|G|$ the order of G. Thus we have the

COROLLARY 1.3: *In a finite group the order of each subgroup divides the order of the group.*

This is commonly known as Lagrange's Theorem, but that is a name, not an attribution. Although it was Lagrange who, in a great memoir published in 1771, started the line of thinking that led to finite group theory, his own ideas were very tentative and fell far short of the result that bears his name. A satisfactory understanding of groups, subgroups and 'Lagrange's Theorem' (which acquired that name only later, $c.1870$) appears to have been achieved first by Galois $c.1830$ (who was then about eighteen years old) and became more widely available in the years 1845–1850.

Our arguments using right cosets could equally well have been carried out for the left cosets defined by

$$aH := \{ah \mid h \in H\}.$$

These are pairwise disjoint and their union is the whole of G. The map $Ha \mapsto a^{-1}H$ can easily be proved to be a one–one correspondence between the set of right cosets and the set of left cosets. (It should not be confused with maps at a lower level, such as the map $ha \mapsto a^{-1}h^{-1}$, that is, the map $g \mapsto g^{-1}$, which is a one–one correspondence between the right coset Ha and the left coset $a^{-1}H$.) Consequently, the index of H in G is the same whether it is defined to be the number of right cosets or the number of left cosets.

Normal subgroups.   The subgroups N with the property that each left coset is a right coset (and vice versa) are called normal subgroups of G and are of particular interest. Since the unique right coset containing the element a is Na, and the unique left coset that contains a is aN, it follows that the subgroup N is normal if and only if $Na = aN$ for all $a \in G$. This is obviously equivalent to the condition that $a^{-1}Na = N$ for all $a \in G$. Following Wielandt it is usual nowadays to write $N \trianglelefteq G$ to mean that N is a normal subgroup of G. In this notation what we have just observed is the

LEMMA 1.4: *If $N \leqslant G$ then $N \trianglelefteq G$ if and only if $a^{-1}Na = N$ for all $a \in G$.*

For subsets A, B of G we define $AB := \{ab \mid a \in A, b \in B\}$, which is obviously another subset of G. Suppose that $N \trianglelefteq G$ and consider the product in this sense of two of its cosets. We find that

$$(Na)(Nb) = N(aN)b = N(Na)b = (NN)(ab) = N(ab),$$

that is, $(Na)(Nb)$ is again a coset, namely $N(ab)$. With this operation the set $\{Ng \mid g \in G\}$ of cosets itself forms a group, known as the quotient group (or sometimes as the factor group) G/N. The identity of G/N is the coset $N1$, that is, N itself, and the inverse of the coset Ng is the coset Ng^{-1}. If $N = G$ then G/N

is the trivial group $\{1\}$, and if $N = \{1\}$ then G/N is essentially the same as G. We talk of non-trivial, proper quotient groups when we wish to exclude these extremes.

Homomorphisms. Most of the importance of normal subgroups and quotient groups comes from their connection with homomorphisms. Recall that, by definition, a homomorphism from a group G to a group H is a map $\theta: G \to H$ (of underlying sets) such that

$$(ab)\theta = a\theta b\theta \qquad \text{for all } a, b \in G. \tag{1.5}$$

In the left side of this formula the concatenation ab indicates the product in G, in the right side $a\theta b\theta$ means the product of $a\theta$ and $b\theta$ in H. Since the definition of a group includes the inverse function and the identity element, and since in principle we want homomorphisms to preserve *all* the relevant structure, we ought to require also that $a^{-1}\theta = (a\theta)^{-1}$ for all $a \in G$ and that $1\theta = 1$. Both of these conditions already follow from (1.5) however:

LEMMA 1.5: *If G, H are groups and $\theta: G \to H$ is a homomorphism then*
(i) $1\theta = 1$ *and*
(ii) $(a^{-1})\theta = (a\theta)^{-1}$ *for all $a \in G$.*

There are some rather obvious examples of homomorphisms. One is the map $a \mapsto 1$ (for all $a \in G$) which is known as the trivial homomorphism. Another is the map $a \mapsto a$ (for all $a \in G$), known as the identity automorphism of G. Only a little less obviously, if $N \trianglelefteq G$ then the map $\theta: G \to G/N$ defined by $a\theta := Na$ is a homomorphism—the fact that $NaNb = Nab$ is precisely the condition that $a\theta b\theta = (ab)\theta$. Since every element of G/N is of the form Ng for some $g \in G$, as a map θ is onto. It is known as the natural (or canonical) surjection (or epimorphism, or projection) of G onto G/N.

If $\theta: G \to H$ is both one–one and onto, that is, if θ is bijective, then $\theta^{-1}: H \to G$ exists as a mapping. If θ is a homomorphism then $\theta^{-1}: H \to G$ will also be a homomorphism, and θ is said to be an isomorphism. If there exists an isomorphism from G to H then we say that G is isomorphic to H and write $G \cong H$. An isomorphism from G to itself is known as an automorphism of G.

For any homomorphism $\theta: G \to H$ one finds that the image of θ, defined by

$$\text{Im } \theta := \{g\theta \mid g \in G\},$$

is a subgroup of H. The so-called kernel of θ is defined by

$$\text{Ker } \theta := \{x \in G \mid x\theta = 1\}.$$

Since $1\theta = 1$ we have that $1 \in \text{Ker } \theta$; if $a \in \text{Ker } \theta$ then, since $a^{-1}\theta = (a\theta)^{-1} = 1^{-1} = 1$, also $a^{-1} \in \text{Ker } \theta$; and if $a, b \in \text{Ker } \theta$ then $(ab)\theta = a\theta b\theta = 1$, so $ab \in \text{Ker } \theta$. This shows that $\text{Ker } \theta$ is a subgroup of G. In fact $\text{Ker } \theta \trianglelefteq G$ because if $x \in \text{Ker}\,\theta$ and $a \in G$ then

$$(a^{-1}xa)\theta = (a\theta)^{-1}(x\theta)(a\theta) = (a\theta)^{-1} . 1 . a\theta = 1,$$

which shows that $a^{-1}xa \in \text{Ker}\,\theta$. Thus we have

LEMMA 1.6: *If G, H are groups and $\theta: G \to H$ is a homomorphism then*
(i)   *Im θ is a subgroup of H and*
(ii) *Ker θ is a normal subgroup of G.*

Now $a\theta = b\theta$ if and only if $(ab^{-1})\theta = 1$, that is, $ab^{-1} \in \text{Ker}\,\theta$. Therefore $a\theta = b\theta$ if and only if a, b lie in the same coset of $\text{Ker}\,\theta$. This may be pictured as in Fig. 1.2. It follows that θ is injective (one–one) if and only if $\text{Ker}\,\theta = \{1\}$, and in this case of course θ is an isomorphism of G with $\text{Im}\,\theta$. In general, if $\psi:(\text{Ker}\,\theta)a \mapsto a\theta$ for $a \in G$, then $\psi: G/\text{Ker}\,\theta \to H$ is well-defined and is an injective homomorphism such that $\theta = \phi\psi$, where ϕ is the canonical surjection, $\phi: G \to G/\text{Ker}\,\theta$.

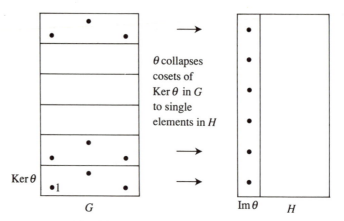

θ collapses cosets of $\text{Ker}\,\theta$ in G to single elements in H

Fig. 1.2. Portrait of a homomorphism.

This gives the famous

THEOREM 1.7 (FIRST ISOMORPHISM THEOREM): *If $\theta: G \to H$ is a homomorphism then $\text{Im}\,\theta \cong G/\text{Ker}\,\theta$.*

Similar considerations give the

THEOREM 1.8 (SECOND ISOMORPHISM THEOREM): *If $H \leqslant G$ and $N \trianglelefteq G$ then $NH \leqslant G$, $H \cap N \trianglelefteq H$, and $NH/N \cong H/(H \cap N)$.*

One can prove this directly by showing that the prescription $Nh \mapsto (H \cap N)h$ defines an isomorphism. Alternatively, consider the canonic epimorphism $\theta: G \to G/N$, and the image H^* of H under θ. On the one hand the kernel of the restriction $\theta{\restriction}H$ is $H \cap N$, so, by the first isomorphism theorem, $H^* \cong H/(H \cap N)$. On the other hand, NH is the set of *all* elements of

G that are mapped into H^*, so $NH \leqslant G$ and, as the kernel of $\theta \upharpoonright NH$ obviously is N, we have $H^* \cong NH/N$.

Suppose that $N \trianglelefteq G$. If $N \leqslant H \leqslant G$, let $H^* := H/N$. Then $H^* \leqslant G^*$ (where $G^* = G/N$), and the idea in the last sentence of the proof we have just sketched shows that every subgroup of G/N is of the form H^* for some subgroup H of G that contains N. Thus we have a one–one correspondence between the set of subgroups of G/N and the set of those subgroups of G that contain N. Furthermore, if $N \leqslant M$ then $M^* \trianglelefteq G^*$ precisely when $M \trianglelefteq G$. In this case we have epimorphisms (surjective homomorphisms) $G \overset{\theta}{\longrightarrow} G^* \overset{\phi}{\longrightarrow} G^{**}$, where $G^{**} := G^*/M^*$. Now $\operatorname{Ker} \theta = N$ and $\operatorname{Ker} \phi = M^* = M\theta$, so the composite $\theta\phi$ is an epimorphism whose kernel can easily be calculated to be M. Consequently,

$$G/M \cong \operatorname{Im}(\theta\phi) = G^{**} = G^*/M^*.$$

This is the third isomorphism theorem.

THEOREM 1.9 (THIRD ISOMORPHISM THEOREM): *If* $N \trianglelefteq G$ *and if* $N \leqslant M \trianglelefteq G$, *then* $M/N \trianglelefteq G/N$ *and* $(G/N)/(M/N) \cong G/M$.

Cyclic groups. Let g be an element of the group G. One defines powers of g inductively by the rules

$$g^0 := 1,$$

$$g^n := g^{(n-1)} \cdot g \qquad \text{for } n := 1, 2, \ldots,$$

$$g^n := (g^{-n})^{-1} \qquad \text{for } n < 0,$$

and the familiar law of indices holds: $g^m \cdot g^n = g^{m+n}$. The set $\{g^r \mid r = 0, \pm 1, \pm 2, \ldots\}$ is a subgroup of G, the cyclic subgroup $\langle g \rangle$ generated by g.

It may happen (if G is finite then it must happen) that two apparently different powers of g coincide. If $g^i = g^j$ where (without loss of generality) $i < j$, then $g^{j-i} = 1$, and so there is a positive power of g which is 1. Then the order of g is defined by

$$\operatorname{ord}(g) := \min\{m > 0 \mid g^m = 1\}.$$

If $\operatorname{ord}(g) = m$ then one finds that

$$\langle g \rangle = \{1, g, g^2, \ldots, g^{m-1}\},$$

and so $|\langle g \rangle| = \operatorname{ord}(g)$. It follows from Lagrange's Theorem that if G is finite then the order of any element divides $|G|$.

If $G = \langle g \rangle$ for some g, then G is said to be cyclic. A simple but significant fact about cyclic groups is that they are abelian. A cyclic group is determined up to isomorphism by its order: if g has order m then the map $r \mapsto g^r$ yields an isomorphism $Z_m \to G$, where Z_m denotes the additive group of integers modulo m. One finds that every subgroup of a cyclic group is cyclic: if $G = \langle g \rangle$ and $H \leqslant G$ then either $H = \{1\}$ or $H = \langle g^s \rangle$ where $s := \min\{r \mid r > 0, g^r \in H\}$.

Conjugates and centralisers.   Again, let g be an element of the group G. If $x \in G$ then $x^{-1}gx$ (sometimes written g^x) is known as the conjugate of g by x. The set of all conjugates of g, that is $\{x^{-1}gx \mid x \in G\}$, is known as the conjugacy class of g in G. We define the centraliser of g in G by

$$C_G(g) := \{z \in G \mid z^{-1}gz = g\}.$$

Thus the centraliser of g consists of all elements that commute with g. Certainly $1 \in C_G(g)$ and if $a, b \in C_G(g)$ then

$$(ab^{-1})^{-1}g(ab^{-1}) = b(a^{-1}ga)b^{-1} = bgb^{-1} = g,$$

which shows that $ab^{-1} \in C_G(g)$. Therefore $C_G(g)$ is a subgroup of G. Consider now a fixed conjugate $x^{-1}gx$ of g in G. We have that

$$
\begin{aligned}
\{y \mid y^{-1}gy = x^{-1}gx\} &= \{y \mid xy^{-1}gyx^{-1} = g\} \\
&= \{y \mid (yx^{-1})^{-1}g(yx^{-1}) = g\} \\
&= \{y \mid yx^{-1} \in C_G(g)\} \\
&= \{y \mid y \in C_G(g)x\} \\
&= C_G(g)x.
\end{aligned}
$$

Thus the conjugates $x^{-1}gx$ and $y^{-1}gy$ are equal if and only if the cosets $C_G(g)x$ and $C_G(g)y$ coincide. Consequently there is a one–one correspondence between the conjugacy class of g in G and the set of right cosets of the centraliser of g, and, in particular, the number of distinct conjugates of g is equal to $|G:C_G(g)|$. From Lagrange's Theorem it follows that if G is finite then the size of every conjugacy class divides the order of G. In later chapters of this book we shall see that this is just a special case of a very general fact about group actions.

Direct products.   The (external) direct product of groups A, B is defined by $A \times B := \{(a, b) \mid a \in A, b \in B\}$ with coordinatewise multiplication $(a_1, b_1)(a_2, b_2) = (a_1a_2, b_1b_2)$. The unit element is $(1, 1)$, the inverse of (a, b) is (a^{-1}, b^{-1}), and it is routine to verify that the axioms defining groups hold. We usually identify A in the obvious way with $\{(a, 1) \mid a \in A\}$, which is a subgroup of $A \times B$ canonically isomorphic with A; similarly we think of B as a subgroup of $A \times B$.

One can recognise when a given group is a direct product in several ways. Here are two of the most important. Suppose that A, B are subgroups of G. If

(i) $ab = ba$ for all $a \in A$, $b \in B$ (that is, A, B commute elementwise); and
(ii) each element g of G has a unique expression in the form $g = ab$ for some $a \in A$, $b \in B$,

or if

(i) $A \trianglelefteq G$, $B \trianglelefteq G$; and
(ii) $G = AB$ and $A \cap B = 1$,

then the map $(a, b) \mapsto ab$ is an isomorphism of $A \times B$ with G.

The symmetric groups.   Let Ω be any set. We define the symmetric group on Ω to be the group of all permutations of Ω: that is,

$$\mathrm{Sym}(\Omega) := \{ f \mid f : \Omega \to \Omega, \ f \text{ bijective} \},$$

with composition of mappings as the group multiplication. The unit is the identity map on Ω, the inverse is the inverse function which always exists for a bijective map, and, since composition of mappings always is associative, this really is a group. If there is a one–one correspondence between the sets Ω and Ω' then $\mathrm{Sym}(\Omega)$ and $\mathrm{Sym}(\Omega')$ are isomorphic. In fact, if $h : \Omega \to \Omega'$ is a bijection then the map $\theta : f \mapsto h^{-1}fh$ is an isomorphism $\theta : \mathrm{Sym}(\Omega) \to \mathrm{Sym}(\Omega')$. As an abstract group therefore $\mathrm{Sym}(\Omega)$ is determined by the cardinality of Ω. In particular, if Ω is finite and has n elements, then we write S_n for $\mathrm{Sym}(\Omega)$ and we talk of the symmetric group of degree n. Of course S_n is finite; its order is $n!$.

By a cycle $(\alpha_1 \alpha_2 \ldots \alpha_a)$ in the symmetric group S_n (or $\mathrm{Sym}(\Omega)$) we mean that permutation which maps α_1 to α_2, α_2 to α_3, ..., α_a to α_1, and any other element of Ω to itself. We get a remarkably economical notation for the elements of S_n by expressing each permutation as a product of disjoint cycles

$$(\alpha_1 \alpha_2 \ldots \alpha_a)(\beta_1 \beta_2 \ldots \beta_b) \ldots .$$

Such an expression is unique up to the cyclic order within each cycle and, since disjoint cycles commute, up to the order in which the cycles are written down. Cycles of length 1, that is, fixed points, are usually (but by no means always) omitted in this notation. A cycle of length 2 is known as a transposition. Now

$$(\alpha_1 \alpha_2 \ldots \alpha_a) = (\alpha_1 \alpha_2)(\alpha_1 \alpha_3) \ldots (\alpha_1 \alpha_a),$$

that is, a cycle of length a can be obtained as a product of $a - 1$ transpositions. So we have

LEMMA 1.10: (i)   *Every permutation in S_n can be expressed as a product of transpositions (that is, the set of transpositions generates S_n);*
(ii)   *A permutation that consists of disjoint cycles of lengths $a_1, a_2, \ldots, a_k$ can be expressed as a product of $a_1 + a_2 + \cdots + a_k - k$ transpositions.*

A permutation that has c_r cycles of length r (for $r := 1, 2, 3, \ldots$) is said to be of cycle type $1^{c_1} . 2^{c_2} . 3^{c_3} \ldots$ (terms where $c_r = 0$ are omitted). Thus, for example, the permutation $(1\ 2\ 3)(4\ 5\ 6\ 7)$ in S_{10} has cycle type $1^3 . 3^1 . 4^1$, and an a-cycle in S_n has cycle type $1^{n-a} . a^1$. Now if $f : \alpha \mapsto \beta$ then $g^{-1}fg : \alpha g \mapsto \beta g$, and so the conjugate $g^{-1}fg$ may be obtained very easily in terms of the cycles of f: if

$$f = (\alpha_1 \alpha_2 \ldots \alpha_a)(\beta_1 \beta_2 \ldots \beta_b) \ldots$$

then

$$g^{-1}fg = (\alpha'_1 \alpha'_2 \ldots \alpha'_a)(\beta'_1 \beta'_2 \ldots \beta'_b) \ldots ,$$

where $g : \alpha_i \mapsto \alpha'_i$, $\beta_i \mapsto \beta'_i$, etc. So we have

LEMMA 1.11: *Permutations $f, f' \in S_n$ are conjugate in S_n if and only if they have the same cycle type.*

The themes of these two lemmas will be developed further in the exercises and in later chapters.

Specimen solutions to selected exercises.

The following specimen solutions, being of the nature of an extension of the text, are placed between the main body of the chapter and the exercises. The wise and conscientious reader will, however, try the exercises unaided before attempting to evaluate what we have offered.

EXERCISE 1.6 (i)   *Let A, B be subgroups of the group G. Show that if $x, y \in G$ then $Ax \cap By$ is either empty or a right coset of $A \cap B$. Deduce that*

$$|G:A \cap B| \leqslant |G:A| . |G:B|$$

[Poincaré].
   Now suppose that A, B are subgroups of finite index m, n respectively in G, where $\mathrm{hcf}(m, n) = 1$.
(ii)   *What can you say about $|G:A \cap B|$?*
(iii)   *Prove that $AB = G$.*

Solution. (i)   Suppose that $Ax \cap By \neq \varnothing$, say $g \in Ax \cap By$. Since $g \in Ax$ we have that $Ag = Ax$; similarly, $Bg = By$. Now for any $h \in Ax \cap By$ there exist $a \in A$, $b \in B$ such that $h = ag$ and $h = bg$. It follows that $a = b \in A \cap B$, and so $h \in (A \cap B)g$. This shows that $Ax \cap By \subseteq (A \cap B)g$. On the other hand, we certainly have that $(A \cap B)g \subseteq Ag = Ax$, and likewise that $(A \cap B)g \subseteq By$, so that $(A \cap B)g \subseteq Ax \cap By$. Thus $Ax \cap By$ is either empty or it is the coset $(A \cap B)g$ of $A \cap B$.

Every coset of $A \cap B$ certainly does arise as the intersection of some coset of A and some coset of B because $(A \cap B)g = Ag \cap Bg$. Thus the number of cosets of $A \cap B$ is at most the number of pairs (Ax, By), that is, $|G:A \cap B| \leqslant |G:A| . |G:B|$.

(ii)   The question is worded to suggest that our group G need not be finite so we should try to give an argument that is valid quite generally. Let $D := A \cap B$. Since each coset of A contains $|A:D|$ cosets of D we have that

$|G:D| = |G:A| \cdot |A:D|$. Similarly, $|G:D| = |G:B| \cdot |B:D|$. Thus $|G:D|$ is divisible by both m and n, and, since these are coprime, $|G:D|$ is a multiple of mn. But we know that $0 < |G:D| \leqslant mn$. Therefore $|G:D| = |G:A \cap B| = mn$.

(iii) Since $|G:A \cap B| = |G:A| \cdot |G:B|$ we know from our proof of the inequality in (i) that every intersection $Ax \cap By$ must be non-empty. In particular, if $g \in G$ then $Ag \cap B \neq \varnothing$. This means that there exist $a \in A$, $b \in B$ such that $ag = b$, and so $g = a^{-1}b \in AB$. Thus $G = AB$ as required.

EXERCISE 1.7 *Prove that the product of any two right cosets of H in G is again a right coset if and only if $H \trianglelefteq G$.*

Solution.   Suppose first that $H \trianglelefteq G$. Let Hx, Hy be any two right cosets of H. Since H is normal we know that $xH = Hx$ and so $HxHy = HHxy = Hxy$, that is, $HxHy$ is the right coset Hxy.

For the converse suppose that the product of any two right cosets is again a right coset of H. In particular, for any $x \in G$ the set $Hx^{-1} \cdot Hx$ is a right coset. Since it contains the element 1 it is the unique coset containing 1, namely H. Thus $Hx^{-1}Hx = H$, from which it follows that $x^{-1}Hx \subseteq H$. Similarly, $Hx \cdot Hx^{-1}$ must be the coset containing 1, namely H, and so $xHx^{-1} \subseteq H$, that is, $H \subseteq x^{-1}Hx$. Thus $H = x^{-1}Hx$ for all $x \in G$ and so H is a normal subgroup, as required.

EXERCISE 1.9 *Prove that in a finite group G the order of any element divides $|G|$. Deduce Fermat's Little Theorem that, if p is a prime number and x is an integer not divisible by p, then $x^{p-1} \equiv 1 \pmod{p}$.*

Solution.   Let $g \in G$. We have defined the order of g to be the least positive integer m such that $g^m = 1$. Define a binary relation $\sim$ on G by specifying that $x \sim y$ if and only if $y = g^r x$ for some integer r. It is easy to prove that $\sim$ is an equivalence relation. Furthermore, the equivalence class containing the element x of G is $\{x, gx, g^2x, \ldots, g^{m-1}x\}$. For certainly these elements are, by definition of $\sim$ equivalent to x. On the other hand, if $y \sim x$ then $y = g^r x$. Using the fact that the integer r can be expressed in the form $qm + r_0$ with $0 \leqslant r_0 \leqslant m-1$ and the fact that $g^m = 1$ we see that $y = g^{r_0}x$ so that y lies in the displayed set. Now the elements $x, gx, \ldots, g^{m-1}x$ must all be different because otherwise we would have an equation $g^ix = g^jx$ with $0 \leqslant i < j \leqslant m-1$, from which we would derive that $g^{j-i} = 1$ contrary to the definition of m. Thus each equivalence class in G has size m and so, since G is the disjoint union of the equivalence classes, m divides $|G|$.

Since $g^n = 1$ for any multiple n of m, it is an immediate corollary that $g^{|G|} = 1$. Let x be an integer not divisible by the prime number p. In the field $\mathbb{Z}_p$ of integers modulo p the element $\bar{x}$ represented by x is non-zero, and so it lies in the multiplicative group $\mathbb{Z}_p^{\times}$. This multiplicative group consists of all the non-zero members of $\mathbb{Z}_p$ and so its order is $p-1$. It follows that $(\bar{x})^{p-1} = 1$ in $\mathbb{Z}_p$. Translating back to an equation in $\mathbb{Z}$ we have that $x^{p-1} \equiv 1 \pmod{p}$ as required.

EXERCISE 1.12   *Let G be a cyclic group of order n.*
(i)   *Show that G has a subgroup of order m if and only if $m \mid n$.*
(ii)   *Suppose that $m \mid n$. Show that there is precisely one subgroup of order m in G.*
(iii)   *Suppose that $n = rs$. Show that $G \cong Z_r \times Z_s$ if and only if $\mathrm{hcf}(r, s) = 1$.*

Solution. (i)   If G has a subgroup of order m then, by Lagrange's Theorem (or its corollary) we know that $m \mid n$. This of course is true for any finite group; the converse depends on the special nature of cyclic groups. Let g be a generator for G. Suppose that $m \mid n$, so that $n = mk$ for some integer k, and define $h := g^k$. Then $h^m = g^n = 1$ and if $1 < i < m$ then $h^i = g^{ki} \neq 1$. Thus the order of h is m, and so it generates a subgroup of order m.

(ii)   We are supposing now that $n = mk$. We have already exhibited one subgroup, namely that generated by g^k, which is of order m. Let H be any subgroup of order m and let x be any element of H. We know on the one hand that $x^m = 1$ (see Exercise 1.9), and on the other hand that $x = g^r$ for some $r \in \mathbb{Z}$. Then $g^{rm} = 1$ and it follows that $n \mid rm$, that is, $km \mid rm$. Thus $k \mid r$ and so x is a power of g^k. This shows that $H \leqslant \langle g^k \rangle$. Since both groups have order m we must have $H = \langle g^k \rangle$, and so $\langle g^k \rangle$ is the unique subgroup of order m.

(iii)   Now we are given that $n = rs$. Let R, S be the unique subgroups of orders r, s respectively in G. Note that $R = \langle g^s \rangle$ and $S = \langle g^r \rangle$ by the proof of (ii) and, in particular, R, S are cyclic, that is, $R \cong Z_r$, $S \cong Z_s$. If $\mathrm{hcf}(r, s) = 1$ then, since $R \cap S$ is a subgroup whose order must (by Lagrange's Theorem) divide both r and s, we have $R \cap S = \{1\}$. Furthermore, by the Euclidean Algorithm there exist integers u, v such that $su + rv = 1$. Then for any integer i we have $g^i = (g^s)^{ui}(g^r)^{vi} \in RS$ and so $G = RS = R \times S$. Thus $G \cong Z_r \times Z_s$.

Suppose finally that $G \cong Z_r \times Z_s$. Then we must have $G = R \times S$ and in particular, $R \cap S = \{1\}$. But R, S both contain the unique subgroup T of order $\mathrm{hcf}(r, s)$. Therefore $\mathrm{hcf}(r, s) = 1$ as required.

EXERCISE 1.16   *We define the 'support' of an element of $\mathrm{Sym}(\Omega)$ by $\mathrm{supp}(f) := \{\omega \in \Omega \mid \omega f \neq \omega\}$. Suppose that f, g are permutations and that $|\mathrm{supp}(f) \cap \mathrm{supp}(g)| = 1$. Prove that the commutator $f^{-1}g^{-1}fg$ is a 3-cycle.*

Solution.   Let α be the single element in $\mathrm{supp}(f) \cap \mathrm{supp}(g)$ and let $\beta := \alpha g$, $\gamma := \alpha f$. Now αf^{-1} lies in $\mathrm{supp}(f)$ but is not equal to α and therefore $\alpha f^{-1} \notin \mathrm{supp}(g)$. Note, however, that $\mathrm{supp}(x) = \mathrm{supp}(x^{-1})$ for any permutation x. Thus $\alpha f^{-1} \notin \mathrm{supp}(g^{-1})$ and so $\alpha f^{-1}g^{-1} = \alpha f^{-1}$. It follows that $\alpha f^{-1}g^{-1}f = \alpha$, and then that $\alpha f^{-1}g^{-1}fg = \alpha g = \beta$. Next, since $\beta \in \mathrm{supp}(g)$ and $\beta \neq \alpha$, we know that $\beta \notin \mathrm{supp}(f) = \mathrm{supp}(f^{-1})$, whence $\beta f^{-1} = \beta$. Therefore $\beta f^{-1}g^{-1}fg = \beta g^{-1}fg = \alpha fg = \gamma g$. But γ lies in $\mathrm{supp}(f)$ and is different from α, so it does not lie in $\mathrm{supp}(g)$, that is $\gamma g = \gamma$. Thus $\beta f^{-1}g^{-1}fg = \gamma$. Similarly, $\gamma f^{-1}g^{-1}fg = \alpha g^{-1}fg = (\alpha g^{-1})g = \alpha$, since αg^{-1} does not lie in $\mathrm{supp}(f)$. Thus $(\alpha \ \beta \ \gamma)$ is one of the cycles of the commutator $f^{-1}g^{-1}fg$.

Suppose now that $\omega \in \Omega \setminus \{\alpha, \beta, \gamma\}$. If $\omega \notin \mathrm{supp}(f) \cup \mathrm{supp}(g)$ then it is fixed by f, f^{-1}, g and g^{-1}, and so certainly $\omega f^{-1}g^{-1}fg = \omega$. If $\omega \in \text{supp}(f)$ then $\omega f^{-1} \notin \text{supp}(g)$ and $\omega f^{-1} \neq \alpha$; therefore $\omega f^{-1}g^{-1}fg = \omega f^{-1}fg = \omega g = \omega$. If $\omega \in \text{supp}(g)$ then $\omega \notin \text{supp}(f)$, and ωg^{-1} must be fixed by f, whence $\omega f^{-1}g^{-1}fg = \omega g^{-1}fg = \omega g^{-1}g = \omega$. Thus all elements of Ω other than α, β, γ are fixed by the commutator and so $f^{-1}g^{-1}fg = (\alpha\,\beta\,\gamma)$ as required.

EXERCISE 1.19   *For which values of n in the range $3 \leqslant n \leqslant 7$ can one find permutations $\sigma, \tau \in S_n$ such that σ has order 2, τ has order 3, and $\sigma\tau$ has order n?*

Solution.   In S_3 any element σ of order 2 is a transposition and hence an odd permutation; on the other hand, elements of order 3 are even permutations, and so the product $\sigma\tau$ would have to be an odd permutation and could not be of order 3.

In S_4, if $\sigma := (1\ 2)$ and $\tau := (2\ 3\ 4)$ then $\sigma\tau = (1\ 3\ 4\ 2)$, so σ, τ are as required.

In S_5 we take $\sigma := (1\ 2)(3\ 4)$ and $\tau := (1\ 3\ 5)$ and find that $\sigma\tau = (1\ 2\ 3\ 4\ 5)$ to solve the problem.

In S_6 we take $\sigma := (1\ 2)$ and $\tau := (3\ 4\ 5)$. Then $\sigma\tau = (1\ 2)(3\ 4\ 5)$, which has order 6.

In S_7 we take $\sigma := (1\ 2)(3\ 4)$ and $\tau := (1\ 3\ 5)(2\ 6\ 7)$ and find that $\sigma\tau = (1\ 6\ 7\ 2\ 3\ 4\ 5)$, which has order 7.

Thus σ, τ can be found as specified if $4 \leqslant n \leqslant 7$ but can not be found if $n = 3$.

Exercises

1.1.   Show that if, in the notation of our formal definition, $(G, \alpha, \beta, 1)$ and $(G, \alpha, \beta', 1')$ are groups, then $\beta' = \beta$ and $1' = 1$.

1.2.   Consider non-empty sets G equipped with an associative binary operation. Of the following propositions:

(A) $\begin{cases} \text{there exist } 1 \in G \text{ and } \beta\colon G \to G \text{ such that axioms (1.2) and} \\ \text{(1.3) hold;} \end{cases}$

(B) $\begin{cases} \text{there exists } 1 \in G \text{ such that } 1a = a = a1 \text{ for all } a \in G, \text{ and} \\ \text{for each } a \in G \text{ there exists } b \in G \text{ such that } ab = 1 = ba; \end{cases}$

(C) $\begin{cases} \text{there exists } e \in G \text{ such that } ae = a \text{ for all } a \in G \text{ and} \\ \text{for each } a \in G \text{ there exists } b \in G \text{ such that } ab = e; \end{cases}$

(D) $\begin{cases} \text{there exists } e \in G \text{ such that } ae = a \text{ for all } a \in G \text{ and} \\ \text{for each } a \in G \text{ there exists } b \in G \text{ such that } ba = e; \end{cases}$

(E) $\begin{cases} \text{for each } a \in G \text{ there exist } e, b \in G \text{ such that } ea = a = ae \text{ and} \\ ab = e = ba; \end{cases}$

(F) $\begin{cases} \text{for all } a, b \in G \text{ there exist unique elements } x, y \in G \text{ such that} \\ ax = b \text{ and } ya = b, \end{cases}$

which can be used to axiomatize group theory?

1.3. An anti-automorphism of a group G is a bijective map $\phi:G \to G$ such that

$$(ab)\phi = b\phi \cdot a\phi \quad \text{for all } a, b \in G.$$

(i) Show that the map $\beta:g \mapsto g^{-1}$ is an anti-automorphism.

(ii) Show that the composite of two anti-automorphisms is an automorphism.

(iii) Deduce that if ϕ is any anti-automorphism of G then there is an automorphism γ such that

$$a\phi = (a\gamma)^{-1} \quad \text{for all } a \in G.$$

1.4. Let H be a subgroup of the group G. Prove that the relation ρ on G defined by the rule

$$a\rho b \quad \text{if and only if } ab^{-1} \in H$$

is an equivalence relation. What are its equivalence classes?

1.5. (i) Let A, B be subgroups of the group G. Show that if $x, y \in G$ then $Ax \cap By$ is either empty or a right coset of $A \cap B$. Deduce that

$$|G:A \cap B| \leqslant |G:A| \cdot |G:B|$$

[Poincaré].

Now suppose that A, B are subgroups of finite index m, n respectively in G, where $\mathrm{hcf}(m, n) = 1$.

(ii) What can you say about $|G:A \cap B|$?

(iii) Prove that $AB = G$.

1.6. (i) Under what conditions is the union $A \cup B$ of subgroups A, B itself a subgroup?

(ii) Under what conditions is the product AB of subgroups itself a subgroup?

1.7. Prove that the product of any two right cosets of H in G is again a right coset if and only if $H \trianglelefteq G$.

1.8. Let G be a finite group of order mn, where $\mathrm{hcf}(m, n) = 1$, and suppose that G has a normal subgroup M of order m. Show that if $H \leqslant G$ and $|H|$ divides m then $H \leqslant M$.

1.9. Prove that in a finite group G the order of any element divides $|G|$.

Deduce Fermat's Little Theorem that, if p is a prime number and x is an integer not divisible by p, then $x^{p-1} \equiv 1 \pmod{p}$.

1.10. Let A be an abelian group of order n whose elements are $a_1, a_2, \ldots, a_n$, of orders $m_1, m_2, \ldots, m_n$ respectively. Let r be the product $m_1 m_2 \ldots m_n$ and let s be the number of representations of 1 as a product

$$a_1^{\alpha_1} a_2^{\alpha_2} \ldots a_n^{\alpha_n} \quad \text{with} \quad 0 \leqslant \alpha_i \leqslant m_i - 1.$$

(i) Show that for any $b \in A$ the number of representations

$$b = a_1^{\beta_1} a_2^{\beta_2} \ldots a_n^{\beta_n} \quad \text{with} \quad 0 \leqslant \beta_i \leqslant m_i - 1$$

is s and deduce that $r = ns$.

(ii) Deduce that if p is a prime divisor of n then A contains an element of order p. [Cauchy's Theorem for abelian groups, as proved by Frobenius in 1884.]

1.11. (i) Let a be an element of order mn in a group G, where $\mathrm{hcf}(m, n) = 1$. Show that there exist $x, y \in G$ such that $a = xy = yx$ and $\mathrm{ord}(x) = m$, $\mathrm{ord}(y) = n$.

(ii) Let A be an abelian group of order mn where $\mathrm{hcf}(m, n) = 1$. Show that if

$$M := \{a \in A \mid a^m = 1\},$$

$$N := \{a \in A \mid a^n = 1\},$$

then M and N are subgroups of A such that $A = M \times N$, $|M| = m$ and $|N| = n$.

(iii) Deduce that if A is an abelian group of order $p_1^{r_1} p_2^{r_2} \ldots p_k^{r_k}$ where $p_1, \ldots, p_k$ are distinct prime numbers then there are subgroups $P_1, \ldots, P_k$ of A such that $|P_i| = p_i^{r_i}$ and $A = P_1 \times P_2 \times \cdots \times P_k$.

1.12. (i) Let G be a cyclic group of order n. Show that G has a subgroup of order m if and only if $m \mid n$.

(ii) Suppose that $m \mid n$. Show that there is precisely one subgroup of order m in G.

(iii) Suppose that $n = rs$. Show that $G \cong Z_r \times Z_s$ if and only if $\mathrm{hcf}(r, s) = 1$.

1.13. Suppose that $H \leqslant G$ and $g \in G$. Show that $g^{-1} Hg$ is a subgroup of G and is isomorphic to H.

1.14. Let H be a subgroup of index 2 in the group G.

(i) Prove that $H \trianglelefteq G$.

(ii) Prove that if $h \in H$ then the conjugacy class $\{x^{-1} hx \mid x \in G\}$ is a union of 1 or 2 conjugacy classes of H.

(iii) Suppose that for every $h \in H \setminus \{1\}$ there exists $g \in H$ that is conjugate to h in G but not in H. Prove that if $h \in H \setminus \{1\}$ then $C_G(h) \leqslant H$; that every element of $G \setminus H$ has order 2; and that H is commutative.

1.15.   (i) Describe the permutation

$$\begin{pmatrix} 0\ 1\ 2\ 3\ 4\ 5\ 6\ 7\ 8\ 9 \\ 0\ 1\ 3\ 6\ 2\ 5\ 4\ 9\ 7\ 8 \end{pmatrix}$$

in cycle notation.

(ii) Calculate $\pi_1 \pi_2$ where

$$\pi_1 = (1\ 2\ 3\ 4)\,(5\ 6\ 7)\,(8\ 9),$$

$$\pi_2 = (1\ 10)\,(2\ 5\ 9)\,(3\ 4\ 6\ 7).$$

1.16. We define the 'support' of an element of $\mathrm{Sym}(\Omega)$ by

$$\mathrm{supp}(f) := \{\omega \in \Omega \mid \omega f \neq \omega\}.$$

Suppose that f, g are permutations such that

$$|\mathrm{supp}(f) \cap \mathrm{supp}(g)| = 1.$$

Prove that the commutator $f^{-1}g^{-1}fg$ is a 3-cycle.

1.17. Prove that the order of a permutation is the least common multiple of the lengths of its cycles.

1.18. We know from Lemma 1.11 that conjugacy classes in S_n are determined by cycle types. For the cycle type $1^{c_1}.2^{c_2}.3^{c_3}.\ldots$ calculate the size of the corresponding conjugacy class and the order of the centraliser of a permutation of that type.

1.19. For which values of n in the range $3 \leqslant n \leqslant 7$ can one find permutations $\sigma, \tau \in S_n$ such that σ has order 2, τ has order 3 and $\sigma\tau$ has order n?

1.20. For $\pi \in S_n$ we define $c(\pi)$ to be the number of cycles (including those of length 1) in the cycle decomposition of π. Show that if π can be expressed as a product of m transpositions then $c(\pi) \geqslant n - m$. Hence show that if the index $\mathrm{ind}(\pi)$ is defined to be the least number of transpositions whose product is π then $\mathrm{ind}(\pi) = n - c(\pi)$.

Chapter 2

A menagerie of groups

In this chapter we introduce some of the more interesting groups that occur in applications of the theory. We restrict ourselves to a description of one of the most concrete forms of each example. In later chapters you will meet the groups again in other guises.

We have already discussed one important class of examples, the finite symmetric groups S_n. Our first exhibit is closely related to these.

Example 2.1 The alternating groups. Let Ω be a finite set with n elements. A permutation f in $\mathrm{Sym}(\Omega)$ is said to be even if it can be expressed as a product of an even number of transpositions, and odd if it can be expressed as a product of an odd number of transpositions. By Lemma 1.10, every permutation is either even or odd (or perhaps both). Thus, for example, as an r-cycle can be expressed as a product of $r - 1$ transpositions, it is an even permutation if it has odd length, and is odd if r is even. We define the alternating group by

$$\mathrm{Alt}(\Omega) := \{ f \in \mathrm{Sym}(\Omega) \mid f \text{ is even} \}.$$

Certainly, the product of two even permutations is even. Also, if $f = h_1 h_2 \ldots h_r$, where the h_i are transpositions, then $f^{-1} = h_r \ldots h_2 h_1$, and so if f is even then also f^{-1} is even. Thus $\mathrm{Alt}(\Omega)$ is a subgroup of $\mathrm{Sym}(\Omega)$. As with $\mathrm{Sym}(\Omega)$, the group $\mathrm{Alt}(\Omega)$ really depends more on n than on the particular set Ω, and so we usually denote it A_n, and speak of the alternating group of degree n.

A moment ago we left open the question whether a permutation can be simultaneously even and odd. As we expect the reader to know, this is not possible. The alternating group A_n consists of exactly half the permutations in S_n when $n \geqslant 2$. The other half are the odd permutations. This was one of the discoveries of A. L. Cauchy who, in 1815, showed that if $h_1, h_2, \ldots, h_r$ are transpositions, and if the product $h_1 h_2 \ldots h_r$ has c cycles in its cycle decomposition, then $r \equiv n - c \pmod 2$. The proof uses induction on r. The statement is certainly true if $r = 0$ since, by convention, we take a product that has no factors to be 1. Suppose, then, that $r > 0$ and let $f := h_1 h_2 \ldots h_{r-1} h_r$, so that c is the number of cycles of the permutation f. Let $g := h_1 h_2 \ldots h_{r-1}$, and suppose that, in cycle notation,

$$g = (\alpha_1 \ldots \alpha_{b_1})(\beta_1 \ldots \beta_{b_2}) \ldots (\lambda_1 \ldots \lambda_{b_d}).$$

Thus g has d cycles, and as inductive hypothesis we can suppose that $r - 1 \equiv n - d \pmod 2$. Since the cycles of g may be written in any order, and since we can write each cycle starting at any point, we may suppose that

$$h_r = (\alpha_1 \; \alpha_{s+1}) \quad \text{(if both entries of } h_r \text{ lie in the same cycle of } g),$$

or

$$h_r = (\alpha_1 \; \beta_1) \quad \text{(if the entries of } h_r \text{ lie in different cycles of } g).$$

In the former case we have

$$f = gh_r = (\alpha_1 \; \ldots \; \alpha_s)(\alpha_{s+1} \; \ldots \; \alpha_{b_1})(\beta_1 \; \ldots \; \beta_{b_2}) \ldots (\lambda_1 \; \ldots \; \lambda_{b_d})$$

and so $c = d+1$. Thus

$$r \equiv n-d+1 \equiv n-c \pmod 2.$$

And in the latter case

$$f = gh_r = (\alpha_1 \; \ldots \; \alpha_{b_1} \; \beta_1 \; \ldots \; \beta_{b_2})(\gamma_1 \; \ldots \; \gamma_{b_3}) \ldots (\lambda_1 \; \ldots \; \lambda_{b_d}),$$

so that $c = d-1$ and again

$$r \equiv n-d+1 \equiv n-c \pmod 2.$$

Induction completes the proof. It is worth noting the heart of the proof, which is the fact that multiplying a permutation by a transposition increases or decreases the number of cycles by exactly 1.

Now let f be any permutation in S_n. If f can be expressed as a product of r transpositions and also as a product of s transpositions, then

$$r \equiv n-c \equiv s \pmod 2,$$

where c is the number of cycles of f, and so r and s are both even or both odd. If h is any odd permutation then every element of the coset $A_n h$ is odd; furthermore, if g is odd then gh^{-1} is even, and so $g = (gh^{-1})h \in A_n h$. Thus if $n \geqslant 2$ then $|S_n : A_n| = 2$, and the set of odd permutations is the second coset of A_n in S_n.

Example 2.2 Linear groups.   Let F be a field and n a positive integer. We write $M_n(F)$ for the set of $n \times n$ matrices with coefficients from F and define

$$\mathrm{GL}(n, F) := \{A \in M_n(F) \,|\, A \text{ is invertible}\}$$

$$= \{A \in M_n(F) \,|\, \det(A) \neq 0\}.$$

Since a product of invertible matrices is invertible, $\mathrm{GL}(n, F)$ is a group under matrix multiplication (the group of units of the ring $M_n(F)$). It is known as the n-dimensional general linear group over F.

The n-dimensional special linear group is defined by

$$\mathrm{SL}(n, F) := \{A \in M_n(F) \,|\, \det(A) = 1\}.$$

It is a subgroup, in fact a normal subgroup of $\mathrm{GL}(n, F)$. Since the determinant map det: $GL(n, F) \to F^*$ (where F^* denotes the multiplicative group of non-zero elements of F) is a group homomorphism which, as one sees quite easily, is surjective and has $SL(n, F)$ as its kernel, we have $GL(n, F)/SL(n, F) \cong F^*$.

If F is a finite field and has q elements one often writes $GL(n, q)$, $SL(n, q)$ for $GL(n, F)$, $SL(n, F)$. (One can prove that up to isomorphism there is at most one finite field with q elements—if there are any at all—so this is not as misguided as it may appear.) These are finite groups and we can compute their orders as follows. A square matrix is non-singular if and only if its rows are linearly independent. If we write our matrix in the usual way, writing line by line from top to bottom; then when it comes to the $(r+1)$th row we simply have to be sure that we choose one that is linearly independent of its r predecessors. This means that, of the q^n possibilities, we must avoid the q^r linear combinations of the rows that have already been written down. Therefore there are

$$q^n - 1 \qquad \text{choices for the first row,}$$
$$\vdots$$
$$q^n - q^r \qquad \text{choices for the } (r+1)\text{th row,}$$
$$\vdots$$
$$q^n - q^{n-1} \quad \text{choices for the last row.}$$

Consequently,

$$|GL(n, q)| = (q^n - 1)(q^n - q) \ldots (q^n - q^{n-1})$$
$$= q^{\frac{1}{2}n(n-1)} \cdot (q^n - 1)(q^{n-1} - 1) \ldots (q - 1).$$

And of course

$$|SL(n, q)| = \frac{1}{q-1} |GL(n, q)|.$$

Example 2.3 Projective groups. As in Example 2.2, let F be a field and n a positive integer. The set $Z := \{aI_n \mid a \in F^*\}$ of scalar matrices is a normal subgroup in $GL(n, F)$ (after all, $X^{-1}AX = A$ for all $A \in Z$ and all $X \in GL(n, F)$). The projective linear groups over F are defined by

$$PGL(n, F) := GL(n, F)/Z$$
$$PSL(n, F) := SL(n, F)/(SL(n, F) \cap Z) \cong SL(n, F) \cdot Z/Z.$$

The projective groups are obtained from the corresponding ordinary linear groups by identifying matrices that are scalar multiples of each other.

Example 2.4 The Möbius group. One of the groups of Example 2.3 is so important that, like a large railway engine, it deserves a romantic name and individual treatment. This is the group M of Möbius transformations

(fractional linear transformations)

$$T_{abcd} : z \mapsto \frac{az+b}{cz+d} \quad (ad-bc \neq 0)$$

of the extended complex field $\mathbb{C} \cup \{\infty\}$. You will recall the convention that, for fixed complex numbers a, b, c, d with $ad-bc \neq 0$, the fraction $(az+b)/(cz+d)$ is deemed to be

$$\left. \begin{array}{ll} \dfrac{az+b}{cz+d} & \text{if } z \neq \infty, -d/c \\[2mm] a/c & \text{if } z = \infty \\[2mm] \infty & \text{if } z = -d/c \end{array} \right\} \quad \text{if } c \neq 0,$$

$$\left. \begin{array}{ll} (a/d)z + (b/d) & \text{if } z \neq \infty \\[4mm] \infty & \text{if } z = \infty \end{array} \right\} \quad \text{if } c = 0.$$

Since the composite of T_{abcd} and $T_{a'b'c'd'}$ is the Möbius transformation whose coefficients are given by the matrix product

$$\begin{pmatrix} a & c \\ b & d \end{pmatrix} \begin{pmatrix} a' & c' \\ b' & d' \end{pmatrix}$$

the map

$$\begin{pmatrix} a & c \\ b & d \end{pmatrix} \mapsto T_{abcd}$$

is a surjective homomorphism from $GL(2, \mathbb{C})$ to M. It is not hard to compute the kernel (see Exercise 2.12), which turns out to be the set of scalar matrices (as one might expect from the fact that $T_{abcd} = T_{\lambda a \lambda b \lambda c \lambda d}$ for any $\lambda \in \mathbb{C}^*$). Consequently

$$M \cong GL(2, \mathbb{C})/\mathbb{C}^* = PGL(2, \mathbb{C}).$$

Notice that, if $\lambda := (ad-bc)^{-1/2}$ and if $a' := \lambda a$, $b' := \lambda b$, $c' := \lambda c$, $d' := \lambda d$, then $T_{abcd} = T_{a'b'c'd'}$, where $a'd' - b'c' = 1$. Therefore

$$M = \{T_{abcd} \mid ad-bc = 1\},$$

and so

$$M \cong PGL(2, \mathbb{C}) \cong PSL(2, \mathbb{C}) \cong SL(2, \mathbb{C})/\{\pm I\}.$$

One part of this discussion has used a particular property of $\mathbb{C}$ (that $PGL(2, \mathbb{C}) = PSL(2, \mathbb{C})$ depends on the fact that every complex number has a square root) but the rest does not. You will find that for any field F the group

PGL(2, F) is naturally isomorphic with the group of all transformations

$$T_{abcd} : z \mapsto \frac{az+b}{cz+d}$$

of $F \cup \{\infty\}$. But more detailed analysis of the Möbius group in terms of the geometry of the Riemann Sphere (see, for example, Ahlfors, *Complex Analysis* (second edition), pp. 76–89, or Chapter 17 of this book) depends heavily on the special nature of $\mathbb{C}$.

Example 2.5 The orthogonal groups.   An $n \times n$ matrix U with real coefficients is said to be orthogonal if its inverse is its transpose, that is, if $UU' = I$. If U is orthogonal then so is U^{-1}, and if U, V are orthogonal then so is UV, as the calculation

$$(UV)(UV)' = UVV'U' = UIU' = UU' = I$$

shows. Thus the set

$$O(n) := O(n, \mathbb{R}) := \{U \in M_n(\mathbb{R}) \mid UU' = I\}$$

is a subgroup of GL($n, \mathbb{R}$). It is known as the n-dimensional (real) orthogonal group.

If $U \in O(n)$ then

$$(\det(U))^2 = \det(UU') = \det(I) = 1,$$

so $\det(U) = \pm 1$. We define the special orthogonal group by

$$SO(n) := SO(n, \mathbb{R}) := \{U \in M_n(\mathbb{R}) \mid UU' = I, \det(U) = 1\},$$

Since there are orthogonal matrices, such as

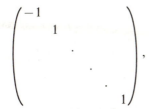

of determinant -1, the group SO(n) is a subgroup of index 2 in O(n).

The great importance of orthogonal matrices derives from their connection with rotations and reflections of euclidean n-space. It is for this reason that they occur not only in linear algebra and geometry, but also in mechanics, physics and chemistry. One can define the orthogonal groups O($n, \mathbb{C}$) and SO($n, \mathbb{C}$) over the field of complex numbers (and indeed, analogous groups O(n, F), SO(n, F) over any field whatsoever) but these have less geometrical importance—although they are still of great interest to algebraists. Over $\mathbb{C}$ the more important groups are the unitary groups that we introduce to you in Exercise 2.15.

Example 2.6 The affine groups.   Let F be an arbitrary field again, and let V be F^n, the vector space of $1 \times n$ row vectors over F. Given $A \in GL(n, F)$ and $b \in F^n$ we define a mapping $T_{A,b} \colon V \to V$ by

$$T_{A,b} \colon x \mapsto xA + b.$$

It is not hard to see that $T_{A,b} = T_{A',b'}$ if and only if $A = A'$ and $b = b'$. These mappings are known as the affine transformations of V; they constitute the affine group,

$$AGL(n, F) := \{T_{A,b} \mid A \in GL(n, F), b \in F^n\}.$$

You will find that $T_{A,b}$ is bijective, that

$$T_{A,b} T_{C,d} = T_{AC,bC+d},$$

and that

$$(T_{A,b})^{-1} = T_{A^{-1}, -bA^{-1}},$$

from which it is clear that $AGL(n, F)$ is a group, indeed a subgroup of $Sym(V)$.

The set

$$T(V) := \{T_{I,b} \mid b \in F^n\}$$

of all translations $x \mapsto x + b$ of V is a subgroup of $AGL(n, F)$, and it is normal. Indeed, it is the kernel of the homomorphism

$$AGL(n, F) \to GL(n, F)$$

given by $T_{A,b} \mapsto A$, and so

$$AGL(n, F)/T(V) \cong GL(n, F).$$

Notice that, if we identify $GL(n, F)$ with the group of mappings of the form $T_{A,0}$, then every affine transformation can be uniquely expressed as the composite of a linear transformation followed by a translation.

Example 2.7 The euclidean group.   In the special case where F is $\mathbb{R}$, the field of real numbers, there are subgroups G, G^+ of the affine group $AGL(n, \mathbb{R})$ that are of great importance in geometry and physics:

$$G := \{T_{A,b} \mid A \in O(n, \mathbb{R})\}$$
$$= \{x \mapsto xA + b \mid A \in O(n), b \in \mathbb{R}^n\},$$
$$G^+ := \{T_{A,b} \mid A \in SO(n, \mathbb{R})\}.$$

Distance in $\mathbb{R}^n$, defined by

$$d(x, y) := \left(\sum (x_i - y_i)^2\right)^{1/2} = ((x - y)(x - y)')^{1/2},$$

is preserved by transformations in G. For, if $x, y \in \mathbb{R}^n$ and $g \in G$, then

$$d(xg, yg) = (((xA+b)-(yA+b))\,((xA+b)-(yA+b))')^{1/2}$$

$$= (((x-y)A)\,((x-y)A)')^{1/2}$$

$$= ((x-y)AA'(x-y)')^{1/2}$$

$$= ((x-y)(x-y)')^{1/2} \quad \text{since} \quad A \in O(n)$$

$$= d(x, y).$$

In fact (see Exercise 2.19 for the 1-dimensional case, and Chapter 10) one can prove that G contains all transformations of $\mathbb{R}^n$ that preserve euclidean distance; that is, G is the group of 'isometries' of $\mathbb{R}^n$. It is sometimes called the euclidean group for this reason. The subgroup G^+ consists of those isometries that also preserve the 'orientation' of space. Its importance is that it consists of those transformations that are physically realisable by, for example, the motion of a rigid body.

Specimen solutions to selected exercises

The following specimen solutions, being of the nature of an extension of the text, are placed between the main body of the chapter and the exercises. The wise and conscientious reader will, however, try the exercises unaided before attempting to evaluate what we have offered.

EXERCISE 2.3   *Suppose that $n \geqslant 3$. Prove that every element of A_n can be written as a product of 3-cycles.*

Solution. By definition, every element of A_n can be written as a product $h_1 h_2 \ldots h_{2s}$ of an even number of transpositions h_i. Therefore it is sufficient to show that a product of two transpositions can be expressed as a product of 3-cycles. So let h_1, h_2 be transpositions. If $h_1 = h_2$ then $h_1 h_2 = 1$, which certainly is a product of 3-cycles. If h_1, h_2 have just one element in common then they may be taken to be (ab), (bc) with a, b, c distinct, and then $h_1 h_2 = (ab)(bc) = (acb)$. If h_1, h_2 have no elements in common then they are of the form $(ab), (cd)$ with a, b, c, d all different, and then $(ab)(cd) = (abc)(adc)$. Thus in all cases $h_1 h_2$ is a product of 3-cycles, as required.

EXERCISE 2.4   *The centre of a group G is defined by*

$$Z(G) := \{z \in G \mid zg = gz \text{ for all } g \in G\}.$$

(i)   *Find the centre of S_n.*
(ii)   *Find the centre of A_n.*

Solution. (i)   Since S_2 is cyclic of order 2 we have $Z(S_2) = S_2$. We shall show that $Z(S_n) = \{1\}$ in all other cases. This is obvious if $n = 1$, so we suppose that $n \geqslant 3$. Let x be a non-trivial permutation in S_n. Choose α such that $\alpha x \neq \alpha$, choose β different from both α and αx, and let h be the transposition $(\alpha\,\beta)$. Now xh maps α to αx, whereas hx maps α to βx which is certainly different from αx. Thus $xh \neq hx$ and so $x \notin Z(S_n)$. Therefore $Z(S_n) = \{1\}$ if $n \geqslant 3$.

(ii)   The alternating group A_n is trivial if $n = 1, 2$ and it is cyclic of order 3 if $n = 3$. Thus $Z(A_n) = A_n$ if $n \leqslant 3$. We shall show that $Z(A_n) = \{1\}$ if $n \geqslant 4$. Suppose that $n \geqslant 4$ and let x be a non-trivial element of A_n. Choose α such that $\alpha x \neq \alpha$. Let $\beta := \alpha x$ and choose γ different from α, β, and βx. Take h to be the 3-cycle $(\alpha\,\beta\,\gamma)$. Then $xh\colon \alpha \mapsto \gamma$, whereas $hx\colon \alpha \mapsto \beta x$, and so certainly $hx \neq xh$. Thus $x \notin Z(A_n)$ and, since this is true for every non-identity permutation $x \in A_n$, we see that $Z(A_n) = \{1\}$ for $n \geqslant 4$, as claimed.

EXERCISE 2.8   *Find the centre (see Exercise 2.4) of* GL(n, F).

Solution.   The general linear group GL(n, F) consists of all $n \times n$ invertible matrices over the field F. We will show that its centre is the group Λ consisting of all non-zero scalar matrices λI_n. Certainly a scalar matrix λI_n commutes with any matrix and so $\Lambda \leqslant Z(\mathrm{GL}(n, F))$. The crux of the matter therefore is to show that a non-scalar invertible matrix does not lie in $Z(\mathrm{GL}(n, F))$. So let $A = (a_{ij})$ and suppose that A is invertible but not scalar. Suppose first that some off-diagonal coefficient a_{rs} of A is non-zero. If E_{sr} is the matrix in which all the coefficients are 0 except the (s, r)-coefficient which is 1 then $X := I + E_{sr} \in \mathrm{GL}(n, F)$. The (s, s)-coefficient of AX is a_{ss}, but the (s, s)-coefficient of XA is $a_{ss} + a_{rs}$. Thus certainly $A \notin Z(\mathrm{GL}(n, F))$ in this case. Suppose then that A is a diagonal matrix but is not scalar. This means that there must exist r, s such that $a_{rr} \neq a_{ss}$. Let X be the permutation matrix corresponding to the transposition $(r\,s)$. Thus $x_{ii} = 1$ if $i \neq r, s$, $x_{rs} = x_{sr} = 1$, and $x_{ij} = 0$ for all other pairs i, j. We find that $X^{-1}AX$ is the same as A except that the (r, r) and (s, s)-coefficients have been interchanged. Thus $XA \neq AX$ and so $A \notin Z(\mathrm{GL}(n, F))$. Therefore $Z(\mathrm{GL}(n, F)) = \Lambda$, as claimed.

EXERCISE 2.17   *Recall that the affine group* AGL(n, F) *was defined as a subgroup of* Sym(F^n). *Prove*
(i)   *that* AGL$(1, 2) \cong S_2$;
(ii)   *that* AGL$(1, 3) \cong S_3$;
(iii)   *that* AGL$(1, 4) \cong A_4$;
(iv)   *that* AGL$(2, 2) \cong S_4$.

Solution.   Recall that AGL(n, F) is defined to be the group of all transformations $T_{A,b}$ of the space F^n of $1 \times n$ row vectors over the field F, where $A \in \mathrm{GL}(n, F)$, $b \in F^n$, and $T_{A,b}\colon x \mapsto xA + b$ for all $x \in F^n$. We write AGL(n, q) to denote AGL(n, F) when F is a finite field with q elements. Note that

$AGL(1, F)$ is the group of all transformations of F itself of the form $x \mapsto xa + b$ with $a, b \in F$ and $a \neq 0$, and therefore $|AGL(1, q)| = (q-1)q$.

(i) As we have just observed, $AGL(1, 2)$ is a subgroup of S_2 of order 2. Therefore $AGL(1, 2) = S_2$.

(ii) Similarly, $AGL(1, 3)$ is a subgroup of S_3 of order 6 and so $AGL(1, 3) = S_3$.

(iii) Now $AGL(1, 4)$ is a subgroup of S_4 of order 12. To see that $AGL(1, 4) = A_4$ we either quote the fact that A_4 is the only subgroup of S_4 of order 12, or we examine each member of $AGL(1, 4)$ and check that it is an even permutation. Knowing that $AGL(1, 4) \leqslant A_4$ tells us that equality holds since the orders are equal.

(iv) The order of $GL(2, 2)$ is easily found to be 6. Thus there are 6 possibilities for A and 4 possibilities for b, therefore 24 transformations $T_{A,b}$ in $AGL(2, 2)$. Consequently $AGL(2, 2)$ is a subgroup of S_4 of order 24, that is, $AGL(2, 2) = S_4$ as required.

EXERCISE 2.20   *If R is any commutative ring we define $GL(n, R)$ to be the group of units of the matrix ring $M_n(R)$. In particular, $GL(n, \mathbb{Z})$ is the group of $n \times n$ matrices with integer coefficients whose inverses also have integer coefficients.*

(i)   *Show that $GL(2, \mathbb{Z}) = \{A \in M_2(\mathbb{Z}) \mid \det(A) = \pm 1\}$. (More generally: $GL(n, \mathbb{Z}) = \{A \in M_n(\mathbb{Z}) \mid \det(A) = \pm 1\}$.)*

(ii)   *Let A be an element of finite order in $GL(2, \mathbb{Z})$. Show that $\operatorname{ord}(A)$ is $1, 2, 3, 4$, or 6. [You might find it helpful to consider the eigenvalues and the characteristic polynomial of A.]*

(iii)   *Exhibit elements of $GL(2, \mathbb{Z})$ that have orders $1, 2, 3, 4$, and 6, respectively.*

Solution. (i) If $A \in GL(n, \mathbb{Z})$ then there exists $B \in GL(n, \mathbb{Z})$ such that $AB = I_n$. Then $\det(A)\det(B) = 1$ and $\det(A), \det(B)$ are integers. It follows that $\det(A) = \det(B) = \pm 1$. Conversely, since $A^{-1} = \det(A)^{-1}\operatorname{adj}(A)$, if $A \in M_n(\mathbb{Z})$ and $\det(A) = \pm 1$ then A^{-1} is an integer matrix and so $A \in GL(n, \mathbb{Z})$.

(ii)   Let A be an element of finite order in $GL(2, \mathbb{Z})$, say

$$A = \begin{pmatrix} a & b \\ c & d \end{pmatrix}$$

and $A^n = I_2$. The eigenvalues λ_1, λ_2 of A satisfy the two equations $x^2 - (a+d)x + (ad - bc) = 0$ and $x^n = 1$. In particular, λ_1, λ_2 are roots of unity and therefore if they are real numbers then they must be ± 1, and $\operatorname{ord}(A)$ will be 1 or 2. If λ_1, λ_2 are non-real complex numbers then, since they are roots of a quadratic equation with real (indeed, integer) coefficients, they must be complex conjugates of each other. In this case therefore $\det(A) = 1$ and $\lambda_1 + \lambda_2 = 2\operatorname{Re}(\lambda_i) = a+d$. Since $\lambda_1 + \lambda_2$ is an integer but is also a sum of two complex roots of unity, it is 1, 0, or -1. If $\lambda_1 + \lambda_2 = 0$ then $\lambda_1 = -\lambda_2 = \pm i$ and so $\mathrm{ord}(A) = 4$. If $\lambda_1 + \lambda_2 = -1$ then λ_1, λ_2 are the complex cube roots of unity (this is easily visible from the geometry of the unit circle in the complex plane, or from the fact that the characteristic polynomial of A is $x^2 + x + 1$ in this case), and $\mathrm{ord}(A) = 3$. If $\lambda_1 + \lambda_2 = 1$ then λ_1, λ_2 are the complex 6th roots of unity $\frac{1}{2}(1 \pm \sqrt{-3})$, and $\mathrm{ord}(A) = 6$. Thus $\mathrm{ord}(A)$ is 1, 2, 3, 4, or 6, as required.

(iii)   Guided by the above analysis, which tells us the determinant $ad - bc$ and the trace $a + d$ of a matrix of each of the given orders we find that

$$\begin{pmatrix} 1 & 0 \\ 0 & 1 \end{pmatrix} \quad \text{has order 1;}$$

$$\begin{pmatrix} -1 & 0 \\ 0 & -1 \end{pmatrix} \quad \text{has order 2;}$$

$$\begin{pmatrix} 0 & 1 \\ -1 & 0 \end{pmatrix} \quad \text{has order 4;}$$

$$\begin{pmatrix} 0 & 1 \\ -1 & -1 \end{pmatrix} \quad \text{has order 3;}$$

$$\begin{pmatrix} 0 & 1 \\ -1 & 1 \end{pmatrix} \quad \text{has order 6;}$$

as required.

Exercises

2.1.   Recall from Exercise 1.16 the definition of support of a permutation: $\mathrm{supp}(f) := \{\omega \mid \omega f \neq \omega\}$. Prove
 (i) that $\mathrm{supp}(fg) \subseteq \mathrm{supp}(f) \cup \mathrm{supp}(g)$;
 (ii) that $\mathrm{supp}(x^{-1}fx) = (\mathrm{supp}(f))x$.
Deduce that the set

$$\mathrm{SF}(\Omega) := \{f \in \mathrm{Sym}(\Omega) \mid \mathrm{supp}(f) \text{ is finite}\}$$

of finitary permutations is a normal subgroup of $\mathrm{Sym}(\Omega)$.

2.2.   Let X be the set of permutations in S_n that can be expressed as a product of r transpositions, where r is a multiple of 3. Show that $X = S_n$. What is the analogous result when 3 is replaced by some other positive integer t?

2.3.   Suppose that $n \geqslant 3$. Prove that every element of A_n can be expressed as a product of 3-cycles.

2.4. The centre of a group G is defined by

$$Z(G) := \{z \in G \mid zg = gz \quad \text{for all} \quad g \in G\}.$$

   (i) Find the centre of S_n.
   (ii) Find the centre of A_n.

2.5. Let A be an abelian group. We define the corresponding generalized dihedral group to be the set of pairs

$$D(A) := \{(a, \varepsilon) \mid a \in A, \varepsilon = \pm 1\}$$

with multiplication given by

$$(a, \varepsilon)(b, \eta) = (ab^\varepsilon, \varepsilon\eta).$$

The unit is $(1, 1)$ and the inverse of (a, ε) is $(a^{-\varepsilon}, \varepsilon)$.
   (i) Prove that $D(A)$ is a group.
   (ii) We identify A with the set of all pairs in which $\varepsilon = +1$. Show that $A \triangleleft D(A)$. What is the index $|D(A):A|$?
   (iii) Show that every element of $D(A)$ not in A has order 2.
   (iv) What condition on A will ensure that $D(A)$ is not abelian?
   (v) What is the centre of $D(A)$?

2.6. The dihedral group D_n is defined to be $D(Z_n)$ (the group defined in Question 2.5 starting with A cyclic of order n). Show that $G \cong D_n$ if and only if G can be generated by elements a, b, where the order of a is n, the order of b is 2, and $b^{-1}ab = a^{-1}$.

2.7. (i) Prove that $S_3 \cong D_3$.
   (ii) Prove that the subgroup of S_4 generated by the permutations $(1\ 2\ 3\ 4)$ and $(1\ 3)$ is isomorphic to D_4.

2.8. Find the centre (see Question 2.4) of $GL(n, F)$.

2.9. Let $T_n := \{\theta \in F \mid \theta^n = 1\}$, the multiplicative group of nth roots of unity in the field F. Show that the group $Z \cap SL(n, F)$ of scalar matrices in the special linear group is isomorphic to T_n.

2.10. Exhibit a subgroup of $SL(n, q)$ whose order is $q^{n(n-1)/2}$.

2.11. (i) Prove that $PGL(n, \mathbb{C}) = PSL(n, \mathbb{C})$ for every n.
   (ii) Prove that $PGL(n, \mathbb{R}) \cong PSL(n, \mathbb{R})$ if and only if n is odd.
   (iii) Prove that if $n > 1$ then $PGL(n, \mathbb{Q}) \ncong PSL(n, \mathbb{Q})$.

2.12. Show that the Möbius transformation T_{abcd}, where $ad - bc \neq 0$, is the identity mapping of $\mathbb{C} \cup \{\infty\}$ if and only if $b = c = 0$ and $a = d$.

2.13. Show that a matrix U in $M_n(\mathbb{R})$ is orthogonal if and only if its rows form an orthonormal basis for $\mathbb{R}^n$.

2.14. (i) Show that $SO(2) = \left\{ \begin{pmatrix} \cos\theta & \sin\theta \\ -\sin\theta & \cos\theta \end{pmatrix} \middle| \theta \in \mathbb{R} \right\}$.

(ii) The so-called circle group is defined by

$$S^1 := \{z \in \mathbb{C} \mid |z| = 1\},$$

with multiplication of complex numbers as the group operation. Show that $SO(2) \cong S^1 \cong R/Z$, where R here denotes the additive group of real numbers and Z the subgroup of integers.

(iii) Show that

$$O(2) = \left\{ \begin{pmatrix} \cos\theta & \sin\theta \\ -\sin\theta & \cos\theta \end{pmatrix} \middle| \theta \in \mathbb{R} \right\} \cup \left\{ \begin{pmatrix} \cos\theta & \sin\theta \\ \sin\theta & -\cos\theta \end{pmatrix} \middle| \theta \in \mathbb{R} \right\}.$$

(iv) Show that $O(2) \cong D(S^1)$ (generalized dihedral group as defined in Question 2.5).

2.15. A matrix U in $M_n(\mathbb{C})$ is said to be unitary if it is invertible and its inverse is the transpose of its complex conjugate: $U\bar{U}' = I_n$. We define

$$U(n) := \{U \in M_n(\mathbb{C}) \mid U\bar{U}' = I_n\},$$

$$SU(n) := \{U \in U(n) \mid \det(U) = 1\}.$$

(i) Show that $U(n)$ is a subgroup of $GL(n, \mathbb{C})$ and that $SU(n)$ is a normal subgroup of $U(n)$. These are the unitary and the special unitary groups.

(ii) Prove that if $U \in U(n)$ then $\det(U)$ is a complex number of absolute value 1, and deduce that $U(n)/SU(n) \cong S^1$.

2.16. Prove the assertion made in the text, that affine transformations $T_{A,u}$, $T_{A',u'}$ are equal if and only if $A = A'$ and $u = u'$.

2.17. Recall that the affine group $AGL(n, F)$ was defined as a subgroup of $Sym(F^n)$. Prove
(i) that $AGL(1, 2) = S_2$;
(ii) that $AGL(1, 3) = S_3$;
(iii) that $AGL(1, 4) = A_4$;
(iv) that $AGL(2, 2) = S_4$.

2.18. Let $A := \{(a_{ij}) \in GL(n+1, F) \mid a_{11} = 1, a_{i1} = 0 \text{ for } i > 1\}$. Show that A is a subgroup of $GL(n+1, F)$ isomorphic to $AGL(n, F)$.

2.19. Show that the isometries of $\mathbb{R}^1$ are precisely the maps

$$x \mapsto \varepsilon x + a \quad \text{for all} \quad x \in \mathbb{R}^1$$

where $\varepsilon = \pm 1$ and $a \in \mathbb{R}$.

2.20. If R is any commutative ring we define $GL(n, R)$ to be the group of units of the matrix ring $M_n(R)$. In particular, $GL(n, \mathbb{Z})$ is the group of $n \times n$

matrices with integer coefficients whose inverses also have integer coefficients.

(i) Show that $GL(2, \mathbb{Z}) = \{A \in M_2(\mathbb{Z}) \mid \det(A) = \pm 1\}$. (More generally: $GL(n, \mathbb{Z}) = \{A \in M_n(\mathbb{Z}) \mid \det(A) = \pm 1\}$.)

(ii) Let A be an element of finite order in $GL(2, \mathbb{Z})$. Show that $\mathrm{ord}(A)$ is 1, 2, 3, 4, or 6. [You might find it helpful to consider the eigenvalues and the characteristic polynomial of A.]

(iii) Exhibit elements of $GL(2, \mathbb{Z})$ that have orders 1, 2, 3, 4, and 6, respectively.

Chapter 3

Actions of groups

We come now to the proper subject of the first part of this book: in this chapter we define exactly what we mean by an action of a group on a set and we introduce some of the appropriate language.

Let G be a group and Ω a set. By an *action* of G on Ω we mean a function $\mu: \Omega \times G \to \Omega$ such that for all $\omega \in \Omega$ and all $g, h \in G$

$$((\omega, g)\mu, h)\mu = (\omega, gh)\mu \qquad (3.1)$$

and for all $\omega \in \Omega$

$$(\omega, 1)\mu = \omega. \qquad (3.2)$$

In practice one finds it convenient to do without an explicit name for the action μ and we write ω^g for $(\omega, g)\mu$ whenever there is no risk of confusion. In this notation the axioms look rather simpler:

$$(\omega^g)^h = \omega^{gh}$$

$$\omega^1 = \omega$$

for all $\omega \in \Omega$ and all $g, h \in G$.

Much of group theory, particularly that part which deals with finite groups, originated from the study of permutation groups between about 1845 and 1900. (Although Galois had already understood a great deal about groups when he died, aged 20, in 1832, he seems to have been unable to communicate his discoveries very effectively at that early time. Some of the ideas can be traced even further back, to Cauchy in 1815 and to Ruffini, who worked between 1799 and 1814 expanding on suggestions made by Lagrange in 1770.) By a *permutation group* on the set Ω we mean simply a subgroup of $\text{Sym}(\Omega)$. Such a group G has a *natural action* μ on Ω defined by

$$(\omega, g)\mu := \omega g$$

where the right-hand side of the definition means, of course, the image of ω under the permutation g (bijective mapping on Ω to Ω). Our axioms for a group action then become the statements

$$(\omega g)h = \omega(gh),$$

$$\omega 1 = \omega,$$

and these are certainly true: the former by definition of composition of mappings (the group multiplication in G), and the latter by definition of the identity function on Ω.

More generally, any homomorphism $\rho: G \to \mathrm{Sym}(\Omega)$ gives rise to an action μ of G on Ω defined by

$$\omega^g := (\omega, g)\mu := \omega(g\rho)$$

for all $\omega \in \Omega$ and all $g \in G$. This really is an action because

$$(\omega^g)^h = (\omega(g\rho))(h\rho) = \omega((g\rho)(h\rho)) = \omega((gh)\rho) = \omega^{gh}$$

for all $\omega \in \Omega$ and all $g, h \in G$, and

$$\omega^1 = \omega(1\rho) = \omega^1 = \omega.$$

Conversely, suppose that μ is an action of G on Ω. For a fixed member g of G consider the mapping $\rho_g : \Omega \to \Omega$ defined by

$$\rho_g : \omega \mapsto \omega^g.$$

This is bijective: it has a two-sided inverse, namely $\rho_{g^{-1}}$, because, for all $\omega \in \Omega$,

$$\omega \rho_g \rho_{g^{-1}} = (\omega^g)^{g^{-1}} = \omega^{(gg^{-1})} = \omega^1 = \omega,$$

and similarly $\omega \rho_{g^{-1}} \rho_g = \omega_g$, which shows that

$$\rho_g \rho_{g^{-1}} = \rho_{g^{-1}} \rho_g = 1.$$

In this way each element of G acts as a permutation of Ω. Furthermore, the map

$$\rho: G \to \mathrm{Sym}(\Omega), \qquad \rho: g \mapsto \rho_g$$

is a homomorphism. This is the substance of our first axiom describing actions: for all $\omega \in \Omega$ we have

$$\omega \rho_{gh} = \omega^{gh} = (\omega^g)^h = (\omega\rho_g)\rho_h = \omega(\rho_g\rho_h),$$

and so for all $g, h \in G$ we have

$$\rho_{gh} = \rho_g\rho_h.$$

We call a homomorphism $\rho: G \to \mathrm{Sym}(\Omega)$ a *permutation representation* of G, or a *representation of G as a group of transformations* of Ω. What we have just shown is that every such representation gives rise to an action of G on Ω and that conversely every action gives rise to a permutation representation. We leave it to the reader to check that the passage from μ to ρ and back to an action returns the original action μ; and that passage from ρ to μ and then from μ to its associated permutation representation returns the original homomorphism ρ. To summarise, we have

LEMMA 3.1: *There is a one–one correspondence between actions of the group G on the set Ω and representations of G by permutations of Ω.*

The permutation representation ρ, or the action μ, is said to be *faithful* if Ker $\rho = \{1\}$, that is, if ρ is a monomorphism. In this case we say that G acts faithfully on Ω, we think of G as being identified with its image under ρ, a subgroup of Sym(Ω), and we are essentially back with the important special case of permutation groups. At the other extreme we have the so-called *trivial* action of G on Ω where $\omega^g = \omega$ for all $\omega \in \Omega$ and all $g \in G$. Here of course $G\rho = \{1\} \leqslant$ Sym(Ω) and Ker $\rho = G$.

A set Ω with an action of some group on it is known generally as a *G-space* or a *G-set*. Here 'G-space' is a generic term, really an abbreviation of 'group-space'. It means a triple (Ω, X, μ) where Ω is a set, X is a group and μ is an action of X on Ω. But the term 'G-space' is also often used in a more specific sense: if we are working in a context where a group G is given then by a *G-space* we mean a pair (Ω, μ) where Ω is a set and μ an action of G on Ω. When the term is used in this sense, G is used as a variable, and so we can talk of an *X-space* if the group is X, and so on. As usual, we use the full form (Ω, G, μ) of the notation for a G-space only when precision is necessary. Most of the time it is briefer and clearer to talk of the G-space Ω (if the action μ is implicitly known), or of the *transformation group* (Ω, G) (when the action μ is a natural action of G on Ω, as it always will be in geometrical contexts).

One wants to think of the G-spaces (Ω, G, μ) and (Ω', G', μ') as being essentially the same if Ω can be identified with Ω' and G with G' in such a way that μ and μ' become the same. Formally, we define an *equivalence* of (Ω, G, μ) with (Ω', G', μ') to be a pair (θ, ϕ) of functions where $\theta: \Omega \to \Omega'$ is a bijection, $\phi: G \to G'$ is an isomorphism, and

$$(\omega^g)\theta = (\omega\theta)^{g\phi}$$

for all $\omega \in \Omega$ and all $g \in G$. One checks very easily that $(1, 1)$ is an equivalence of (Ω, G, μ) with itself; that if (θ, ϕ) is an equivalence then so is (θ^{-1}, ϕ^{-1}); and that if

$$(\theta, \phi): (\Omega, G, \mu) \to (\Omega', G', \mu')$$

$$(\theta', \phi'): (\Omega', G', \mu') \to (\Omega'', G'', \mu'')$$

are equivalences then also

$$(\theta\theta', \phi\phi'): (\Omega, G, \mu) \to (\Omega'', G'', \mu'')$$

is an equivalence. Consequently equivalence is an equivalence relation in the usual sense on the class of all G-spaces.

If we are working with actions of the fixed group G then an equivalence of Ω with Ω' is a pair (θ, ϕ) as above, where now ϕ is an automorphism of G. Here is a particularly useful example of an equivalence.

LEMMA 3.2:   *Let Ω be a G-space and let t be a fixed element of G. The pair*
(θ, ϕ) *where*

$$\theta: \omega \mapsto \omega^t,$$

$$\phi: g \mapsto t^{-1}gt,$$

is an equivalence of Ω with itself.

We shall call these *inner self-equivalences* of Ω.

Proof.   Certainly θ is a bijection on Ω to Ω. Likewise ϕ is an automorphism of
G (it is the so-called 'inner' automorphism induced by t) because

$$(t^{-1}gt)(t^{-1}ht) = t^{-1}ght.$$

The main condition to be checked is that $(\omega^g)\theta = (\omega\theta)^{g\phi}$ for all $\omega \in \Omega$, $g \in G$,
and this is embodied in the equation

$$(\omega^g)^t = (\omega^t)^{t^{-1}gt},$$

which proves the result.

In this context of actions of the fixed group G a certain special kind of
equivalence in which the group automorphism is the identity is of particular
importance and is known as a *G-isomorphism*. Thus $\theta: \Omega \to \Omega'$ is a *G-
isomorphism* if it is a bijection such that

$$(\omega^g)\theta = (\omega\theta)^g$$

for all $\omega \in \Omega$ and $g \in G$. Again, we leave to the reader the very easy task of
checking that $1: \Omega \to \Omega$ is always a G-isomorphism, that if $\theta: \Omega \to \Omega'$ is a G-
isomorphism then so is $\theta^{-1}: \Omega' \to \Omega$, and that a composite of G-isomorphisms
is a G-isomorphism.

More generally, one defines a *G-morphism* to be a function $\theta: \Omega \to \Omega'$ for
which

$$(\omega^g)\theta = (\omega\theta)^g$$

for all $\omega \in \Omega$, $g \in G$. This is the appropriate notion of morphism when we
consider G-spaces (still for the fixed group G, remember) as being a kind of
algebraic structure in which the elements of G are thought of as unary
operators (rather like scalar multiplications in vector-space theory). From this
point of view the theory of G-spaces has much in common with other algebraic
theories. For example, one defines a *G-subspace* of Ω to be a subset Ω' that is
closed under the operations in G: so, if $\omega \in \Omega'$ then $\omega^g \in \Omega'$ for all $g \in G$. In
Chapters 5, 6, 7, where we study G-spaces in some detail, our development of
the theory is guided partly by this philosophy.

Specimen solutions to selected exercises

The following specimen solutions, being of the nature of an extension of the text, are placed between the main body of the chapter and the exercises. The wise and conscientious reader will, however, try the exercises unaided before attempting to evaluate what we have offered.

EXERCISE 3.4 (i)   *Let $\rho:G \to \mathrm{Sym}(\Omega)$ and $\rho':G' \to \mathrm{Sym}(\Omega')$ be homomorphisms (i.e. permutation representations). Show that the corresponding actions are equivalent if and only if there exist a bijection $\theta:\Omega \to \Omega'$ and an isomorphism $\phi:G \to G'$ such that $g\phi\rho' = \theta^{-1}(g\rho)\theta$ for all g in G.*

(ii)   *Suppose now that $G,\, H \leqslant \mathrm{Sym}(\Omega)$. Show that the corresponding spaces (Ω, G) and (Ω, H) are equivalent if and only if H is conjugate to G in $\mathrm{Sym}(\Omega)$.*

Solution. (i)   The G-space corresponding to the permutation representation ρ is (Ω, G, μ) where μ is defined by $(\omega, g)\mu := \omega^g := \omega(g\rho)$. Let μ' be the action of G' on Ω' corresponding to the permutation representation ρ'. By definition the spaces (Ω, G, μ), (Ω', G', μ') are equivalent if and only if there exist bijections $\theta:\Omega \to \Omega'$, $\phi:G \to G'$ such that ϕ is a group isomorphism and $(\omega^g)\theta = (\omega\theta)^{g\phi}$ for all $\omega \in \Omega$ and all $g \in G$. Now the following are equivalent:

$(\omega^g)\theta = (\omega\theta)^{g\phi}$ for all $\omega \in \Omega$ and all $g \in G$;

$(\omega(g\rho))\theta = ((\omega\theta)(g\phi\rho'))$ for all $\omega \in \Omega$ and all $g \in G$ [by definition of the actions];

$\omega((g\rho)\theta) = \omega(\theta(g\phi\rho'))$ for all $\omega \in \Omega$ and all $g \in G$;

$(g\rho)\theta = \theta(g\phi\rho')$ for all $g \in G$ [by definition of equality of functions];

$\theta^{-1}(g\rho)\theta = g\phi\rho'$ for all $g \in G$.

Thus the spaces (Ω, G, μ), (Ω', G', μ') are equivalent if and only if there exist a bijection $\theta:\Omega \to \Omega'$ and an isomorphism $\phi:G \to G'$ such that $g\phi\rho' = \theta^{-1}(g\rho)\theta$ for all g in G, as required.

(ii) Now we have $G,\, H \leqslant \mathrm{Sym}(\Omega)$, so that in this case the permutation representations ρ, ρ' are simply the inclusion maps $G \to \mathrm{Sym}(\Omega)$, $H \to \mathrm{Sym}(\Omega)$. Therefore the spaces (Ω, G), (Ω, H) are equivalent if and only if there exist a bijection $\theta:\Omega \to \Omega$ and an isomorphism $\phi:G \to H$ such that $g\phi = \theta^{-1}g\theta$ for all $g \in G$. Being a bijection θ is in fact a member of $\mathrm{Sym}(\Omega)$, and the equation says that ϕ is conjugation by θ (restricted to G). Thus (Ω, G), (Ω, H) are equivalent if and only if $G,\, H$ are conjugate subgroups of $\mathrm{Sym}(\Omega)$, as claimed.

EXERCISE 3.7   *Let $\Omega := \{1, 2, 3\}$ and $G := Z_2$.*
(i)   *How many different actions are there of G on Ω?*
(ii)   *Up to isomorphism how many different G-spaces do these give?*

Solution. (i) We know that each action gives rise to a permutation representation $\rho: G \to \text{Sym}(\Omega)$ and conversely each permutation representation gives rise to a unique action. Now $Z_2 = \{0, 1\}$ with the operation of addition modulo 2 and any homorphism $\rho: Z_2 \to \text{Sym}(\Omega)$ is determined by the image of the generator 1. Since 1 has order 2 as element of Z_2 it must be mapped by a homomorphism ρ to an element of order 1 or 2; conversely, given any element of the order 1 or 2 in $\text{Sym}(\Omega)$ there is a homomorphism $\rho: Z_2 \to \text{Sym}(\Omega)$ mapping 1 to that element. The relevant elements of $\text{Sym}(\Omega)$ when $\Omega = \{1, 2, 3\}$ are the identity and the three transpositions (1 2), (2 3), (1 3). Therefore there are 4 different actions of Z_2 on Ω.

(ii) It is easy to see that the action in which $\omega^g = \omega$ for all $\omega \in \Omega$ and all $g \in G$ (known as the trivial action) gives a G-space that is not isomorphic to any other G-space. The other three actions, however, all give isomorphic Z_2-spaces. If the homomorphisms ρ_1, ρ_2, ρ_3 map the generator 1 of Z_2 to (1 2), (2 3), (1 3), respectively, giving rise to actions μ_1, μ_2, μ_3, and $\theta: \Omega \to \Omega$ is the permutation (1 2 3) then, as one checks easily from first principles, θ is an isomorphism $(\Omega, \mu_1) \to (\Omega, \mu_2)$ and an isomorphism $(\Omega, \mu_2) \to (\Omega, \mu_3)$. Thus up to isomorphism there are just 2 different Z_2-spaces of size 3.

EXERCISE 3.10   *Let G be a group and let Ω_1, Ω_2 be G-spaces (with actions μ_1, μ_2). We define*

$$\Omega_1 + \Omega_2 := (\Omega_1 \times \{1\}) \cup (\Omega_2 \times \{2\})$$

and μ by

$$\mu: ((\omega, i), g) \mapsto ((\omega, g)\mu_i, i)$$

(for $i := 1, 2, \omega \in \Omega_i$ and $g \in G$). Show that this makes Ω into a G-space (the sum of Ω_1 and Ω_2).

Similarly, we define $\mu: (\Omega_1 \times \Omega_2) \times G \to \Omega_1 \times \Omega_2$ by $(\omega_1, \omega_2)^g := (\omega_1^g, \omega_2^g)$. Show that this is an action (giving the product space).

Solution.   Certainly μ is a map $(\Omega_1 + \Omega_2) \times G \to \Omega_1 + \Omega_2$. We need to check the conditions specifying an action. Since μ_i is an action we know that $(\omega, 1)\mu_i = \omega$ for all $\omega \in \Omega_i$, and that $((\omega, g)\mu_i, h)\mu_i = (\omega, gh)\mu_i$ for all $\omega \in \Omega_i$, $g, h \in G$. Elements of $\Omega_1 + \Omega_2$ are of one and only one of the forms $(\omega, 1)$ with $\omega \in \Omega_1$, $(\omega, 2)$ with $\omega \in \Omega_2$. Now $((\omega, i), 1)\mu = ((\omega, 1)\mu_i, i) = (\omega, i)$ (where $i \in \{1, 2\}$ and $\omega \in \Omega_i$), so the first condition for a G-space is satisfied. And if $g, h \in G$ then

$$\begin{aligned} (((\omega, i), g)\mu, h)\mu &= (((\omega, g)\mu_i, i), h)\mu && \text{[by definition of } \mu\text{]} \\ &= (((\omega, g)\mu_i, h)\mu_i, i) && \text{[by definition of } \mu\text{]} \\ &= ((\omega, gh)\mu_i, i) && \text{[since } \mu_i \text{ is an action]} \\ &= ((\omega, i), gh)\mu && \text{[by definition of } \mu\text{]} \end{aligned}$$

Thus μ is indeed an action.

Similarly, for the product space we have

$$(\omega_1, \omega_2)^1 = (\omega_1^1, \omega_2^1) = (\omega_1, \omega_2)$$

for all $(\omega_1, \omega_2) \in \Omega_1 \times \Omega_2$, and

$$
\begin{aligned}
((\omega_1, \omega_2)^g)^h &= (\omega_1^g, \omega_2^g)^h && \text{[by definition of } \mu] \\
&= ((\omega_1^g)^h, (\omega_2^g)^h) && \text{[by definition of } \mu] \\
&= (\omega_1^{(gh)}, \omega_2^{(gh)}) && \text{[since } \mu_1, \mu_2 \text{ are actions]} \\
&= (\omega_1, \omega_2)^{(gh)} && \text{[by definition of } \mu]
\end{aligned}
$$

for all $(\omega_1, \omega_2) \in \Omega_1 \times \Omega_2$ and all $g, h \in G$. Thus μ as given does satisfy the conditions for an action, as required.

Exercises

3.1.   In the definition of an action does axiom (3.2) follow from axiom (3.1)?

3.2.   Prove from the axioms for an action that if $\alpha^g = \beta$ then $\alpha = \beta^{g^{-1}}$.

3.3.   (i) Show how a one–one correspondence between sets Ω and Ω' gives rise to an equivalence $(\Omega, \mathrm{Sym}(\Omega))$ with $(\Omega', \mathrm{Sym}(\Omega'))$.
   (ii) Do the same when Ω, Ω' are finite for the permutation groups $(\Omega, \mathrm{Alt}(\Omega))$ and $(\Omega', \mathrm{Alt}(\Omega'))$.

3.4.   (i) Let $\rho: G \to \mathrm{Sym}(\Omega)$ and $\rho': G' \to \mathrm{Sym}(\Omega')$ be homomorphisms (i.e. permutation representations). Show that the corresponding actions are equivalent if and only if there exist a bijection $\theta: \Omega \to \Omega'$ and an isomorphism $\phi: G \to G'$ such that

$$g\phi\rho' = \theta^{-1}(g\rho)\theta$$

for all g in G.
   (ii) Suppose now that $G, H \leqslant \mathrm{Sym}(\Omega)$. Show that the corresponding spaces (Ω, G) and (Ω, H) are equivalent if and only if H is conjugate to G in $\mathrm{Sym}(\Omega)$.

3.5.   Suppose that $G \leqslant \mathrm{Sym}(\Omega)$.
   (i) Show that $\mathrm{Aut}(\Omega)$, the group of G-automorphisms of Ω, is the centraliser of G in $\mathrm{Sym}(\Omega)$. (If X is a subgroup of Y the centraliser of X in Y is defined by

$$C_Y(X) := \{y \in Y \mid xy = yx \text{ for all } x \in X\}.)$$

   (ii) Identify the group of 'self-equivalences' of Ω as G-space in similar terms.

3.6.   We speak of the kernel of an action or of a G-space Ω meaning the kernel of the corresponding permutation representation $\rho: G \to \mathrm{Sym}(\Omega)$.
   (i) Show that isomorphic G-spaces have the same kernel.
   (ii) Is it true that equivalent G-spaces always have the same kernel?

3.7. Let $\Omega := \{1, 2, 3\}$ and $G := Z_2$.
  (i) How many different actions are there of G on Ω?
  (ii) Up to isomorphism how many different G-spaces do these give?

3.8. Let $\Omega := \{1, 2\}$ and $G := Z_2 \times Z_2$.
  (i) Show that there are 4 different actions of G on Ω.
  (ii) One of these is the trivial action. The other 3 give G-spaces $\Omega_1, \Omega_2,$
    Ω_3, say. Show that no two of them are isomorphic, but that $\Omega_1, \Omega_2,$
    Ω_3 are all equivalent.

3.9. Let G_1, G_2 be groups, let Ω_1 be a G_1-space and Ω_2 a G_2-space, and let
  $G := G_1 \times G_2$.
  (i) Define

$$\Omega := (\Omega_1 \times \{1\}) \cup (\Omega_2 \times \{2\})$$

  and $\mu : \Omega \times G \to \Omega$ by

$$(\omega, i)^{(g_1, g_2)} := (\omega^{g_i}, i).$$

  Show that this gives an action of G on Ω.
  (ii) Define $\Omega := \Omega_1 \times \Omega_2$ and $\mu : \Omega \times G \to \Omega$ by

$$(\omega_1, \omega_2)^{(g_1, g_2)} := (\omega_1^{g_1}, \omega_2^{g_2}).$$

  Show that this defines an action.
  [These spaces are known as the *disjoint union* (or *sum*) and the *product*
  space, respectively. Notice particularly that they are G-spaces for the
  direct product $G_1 \times G_2$.]

3.10. Let G be a group and let Ω_1, Ω_2 be G-spaces (with actions μ_1, μ_2). We
  define

$$\Omega_1 + \Omega_2 := (\Omega_1 \times \{1\}) \cup (\Omega_2 \times \{2\})$$

  and μ by

$$\mu : ((\omega, i), g) \mapsto ((\omega, g)\mu_i, i)$$

  (for $i := 1, 2$, $\omega \in \Omega_i$ and $g \in G$). Show that this makes $\Omega_1 + \Omega_2$ into a
  G-space (the *sum* of Ω_1 and Ω_2).
  Similarly, we define $\mu : (\Omega_1 \times \Omega_2) \times G \to \Omega_1 \times \Omega_2$ by

$$(\omega_1, \omega_2)^g := (\omega_1^g, \omega_2^g).$$

  Show that this is an action (giving the *product* space).

3.11. Let Ω be a G-space. We define

$$\Omega^k := \{(\omega_1, \omega_2, \ldots, \omega_k) \mid \omega_i \in \Omega\},$$
$$\Omega^{(k)} := \{(\omega_1, \omega_2, \ldots, \omega_k) \mid \omega_i \in \Omega, \text{ all different}\},$$
$$\Omega^{\{k\}} := \{E \mid E \subseteq \Omega, |E| = k\},$$
$$\mathscr{P}(\Omega) := \{E \mid E \subseteq \Omega\}.$$

  Show how each of these is a G-space in a natural way. Show also that $\Omega^{(k)}$
  is a G-subspace of Ω^k and that $\Omega^{\{k\}}$ is a G-subspace of $\mathscr{P}(\Omega)$.

Chapter 4

A garden of *G*-spaces

This chapter is intended to exhibit a variety of group actions that illustrate the concepts introduced in Chapter 3, and to sow some seeds that will later provide food for thought. Although we limit ourselves here to little more than sketches, each of the examples is of some importance in some branch of mathematics, and will be treated in greater detail in later chapters.

Example 4.1 The regular representation. Let G be any group, let $\Omega := G$, and define $\mu: \Omega \times G \to \Omega$ by

$$\mu: (\omega, g) \mapsto \omega g$$

where ωg on the right of this formula denotes the product of ω and g as group elements. That this prescription does give an action of G is an immediate consequence of the first two group axioms. This is the action of G on itself by what one often calls 'right translation'. The G-space Ω is known as the (right) regular space, and the corresponding homomorphism $G \to \mathrm{Sym}(G)$ as the regular permutation representation of G, or, sometimes, as the Cayley representation. Notice that this is a faithful action. Consequently G is isomorphic with a subgroup of $\mathrm{Sym}(G)$, and we have Cayley's theorem that every finite group with n elements is isomorphic to a subgroup of S_n.

Example 4.2 Coset spaces. Let G be any group. For a subgroup H of G we define

$$\cos(G:H) := \{Hx \mid x \in G\}$$

and

$$(Hx)^g := Hxg$$

for $g \in G$. It is straightforward to check that this does define an action of G on the set of cosets, an action that we again call right translation. The G-space $\cos(G:H)$ is known as the coset space of H in G.

For later use we work out an item of information about $\cos(G:H)$.

LEMMA 4.2.1: *Let K be the kernel of the permutation representation associated with the action of G on $\cos(G:H)$. Then*
(i)   *K is the greatest normal subgroup of G that is contained in H; and*
(ii) $K = \bigcap_{x \in G} x^{-1}Hx$.

Notice that, even if G and H are infinite, the description 'greatest normal subgroup of G contained in H' makes sense: the subgroup of G generated by the union of all the normal subgroups that lie in H is obviously contained in H, is normal in G, and, by construction, contains every normal subgroup of G that is contained in H. We shall, however, also prove as part of the lemma that K has these properties.

Proof. From its definition,

$$K = \{g \in G \mid Hxg = Hx \text{ for all } x \in G\}.$$

Now, as K is the kernel of a homomorphism $\rho: G \to \text{Sym}(\Omega)$ (where $\Omega := \cos(G\!:\!H)$), we know that $K \trianglelefteq G$. Also, if $g \in K$ then, in particular, $Hg = H$, so $g \in H$: thus $K \leqslant H$. And, if $L \trianglelefteq G$ and $L \leqslant H$ then, for all $g \in L$ we have

$$Hxg = H(xgx^{-1})x = Hx$$

(since $xgx^{-1} \in L$), and so $L \leqslant K$. We have shown that K contains every normal subgroup of G that is contained in H, and this proves (i).

Statement (ii) can be proved from (i), or directly as follows:

$$
\begin{aligned}
K &= \{g \in G \mid Hxg = Hx \text{ for all } x \in G\} \\
&= \{g \in G \mid Hxgx^{-1} = H \text{ for all } x \in G\} \\
&= \{g \in G \mid xgx^{-1} \in H \text{ for all } x \in G\} \\
&= \{g \in G \mid g \in x^{-1}Hx \text{ for all } x \in G\} \\
&= \bigcap_{x \in G} x^{-1}Hx.
\end{aligned}
$$

The following special case of Lemma 4.2.1 is often useful.

COROLLARY 4.2.2: *Let G be a group, $H \leqslant G$ and $|G\!:\!H| = n$ (finite). There is a normal subgroup K of G such that $K \leqslant H$ and $|G\!:\!K|$ divides $n!$.*

For, G/K in the lemma is isomorphic to a subgroup of S_n. Therefore it is finite and, by Lagrange's Theorem, its order divides $n!$.

We hope it is not already too late to offer some words of warning about notation and terminology. First, remember that, although the description $\{Hx \mid x \in G\}$ seems to suggest that $\cos(G\!:\!H)$ contains as many members as there are elements in G, that is only exceptionally the case: cosets Hx, Hy may be the same even when x, y are different. Figure 4.1 should remind you what happens.

Secondly, if we had to qualify the term 'coset space' we should call $\cos(G\!:\!H)$ as defined above a 'right coset space' because G acts by right translation. Also, this is consistent with our terminology for cosets—indeed, it is one reason why

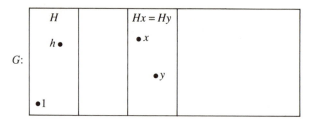

Fig. 4.1. Cosets of H in G.

we call cosets Hx right cosets. (Authors appear to have been almost equally divided up to now: some call these right cosets because the coset representative appears on the right, others call them left cosets because the relevant subgroup appears on the left.) There is a reasonably natural definition of the left coset space of H in G (see Exercise 4.4) but this gives nothing really new since the map $Hx \mapsto x^{-1}H$ is a G-isomorphism of the right coset space to the left coset space.

Third, we have to confess that our notation $\cos(G:H)$ is by no means standard. But we do not know of a notation that *is* generally accepted. In some contexts one sees G/H used, but as this is universally accepted as denoting the quotient group (and carries the suggestion that $H \trianglelefteq G$), and is frequently needed in this sense when coset spaces are also under discussion, we prefer to avoid it. The notations $H \setminus G$, $G:H$ and $(G:H)$ have also been used. But each has disadvantages. We have chosen $\cos(G:H)$ because it is still quite unambiguous and because it seems adequately suggestive.

Example 4.3 The symmetric group acting on $F[x_1, \ldots, x_n]$. The symmetric group S_n, thought of as the group of all permutations of $\{1, 2, \ldots, n\}$, acts on the set $F[x_1, x_2, \ldots, x_n]$ of polynomials with coefficients from a field F by permuting the variables: for a polynomial p and permutation σ we define p^σ by

$$p^\sigma(x_1, x_2, \ldots, x_n) := p(x_{1\sigma^{-1}}, x_{2\sigma^{-1}}, \ldots, x_{n\sigma^{-1}}).$$

Note that σ^{-1} is needed in the subscripts to ensure that $(p^\sigma)^\tau = p^{\sigma\tau}$. Of course $F[x_1, x_2, \ldots, x_n]$ is a ring, and S_n acts as a group of ring automorphisms:

$$(p_1 + p_2)^\sigma = p_1^\sigma + p_2^\sigma$$
$$(p_1 p_2)^\sigma = p_1^\sigma p_2^\sigma$$

for any polynomials p_1, p_2 and any permutation σ. Between 1845 and about 1870 finite groups were thought of almost exclusively as groups of 'substitutions' of variables in this way.

Example 4.4 Linear groups and vector spaces. Let F be a field and let F^n be the vector space of all $1 \times n$ row vectors over F. The group $\mathrm{GL}(n, F)$ of all $n \times n$

invertible matrices over F acts on F^n by right multiplication:

$$(x, A) \mapsto xA$$

for $x \in F^n$, $A \in \mathrm{GL}(n, F)$. Axiom (3.1) defining group actions is true because matrix multiplication is associative, and axiom (3.2) holds because the unit matrix I is an identity element for matrix multiplication.

   In this example the group $\mathrm{GL}(n, F)$ acts as vector space automorphisms (i.e. invertible linear transformations) as the equation

$$(\lambda x + \mu y)A = \lambda xA + \mu yA$$

(for $\lambda, \mu \in F$, $x, y \in F^n$, $A \in \mathrm{GL}(n, F)$) says, and in fact all vector space automorphisms of F^n are obtained this way. As we hope the reader knows, this example is canonical; it provides a standard model for n-dimensional vector spaces over F and their automorphism groups. Let V be any n-dimensional vector space over F. Choose a basis $v_1, v_2, \ldots, v_n$ for V. With respect to this basis each element $v \in V$ has a unique expression as a linear combination

$$v = x_1 v_1 + x_2 v_2 + \cdots + x_n v_n$$

for suitable scalars $x_1, x_2, \ldots, x_n$ in F. Consequently the map

$$\theta: V \to F^n, \quad v \mapsto (x_1, x_2, \ldots, x_n)$$

is well-defined, and of course it is a bijection. Moreover, if α is a linear transformation of V then

$$v_i \alpha = \sum_j a_{ij} v_j$$

for uniquely determinable scalars a_{ij}, and if α is invertible then the matrix (a_{ij}) lies in $\mathrm{GL}(n, F)$. The map

$$\phi: \alpha \mapsto (a_{ij})$$

is an isomorphism

$$\phi: \mathrm{Aut}(V) \to \mathrm{GL}(n, F),$$

and the pair (θ, ϕ) is an equivalence of the transformation group $(V, \mathrm{Aut}(V))$ with the G-space $(F^n, \mathrm{GL}(n, F))$. The proof is a routine calculation: if $v \in V$ and $\alpha \in \mathrm{Aut}(V)$, then

$$(v\alpha)\theta = \left(\left(\sum_i x_i v_i \right) \alpha \right) \theta$$

$$= \left(\sum_i x_i (v_i \alpha) \right) \theta$$

$$= \left(\sum_i x_i \left(\sum_j a_{ij} v_j \right) \right) \theta$$

$$= \left(\sum_j \left(\sum_i x_i a_{ij} \right) v_j \right) \theta$$

$$= \left(\sum_i x_i a_{i1}, \sum_i x_i a_{i2}, \ldots, \sum_i x_i a_{in} \right),$$

while

$$(v\theta)(\alpha\phi) = (x_1, x_2, \ldots, x_n) \begin{pmatrix} a_{11} & a_{12} & \cdots & a_{1n} \\ & & \cdot & \\ & \cdot & \cdot & \cdot \\ & & \cdot & \\ a_{n1} & a_{n2} & \cdots & a_{nn} \end{pmatrix}$$

$$= \left(\sum_i x_i a_{i1}, \sum_i x_i a_{i2}, \ldots, \sum_i x_i a_{in} \right).$$

Of course there is no need to limit oneself to automorphisms of V and invertible matrices. Once a basis has been chosen, the passage to coefficients and matrices provides an equivalence of V and the ring of all linear transformations of it with F^n acted on by $M_n(F)$.

Example 4.5 Projective groups and projective spaces.   Again, let F be any field and n a positive integer. We define

$$\Sigma_k := \{ W \mid W \leqslant F^n \text{ and dim } W = k+1 \}$$

for $0 \leqslant k \leqslant n-1$. Since $GL(n, F)$ acts on F^n as a group of linear transformations elements of $GL(n, F)$ map vector subspaces to vector subspaces and preserve dimension. Therefore $GL(n, F)$ acts on Σ_k in a natural way. Now a scalar matrix aI_n fixes every subspace of F^n (setwise, not necessarily elementwise), and so the group Z of non-zero scalar matrices lies in the kernel of the action. Consequently $PGL(n, F)$ acts naturally on Σ_k, for remember that PGL is defined to be the quotient group GL/Z. Geometrically Σ_k is the set of so-called 'k-flats' in the $(n-1)$-dimensional projective geometry $PG(n-1, F)$. These geometries will reappear from time to time in later chapters of this book. For the moment we consider just the case $n = 2$.

The 'projective line' $PG(1, F)$ is defined to be the G-space consisting of the set of 1-dimensional subspaces of F^2 with the natural action of $PGL(2, F)$ on it. If W is a 1-dimensional subspace then either W is spanned by $(z, 1)$ for some z in F, or $W = \{(a, 0)\,|\,a \in F\}$, in which case it is spanned by $(1, 0)$. So we identify $PG(1, F)$ with $F \cup \{\infty\}$ by the map

$$\{(az, a)\,|\,a \in F\} \longmapsto z$$

$$\{(a, 0)\,|\,a \in F\} \longmapsto \infty,$$

which associates a single 'coordinate' with each point of $PG(1, F)$. The coordinate of the subspace of F^2 spanned by (x, y) is x/y with the convention that $x/0 = \infty$ if $x \neq 0$; we may think of it as $1/m$, where m is the 'slope' of the line through the origin in F^2 that represents our 1-dimensional subspace pictorially. Consider now the action of the non-singular matrix $\begin{pmatrix} a & c \\ b & d \end{pmatrix}$. Since

$$(x, y)\begin{pmatrix} a & c \\ b & d \end{pmatrix} = (ax + by, \; cx + dy),$$

we have

$$x/y \longmapsto \frac{ax + by}{cx + dy} = \frac{a(x/y) + b}{c(x/y) + d}.$$

This is the transformation

$$T_{abcd} \colon z \longmapsto \frac{az + b}{cz + d}$$

of $F \cup \{\infty\}$ that we considered in Example 2.4 of Chapter 2. Thus $F \cup \{\infty\}$ with the fractional linear group acting on it is equivalent to the projective line over F.

Example 4.6 The orthogonal group acting on spheres.   Another example of geometrical interest is the group $O(n, \mathbb{R})$ acting on the $(n-1)$-sphere in $\mathbb{R}^n$ defined by

$$S^{n-1} := \{x \in \mathbb{R}^n\,|\,\|x\| = 1\}.$$

The point is that if $\|x\| = 1$, that is, if $xx' = 1$, and if $A \in O(n, \mathbb{R})$, then

$$(xA)(xA)' = xAA'x' = xx' = 1.$$

Thus if $x \in S^{n-1}$ then also $xA \in S^{n-1}$ and the prescription $(x, A) \longmapsto xA$ describes a group action exactly as it does for $GL(n, \mathbb{R})$ acting on $\mathbb{R}^n$.

Example 4.7 Affine groups and affine spaces.   The group AGL(n, F) was defined in Chapter 2 as the group of bijections

$$x \mapsto xA + b \quad (A \in GL(n, F), b \in F^n)$$

of F^n. Thus F^n is naturally an AGL(n, F)-space: this is known as the n-dimensional affine space over F. Notice that GL(n, F) acting as in Example 4.4 is a subgroup of AGL(n, F); that is, non-singular linear transformations are some of the affine transformations of F^n. Notice also that the group of translations $x \mapsto x + b$, which, recall, is a normal subgroup of AGL(n, F), acts on the affine space F^n regularly in the sense of Example 4.1.

Example 4.8 Actions by conjugation.   Let G be any group. We can take $\Omega := G$ and define

$$\omega^g := g^{-1}\omega g$$

for $\omega \in \Omega$ and $g \in G$. It follows easily from the group axioms that this defines an action of G on itself.

There are several useful variations on this theme: G acts on the set of its subgroups by conjugation; if $N \trianglelefteq G$ then G acts on N, and on the set of all subgroups of N, by conjugation. Examples like these will reappear many times in this book.

Specimen solutions to selected exercises

The following specimen solutions, being of the nature of an extension of the text, are placed between the main body of the chapter and the exercises. The wise and conscientious reader will, however, try the exercises unaided before attempting to evaluate what we have offered.

EXERCISE 4.3   *Let G be a finite group of order $2^r m$ where m is odd and $r \geqslant 1$. Suppose that G contains an element a of order 2^r.*
(i)   *Show that in the regular permutation representation of G the element a acts as an odd permutation.*
(ii)   *Deduce that G has a subgroup of index 2.*
(iii)   *Now show that this subgroup of index 2 contains an element of order 2^{r-1}, and use induction to prove that G has a normal subgroup of order m.*

Solution. (i)   The regular representation is G acting on itself by right multiplication. For any element $g \in G$ the cycle of g that contains a given element x in this permutation representation is $(x \ xg \ xg^2 \ \ldots \ xg^{k-1})$, where k is the least positive integer such that $xg^k = x$. This means that $k = \text{ord}(g)$ and shows that every cycle of g in the regular representation has the same length ord(g). Applying this to a we see that in the regular representation it is a product of m disjoint cycles, each of length 2^r. Since cycles of even length are odd permutations, and since m is odd, in this permutation representation a acts as an odd permutation.

(ii)   Let H be the set of elements of G that act as even permutations in the regular representation of G. Then H is a normal subgroup of G. But if $g \in G \setminus H$ then $ga^{-1} \in H$, and so $G = H \cup Ha$. Thus $|G:H| = 2$.

(iii)   The element a^2 has order 2^{r-1} and certainly lies in H. If $r = 1$ then H is our normal subgroup of order m. Suppose therefore that $r > 1$. As inductive hypothesis we may suppose that H has a normal subgroup M of order m. Now the quotient group H/M has order 2^{r-1} and therefore the order of any of its non-trivial elements is a power of 2, in particular, all its non-trivial elements have even order. It follows that any element of H not in M has even order. On the other hand, the order of any element of M divides m, which is odd. Thus $M = \{x \in H \mid \mathrm{ord}(x) \text{ is odd}\}$. For any $g \in G$ we know that $g^{-1}Hg = H$ and that $g^{-1}xg$ has the same order as x. Therefore $g^{-1}Mg = M$, and M is a normal subgroup of G of order m, as required.

EXERCISE 4.6   *Recall that the group G is said to be simple if it has no non-trivial proper normal subgroups. Show that if G is simple and has a subgroup of index k, where $k > 2$, then $|G|$ divides $k!/2$.*

Solution.   The fact that $|G|$ divides $k!$ is an immediate consequence of Corollary 4.2.2. What is asked for is a refinement of that. Let G be a simple group, let H be a subgroup of index k in G, where $k > 2$, and let Ω be the coset space $\cos(G:H)$. The action of G on Ω gives rise to a homomorphism $\rho: G \to \mathrm{Sym}(\Omega) \cong S_k$, namely the permutation representation associated with the G-space Ω. The kernel of ρ is a normal subgroup of G contained in H. Since G is simple we must have $\mathrm{Ker}(\rho) = \{1\}$, and so ρ may be thought of as an isomorphism of G to a subgroup of S_k. We can identify G with $\mathrm{Im}(\rho)$, and think of G as actually being a subgroup of S_k. Now $G \cap A_k$ is a normal subgroup of index 1 or 2 in G. Since G is simple and of order > 2 the index cannot be 2. Therefore in fact $G \leqslant A_k$ and by Lagrange's Theorem $|G|$ divides $k!/2$, as required.

EXERCISE 4.7   *Let G be a finite group whose order is a power p^a of a prime number p. By considering the kernel of the action of G on the coset space, show that any subgroup of index p is normal.*

Solution.   Again, this is a consequence of the corollary in our discussion of Example 4.2, but we give a self-contained proof. Let H be a subgroup of index p in G and let Ω be the coset space $\cos(G:H)$. The permutation representation associated with the action of G on Ω is a homomorphism $\rho: G \to \mathrm{Sym}(\Omega)$ whose kernel K is contained in H. By the First Isomorphism Theorem $G/K \cong \mathrm{Im}(\rho) \leqslant \mathrm{Sym}(\Omega) \cong S_p$ and so by Lagrange's Theorem $|G/K|$ divides $p!$.

On the other hand, $|G/K|$ divides $|G|$, which is p^a. The highest power of p dividing $p!$ is p and so $|G:K|$ divides p. Therefore, since $K \leqslant H$ we must have $K = H$. But, being the kernel of a homomorphism, K is normal in G. Consequently H is normal in G, as required.

EXERCISE 4.14   *Let Ω be the space of all functions $f: \mathbb{R}^3 \to \mathbb{R}$ all of whose partial derivatives (of any order) exist everywhere. Let G be the euclidean group of $\mathbb{R}^3$, that is, the set of all transformations*

$$T_{U,b}: a \mapsto aU + b \quad (a \in \mathbb{R}^3),$$

where $U \in O(3)$ and $b \in \mathbb{R}^3$. For $g \in G$ and $f \in \Omega$ define f^g by $f^g(x, y, z) :=$ $f((x, y, z)g^{-1})$.
   (i)   *Show that the prescription $(f, g) \mapsto f^g$ gives an action of G on Ω.*
   (ii)   *Let ∇ be the Laplace operator on Ω,*

$$\nabla f := \frac{\partial^2 f}{\partial x^2} + \frac{\partial^2 f}{\partial y^2} + \frac{\partial^2 f}{\partial z^2}.$$

Show that $\nabla(f^g) = (\nabla f)^g$ for all $f \in \Omega$ and all $g \in G$.
   (iii)   *Hence show that euclidean mappings transform harmonic functions to harmonic functions.*

Solution. (i)   The chain rule for partial differentiation tells us that if

$$\phi(x, y, z) := f(a_{11}x + a_{21}y + a_{31}z + b_1, a_{12}x + a_{22}y + a_{32}z + b_2,$$

$$a_{13}x + a_{23}y + a_{33}z + b_3)$$

then

$$\frac{\partial \phi}{\partial x} = a_{11} \frac{\partial f}{\partial x} + a_{12} \frac{\partial f}{\partial y} + a_{13} \frac{\partial f}{\partial z},$$

$$\frac{\partial \phi}{\partial y} = a_{21} \frac{\partial f}{\partial x} + a_{22} \frac{\partial f}{\partial y} + a_{23} \frac{\partial f}{\partial z},$$

$$\frac{\partial \phi}{\partial z} = a_{31} \frac{\partial f}{\partial x} + a_{32} \frac{\partial f}{\partial y} + a_{33} \frac{\partial f}{\partial z}.$$

Therefore if $f \in \Omega$ then certainly $f^g \in \Omega$ and our prescription describes a map $\Omega \times G \to \Omega$. It is clear that $f^1 = f$ for all $f \in \Omega$. And, since

$$(f^g)^h(x, y, z) = f^g((x, y, z)h^{-1}) = f(((x, y, z)h^{-1})g^{-1})$$

$$= f((x, y, z)(h^{-1}g^{-1})) = f((x, y, z)(gh^{-1})) = f^{(gh)}(x, y, z)$$

for any $(x, y, z) \in \mathbb{R}^3$, we also have that $(f^g)^h = f^{(gh)}$ for any $g, h \in G$. Thus our prescription does define an action of G on Ω.

(ii)   It is clear that if g is a translation $(x, y, z) \mapsto (x+b_1, y+b_2, z+b_3)$ then $\nabla(f^g) = (\nabla f)^g$ for any $f \in \Omega$. Therefore we will concentrate on proving that this equation holds when $g:(x, y, z) \mapsto (x, y, z)U$ and U is a 3×3 matrix $(a_{ij}) \in O(3)$. Applying the chain rule once more to the formula in (i) above (which now represents $f((x, y, z)U^{-1})$) we find that if $\phi := f^g$ then

$$\frac{\partial^2 \phi}{\partial x^2} = a_{11}\left(a_{11}\frac{\partial^2 f}{\partial x^2} + a_{12}\frac{\partial^2 f}{\partial x \partial y} + a_{13}\frac{\partial^2 f}{\partial x \partial z}\right)$$

$$+ a_{12}\left(a_{11}\frac{\partial^2 f}{\partial y \partial x} + a_{12}\frac{\partial^2 f}{\partial y^2} + a_{13}\frac{\partial^2 f}{\partial y \partial z}\right)$$

$$+ a_{13}\left(a_{11}\frac{\partial^2 f}{\partial z \partial x} + a_{12}\frac{\partial^2 f}{\partial z \partial y} + a_{13}\frac{\partial^2 f}{\partial z^2}\right).$$

Similar formulae hold for the other partial derivatives of order 2. If we write out ∇f and collect coefficients we find, for example, that

the coefficient of $\dfrac{\partial^2 f}{\partial x^2}$ is $a_{11}^2 + a_{12}^2 + a_{13}^2$;

the coefficient of $\dfrac{\partial^2 f}{\partial x \partial y}$ is $a_{11}a_{21} + a_{12}a_{22} + a_{13}a_{23}$.

Since $U \in O(3)$, its matrix, and therefore also the matrix (a_{ij}), is orthogonal. It follows that terms like $a_{11}^2 + a_{12}^2 + a_{13}^2$ are 1 and terms like $a_{11}a_{21} + a_{12}a_{22} + a_{13}a_{23}$ are 0. Thus

$$\nabla f(a_{11}x + a_{21}y + a_{31}z, a_{12}x + a_{22}y + a_{32}z, a_{13}x + a_{23}y + a_{33}z)$$

$$= \frac{\partial^2}{\partial x^2}f(a_{11}x + a_{21}y + a_{31}z + b_1, a_{12}x + a_{22}y + a_{32}z + b_2,$$

$$a_{13}x + a_{23}y + a_{33}z + b_3)$$

$$+ \frac{\partial^2}{\partial y^2}f(a_{11}x + a_{21}y + a_{31}z + b_1, a_{12}x + a_{22}y + a_{32}z + b_2,$$

$$a_{13}x + a_{23}y + a_{33}z + b_3)$$

$$+ \frac{\partial^2}{\partial z^2}f(a_{11}x + a_{21}y + a_{31}z + b_1, a_{12}x + a_{22}y + a_{32}z + b_2,$$

$$a_{13}x + a_{23}y + a_{33}z + b_3),$$

that is, $\nabla(f^g) = (\nabla F^g)$.

(iii)   A harmonic function is a solution of Laplace's equation $\nabla f = 0$. If f is harmonic then $\nabla(f^g) = (\nabla f)^g = 0$ and so f^g is harmonic as required.

Exercises

4.1.   Let $\Omega := G$ and define $\mu : (\omega, g) \mapsto g^{-1}\omega$.
   (i) Prove that μ is an action of G on Ω. We call Ω the left-regular G-space.
   (ii) Show that it is G-isomorphic to the right regular G-space.
   (iii) For which groups G does straightforward left translation $(\omega, g) \mapsto g\omega$ define an action?

4.2.   Let g be an element of order m in the finite group G. What is the cycle structure of g in the regular permutation representation of G? Use your answer to show that m divides $|G|$.

4.3.   Let G be a finite group of order $2^r m$ where $r \geqslant 1$ and m is odd. Suppose that G contains an element a of order 2^r.
   (i) Show that, in the regular permutation representation of G the element a acts as an odd permutation.
   (ii) Deduce that G has a subgroup of index 2.
   (iii) Now show that this subgroup of index 2 contains an element of order 2^{r-1}, and use induction to prove that G has a normal subgroup of order m.

4.4.   Suppose that $H \leqslant G$ and let $\Omega := \{xH \,|\, x \in G\}$. Define the map μ by $(xH, g) \mapsto g^{-1}xH$.
   (i) Show that μ is an action of G on Ω.
   (ii) Show that this left coset space is G-isomorphic to the right coset space $\cos(G : H)$.

4.5.   Show that if $\Omega := \{xH \,|\, x \in G\}$ and $\mu : (xH, g) \mapsto gxH$ then μ is an action of G on Ω if and only if $H \trianglelefteq G$ and G/H is abelian.

4.6.   Recall that the group G is said to be simple if it has no non-trivial proper normal subgroups. Show that if G is simple and has a subgroup of index k, where $k > 2$, then $|G|$ divides $k!/2$.

4.7.   Let G be a finite group whose order is a power p^a of a prime number p. By considering the kernel of the action of G on the coset space, show that any subgroup of index p is normal.

4.8.   Let p be the smallest prime that divides the order of the finite group G and let N be a normal subgroup of order p. By considering the permutation representation of G acting by conjugation on $N \backslash \{1\}$ show that N is contained in the centre of G.

4.9.   Use the results of Questions 4.7 and 4.8 to show that every group of order p^2, where p is prime, is abelian.

4.10.   Let $\Omega := \{1, 2, 3, 4\}$ and let Π be the set of equivalence relations (or partitions) that divide Ω into two parts each of size 2. Define an action of S_4 on Π and use this to produce a homomorphism $\pi : S_4 \to S_3$. What are the kernel and the image of π, respectively?

4.11. Show that, in their natural permutation representations on the appropriate projective lines,
(i) $PGL(2, 2) = S_3$;
(ii) $PGL(2, 3) = S_4$;
(iii) $PSL(2, 3) = A_4$;
(iv) $PSL(2, 4) = A_5$.

4.12. We defined Σ_k to be the set of $(k+1)$-dimensional subspaces of the vector space F^n.
(i) Prove that the kernel of the action of $GL(n, F)$ on Σ_0 is precisely Z, the set of non-zero scalar matrices. (Consequently $PGL(n, F)$ acts faithfully on the set of points of $PG(n-1, F)$.)
(ii) What is the kernel of the action of $GL(n, F)$ on Σ_k for $1 \leqslant k \leqslant n-1$?

4.13. In affine *n*-space $\mathbb{R}^n$ we define a quadric to be the locus of a quadratic equation $f(x_1, \ldots, x_n) = 0$. Show that affine transformations map quadrics to quadrics (and so $AGL(n, \mathbb{R})$ acts on the set of quadrics).

4.14. Let Ω be the space of all functions $f: \mathbb{R}^3 \to \mathbb{R}$ all of whose partial derivatives (of any order) exist everywhere. Let G be the euclidean group of $\mathbb{R}^3$, that is, the set of all transformations

$$T_{U,b}: a \mapsto aU + b \quad (a \in \mathbb{R}^3),$$

where $U \in O(3)$ and $b \in \mathbb{R}^3$. For $g \in G$ and $f \in \Omega$ define f^g by

$$f^g(x, y, z) := f((x, y, z)g^{-1}).$$

(i) Show that the prescription $(f, g) \mapsto f^g$ gives an action of G on Ω.
(ii) Let ∇ be the Laplace operator on Ω,

$$\nabla f := \frac{\partial^2 f}{\partial x^2} + \frac{\partial^2 f}{\partial y^2} + \frac{\partial^2 f}{\partial z^2}.$$

Show that $\nabla(f^g) = (\nabla f)^g$ for all $f \in \Omega$ and all $g \in G$.
(iii) Hence show that euclidean mappings transform harmonic functions to harmonic functions.

Chapter 5

Transitivity and orbits

We say that Ω is a *transitive G-space*, or that *G acts transitively* on Ω if $\Omega \neq \varnothing$ and for any α, $\beta \in \Omega$ there exists $g \in G$ such that $\alpha^g = \beta$. Of course, such an element g will depend upon α and β. Notice that G is transitive if and only if there is some element $\gamma \in \Omega$ such that for any $\delta \in \Omega$ there exists g (depending on δ) with $\gamma^g = \delta$: for then, given α, β there exist $g_1, g_2 \in G$ such that $\gamma^{g_1} = \alpha$, $\gamma^{g_2} = \beta$, and so, if $g := g_1^{-1} g_2$ then $\alpha^g = \beta$. In symbols our definition of transitivity was

$$(\Omega \neq \varnothing) \text{ and } (\forall \alpha \in \Omega)\,(\forall \beta \in \Omega)\,(\exists g \in G)\,(\alpha^g = \beta),$$

and our reformulation is

$$(\exists \gamma \in \Omega)\,(\forall \delta \in \Omega)\,(\exists g \in G)\,(\gamma^g = \delta).$$

Example 5.1. The regular G-space is transitive. For, if α, $\beta \in \Omega = G$, and if we take $g := \alpha^{-1}\beta$, then $\alpha^g = \beta$.

Example 5.2. Any coset space is transitive. For, if $\Omega := \cos(G{:}H)$, and if α, $\beta \in \Omega$, then $\alpha = Ha$, $\beta = Hb$ for some $a, b \in G$, and if we take $g := a^{-1}b$ then we have

$$\alpha^g = (Ha)a^{-1}b = Hb = \beta.$$

Example 5.3. For any field F the linear group $\mathrm{GL}(n, F)$ is transitive on $F^n \backslash \{0\}$. For, given a non-zero vector u in F^n there certainly exists an invertible $n \times n$ matrix A over F, whose first row is u and then

$$(1, 0, \ldots, 0)A = u.$$

Thus our alternative version of the transitivity condition holds.

Example 5.4. The orthogonal group $\mathrm{O}(n)$ is transitive on the $(n-1)$-sphere in $\mathbb{R}^n$. For, if

$$u \in S^{n-1} = \{x \in \mathbb{R}^n \,|\, \|x\| = (xx')^{1/2} = 1\},$$

then (using, for example, the Gram–Schmidt orthogonalisation process) we can extend $\{u\}$ to an orthonormal basis $\{u, u_2, \ldots, u_n\}$ for $\mathbb{R}^n$; and if U is the

$n \times n$ matrix whose ith row is u_i (where $u_1 = u$), then $UU' = I_n$, so $U \in \mathrm{O}(n)$, and

$$(1, 0, \ldots , 0)U = u.$$

Thus our reformulation of the transitivity condition holds.

Recall (from p. 33 in Chapter 3) that a G-subspace of the G-space Ω is, by definition, a subset Ω' such that if $\omega \in \Omega'$ then $\omega g \in \Omega'$ for all $g \in G$. A transitive G-subspace is known as an *orbit*. The importance of the notion of transitivity stems mainly from the following.

THEOREM 5.1: *Every G-space can be expressed in just one way as the disjoint union of a family of orbits.*

Proof. Let Ω be a G-space. We define a relation of equivalence under G by the rule

$$\alpha \equiv \beta \pmod{G} \text{ if and only if there exists } g \text{ in } G \text{ such that } \alpha^g = \beta.$$

Then certainly

$\alpha \equiv \alpha$ because $\alpha^1 = \alpha$;

$(\alpha \equiv \beta) \Rightarrow (\beta \equiv \alpha)$ because if $\alpha^g = \beta$ then $\beta^{g^{-1}} = \alpha$;

$((\alpha \equiv \beta)$ and $(\beta \equiv \gamma)) \Rightarrow (\alpha \equiv \gamma)$ because if $\alpha^g = \beta$ and $\beta^h = \gamma$ then $\alpha^{(gh)} = \gamma$.

These three facts tell us that $\equiv$ is an equivalence relation. It is notationally convenient to choose a suitable set I and list the equivalence classes of the relation $\equiv$ as a family $(\Omega_i)_{i \in I}$: thus

$$\Omega = \bigcup_{i \in I} \Omega_i,$$

$$\Omega_i \cap \Omega_j = \varnothing \quad \text{if } i \neq j,$$

$$\Omega_i \neq \varnothing \quad \text{for all } i \in I.$$

(Notice that if $\Omega = \varnothing$ then we take, as we must, $I := \varnothing$ also. It is a consequence of the definitions, rather than simple convention, that the union of an empty family of sets is the empty set.) The point is, of course, that each equivalence class Ω_i is a transitive G-subspace of Ω: for, if $\omega \in \Omega_i$ and $g \in G$ then certainly $\omega^g \equiv \omega \pmod{G}$, so $\omega^g \in \Omega_i$; and, if $\alpha, \beta \in \Omega_i$ then $\alpha \equiv \beta \pmod{G}$, and so there exists g such that $\alpha^g = \beta$. Thus we have proved the existence of a decomposition of Ω as the union of a family of orbits.

To prove uniqueness, suppose that

$$\Omega = \bigcup_{j \in J} \Sigma_j$$

where $\{\Sigma_j\}_{j \in J}$ is a family of pairwise disjoint, non-empty, transitive G-subspaces. From the definition of transitivity, if α, β lie in one and the same subset Σ_j then $\alpha^g = \beta$ for some $g \in G$, and so $\alpha \equiv \beta \pmod{G}$. Conversely, if

$\alpha \equiv \beta \pmod{G}$, and if $\alpha \in \Sigma_j$, then from the facts that $\beta = \alpha^g$ and Σ_j is a G-subspace, we must have $\beta \in \Sigma_j$ also. Thus $\alpha \equiv \beta \pmod{G}$ if and only if α, β lie in the same subset Σ_j for some $j \in J$. This means that the subsets Σ_j are just the $\equiv$-classes and so the family $\{\Sigma_j\}_{j \in J}$ is the family $\{\Omega_i\}_{i \in I}$ masquerading under a different name. This completes the proof of the theorem.

Example 5.5.   Let X be a group containing G as a subgroup. Then G acts on X by right multiplication. Here the orbits are the left cosets xG for $x \in X$. Similarly, G acts on X by the rule $(\omega, g) \mapsto g^{-1}\omega$, and in this case its orbits are the right cosets Gx.

Example 5.6.   When G acts on itself by conjugation the orbits are the conjugacy classes that we all know and love.

Example 5.7.   The orbits of $O(n)$ acting on $\mathbb{R}^n$ are the spheres Σ_a (for $a \geqslant 0$), where

$$\Sigma_a := \{(x_1, x_2, \ldots, x_n) \mid x_1^2 + x_2^2 + \ldots + x_n^2 = a^2\}$$

Example 5.8.   If Θ is the set of triangles in the euclidean plane $\mathbb{R}^2$, and G is the euclidean group (Example 2.7 with $n = 2$) acting in the natural way on Θ, then two triangles lie in the same orbit if and only if they are congruent in the sense of school editions of Euclid, *Elements*, Book I, Prop. 4.

We have given the theorem and its proof in a rather abstract form suitable for theoretical investigations. In practice it is useful to have a different description of the orbits. If $\alpha \in \Omega$ we define α^G (or αG when G consists of mappings of Ω) by

$$\alpha^g := \{\alpha^g \mid g \in G\},$$

and it is easy to see that this is the G-orbit that contains α. For, certainly $\alpha \in \alpha^G$. Also, if $\omega \in \alpha^G$ then $\omega = \alpha^h$ for some $h \in G$ and so $\omega^g = \alpha^{(hg)} \in \alpha^G$ for all $g \in G$, which shows that α^G is a subspace. And lastly, G is transitive on α^G as is shown by the equation $(\alpha^g)^{g^{-1}h} = \alpha^h$.

To compute the orbits in a G-space Ω one therefore can begin with any element α_1 of Ω, apply to it the members $1, g_1, g_2, \ldots$ of G, each in turn until one has reason to know that no new elements of Ω result. Then $\{\alpha_1, \alpha_1^{g_1}, \alpha_1^{g_2}, \ldots\}$ is a G-orbit Ω_1. If $\Omega_1 = \Omega$ the process is finished; otherwise we choose α_2 from $\Omega \setminus \Omega_1$ and we list the set $\{\alpha_2, \alpha_2^{g_1}, \alpha_2^{g_2}, \ldots\}$, which will be our second orbit Ω_2. When Ω is exhausted the process is complete. In practice, of course, the details of such a calculation must depend on the description of G and of Ω. There are general-purpose programs available for doing it on computers. Most of these assume that G is a group of permutations of Ω and that generators of it are given. That is, the data consist of certain permutations

$g_1, g_2, \ldots, g_k$ which generate G in the sense that every element can be obtained as some product $g_{i_1}^{\pm 1} g_{i_2}^{\pm 1} \ldots g_{i_m}^{\pm 1}$ of them and their inverses. To find the orbits we start with any element α_1 of Ω; we list

$$\alpha_1, \alpha_1^{g_1}, \ldots, \alpha_1^{g_k}, \alpha_1^{g_1^{-1}}, \ldots, \alpha_1^{g_k^{-1}};$$

then we add the images of each of these elements under $g_1, \ldots, g_k, g_1^{-1}, \ldots,$ g_k^{-1}; and so on until we have a list Ω_1 such that the application of any element g_i or g_i^{-1} to any element of it gives us an element of Ω_1 back. Then Ω_1 is invariant under each of $g_1, \ldots, g_k$, so it follows that it is invariant under the group $\langle g_1, \ldots, g_k \rangle$ that they generate. From its construction Ω_1 then is the G-orbit of α_1. If Ω is finite then every permutation g has finite order so g^{-1} is a power of g, and we do not need to deal explicitly with the inverses $g_1^{-1}, \ldots,$ g_k^{-1}. Here is an illustration of the process at work.

Example 5.9. Let $\Omega := \{1, 2, \ldots, 19\}$, and let $G := \langle g_1, g_2, g_3 \rangle \leqslant \mathrm{Sym}(\Omega)$, where

$$g_1 = (1\ \ 4)(2\ \ 10)(3\ \ 11)(6\ \ 12)(7\ \ 13)(8\ \ 14)(9\ \ 15),$$

$$g_2 = (6\ \ 7)(8\ \ 9)(12\ \ 13)(14\ \ 15)(16\ \ 17)(18\ \ 19),$$

$$g_3 = (1\ \ 14\ \ 15)(2\ \ 8\ \ 18\ \ 19\ \ 9)(3\ \ 17\ \ 7\ \ 6\ \ 16)(4\ \ 10)(5\ \ 12\ \ 13).$$

Here fixed points, that is, cycles of length 1, have, as usual, been deleted from the cycle notation. *The problem is*, what are the G-orbits?

Answer: We list in columns the NEW elements of Ω obtained each time we apply g_1, g_2, g_3 to elements already listed.

Begin	1st application	2nd application	3rd application	4th application	5th application
1	4	10	2	19	
	14	8	9		
		15	18		

So $\Omega_1 := \{1, 2, 4, 8, 9, 10, 14, 15, 18, 19\}$ and we have our first orbit. Now we

Continue	1st application	2nd application	3rd application	4th application	5th application
3	11	16	13	12	
	17	7	6	5	

and put $\Omega_2 := \{3, 5, 6, 7, 11, 12, 13, 16, 17\}$. As the process terminates here we have two orbits, one of size 10, the other of size 9.

Note 1.   Although the algorithm is already very quick one can speed it up still further by being less systematic. For example, if at each step we had adjoined all the members of Ω that appear in the same cycle as one of the preceding elements in one or other of the generating permutations, then in Example 5.9 we would have found each of Ω_1, Ω_2 in three steps instead of four.

Note 2.   The algorithm has given us the orbits even though we know very little about G. In this particular case one can show that $G \cong S_{10} \times S_9$, but in general it is a long and difficult process, finding out what the group generated by a given set of permutations is.

Note 3.   For the record, this particular calculation arose some years ago in a question about what are called free groups. In the free group F on two generators there are 19 normal subgroups R such that $F/R \cong A_5$. The permutations g_1, g_2, g_3 give the action on this set of normal subgroups of certain automorphisms ϕ_1, ϕ_2, ϕ_3 that generate the automorphism group of F.

    For infinite sets procedures like the one we have just discussed will be exhausting but not exhaustive, and the question remains, whether one can describe the orbits satisfactorily and how one is to find them. As so often in mathematics, the only method available is to apply guesswork intelligently. Here this usually involves spotting 'invariants' for G acting on Ω, that is, properties of elements of Ω such that α and β share the property if and only if $\alpha \equiv \beta \pmod{G}$. Example 5.7 illustrates what we mean. Here G is the orthogonal group $O(n)$ acting on $\mathbb{R}^n$, and the length of a vector is a numerical invariant of it. That

$$\|x\| = \|xU\| \quad \text{for all } x \in \mathbb{R}^n, \ U \in O(n)$$

was shown by the calculation on p. 23 (see also p. 109); and the converse, that if $\|x\| = \|y\|$ then there is an orthogonal matrix U such that $xU = y$, follows from the calculation in Example 5.4 by appropriate scaling.

    Here are two less obvious examples.

Example 5.10.   Let $\Gamma := \{1, 2, \ldots, n\}$ where n is a natural number, let $G := S_n := \text{Sym}(\Gamma)$, and let $\Omega := \mathscr{P}(\Gamma) := \{E \mid E \subseteq \Gamma\}$. Thus Ω has 2^n members, and G acts on Ω in a naturally defined way. Here it is easy to spot an invariant of a member of Ω. For if $E \subseteq \Omega$ and $g \in G$ then the size of E is unchanged by g: we have $|Eg| = |E|$. Certainly therefore members of Ω that have different sizes must be in different G-orbits. Conversely, if $|E| = |F|$ then there certainly is a permutation in S_n mapping E to F. For, if

$$E \cap F = \{\alpha_1, \ldots, \alpha_k\},$$

$$E = \{\alpha_1, \ldots, \alpha_k, \beta_1, \ldots, \beta_{m-k}\},$$

and

$$F = \{\alpha_1, \ldots, \alpha_k, \gamma_1, \ldots, \gamma_{m-k}\},$$

then the permutation $(\beta_1 \gamma_1)(\beta_2 \gamma_2) \cdots (\beta_{m-k} \gamma_{m-k})$ does what is wanted.

Thus subsets E, F of Γ are equivalent under S_n if and only if $|E| = |F|$, and so the S_n-orbits in $\mathscr{P}(\Gamma)$ are

$$\Omega_0, \Omega_1, \ldots, \Omega_n,$$

where

$$\Omega_i := \{E \subseteq \Gamma \,|\, |E| = i\}.$$

Example 5.11. Let Ω be the set of quadratic forms $\sum_{ij} a_{ij} x_i x_j$ on $\mathbb{R}^n$. The linear group $GL(n, \mathbb{R})$ acts on Ω in a familiar way as a group of linear substitutions. Here the classical theory of quadratic forms gives us two numerical invariants, the rank and the signature, which together distinguish the orbits: two quadratic forms in n variables over $\mathbb{R}$ are equivalent under linear substitution if and only if they have the same rank and signature. This example also illustrates what is meant by a 'canonical form'. Generally, by a set of canonical forms we mean a choice of one element from each orbit of the G-space Ω, that element being chosen because it can be described in a relatively simple and explicit way. The canonical forms for $GL(n, \mathbb{R})$ acting on quadratic forms are these:

$$x_1^2 + \cdots + x_p^2 - x_{p+1}^2 - \cdots - x_{p+q}^2.$$

Here the rank is $p+q$ (notice that $p+q \leqslant n$), the signature is $p-q$, and the point is that a quadratic form is equivalent to this one if and only if it has rank $p+q$ and signature $p-q$.

Specimen solutions to selected exercises

The following specimen solutions, being of the nature of an extension of the text, are placed between the main body of the chapter and the exercises. The wise and conscientious reader will, however, try the exercises unaided before attempting to evaluate what we have offered.

EXERCISE 5.1   *Let $\phi: \Omega \to \Omega'$ be a G-morphism. Show that if Ω_i is an orbit in Ω then $\Omega_i \phi$ is an orbit in Ω'.*

Solution.   We need to show that $\Omega_i \phi$ is a transitive G-subspace of Ω'. Let ω' be any element of $\Omega_i \phi$. Then there exists $\omega \in \Omega_i$ such that $\omega' = \omega \phi$ and for any $g \in G$ we have that $\omega'g = (\omega \phi)g = (\omega g)\phi$ (because ϕ is a G-morphism). But since Ω_i is an orbit it is a G-subspace of Ω, and so $\omega g \in \Omega_i$, whence $\omega'g = (\omega g)\phi \in \Omega_i \phi$. Therefore $\Omega_i \phi$ is a G-subspace of Ω'. To see that G acts transitively on $\Omega_i \phi$ take elements $\omega'_1, \omega'_2 \in \Omega_i \phi$. Choose $\omega_1, \omega_2 \in \Omega_i$ such that $\omega_1 \phi = \omega'_1$, $\omega_2 \phi = \omega'_2$. Since Ω_i is an orbit in Ω there exists $g \in G$ such that $\omega_1 g = \omega_2$. Then $\omega'_1 g = \omega_1 \phi g = \omega_1 g \phi = \omega_2 \phi = \omega'_2$. This shows that G is transitive on $\Omega_i \phi$, and therefore that $\Omega_i \phi$ is a G-orbit, as required.

EXERCISE 5.4   *For which n is the special linear group* SL(n, F) *transitive on* $F^n \setminus \{0\}$?

Solution.   The group SL(1, F) is the trivial group and this is transitive on $F \setminus \{0\}$ if and only if $|F| = 2$. Suppose therefore that $n \geqslant 2$. Let u, v be non-zero vectors in F^n. There exist bases $u_1, \ldots, u_n$ and $v_1, \ldots, v_n$ of F^n such that $u_1 = u, v_1 = v$. Let A be the $n \times n$ matrix such that $u_i A = v_i$ for all relevant i. (If U is the matrix with rows $u_1, \ldots, u_n$ and V is the matrix with rows $v_1, \ldots, v_n$ then $A = U^{-1}V$.) Now A is invertible but its determinant need not be 1. Let $d := \det(A)$, and let $w_1, \ldots, w_{n-1}, w_n$ be the basis $v_1, \ldots, v_{n-1}, d^{-1}v_n$ of F^n. If B is the $n \times n$ matrix such that $u_i B = w_i$ for all i then $uB = v$ and $\det(B) = 1$. Thus if $n \geqslant 2$ then SL(n, F) is transitive on F^n.

EXERCISE 5.10   *Let* α_1, α_2, α_3 *and* β_1, β_2, β_3 *be triples of distinct points of* PG(1, F) (*or of* $F \cup \{\infty\}$ [*see p. 43*]). *Show that there is exactly one element g of* PGL(2, F) (*or fractional linear transformation over F respectively*) *such that* $\alpha_i g = \beta_i$ *for* $1 \leqslant i \leqslant 3$.

Solution.   Points of PG(1, F) are 1-dimensional subspaces of F^2, that is, they may be represented by non-zero vectors in F^2 with the convention that vectors u, u' represent the same point if and only if they are scalar multiples of each other. Let u_1, u_2, u_3 represent the points $\alpha_1, \alpha_2, \alpha_3$ respectively. Then u_1, u_2 are linearly independent and so we may write u_3 as $\lambda_1 u_1 + \lambda_2 u_2$. Notice that both λ_1 and λ_2 must be non-zero. Now we can choose vectors v_1, v_2, v_3 to represent the points $\beta_1, \beta_2, \beta_2$ respectively and we will have scalars μ_1, μ_2 such that $v_3 = \mu_1 v_1 + \mu_2 v_2$. We seek a matrix $A \in$ GL(2, F) such that $u_1 A$ is a scalar multiple of v_1, $u_2 A$ is a scalar multiple of v_2 and $u_3 A$ is a scalar multiple of v_3. This requires $u_1 A = \xi v_1$, $u_2 A = \eta v_2$ and $(\lambda_1 u_1 + \lambda_2 u_2)A = \zeta(\mu_1 v_1 + \mu_2 v_2)$, and is possible if and only if

$$\lambda_1 \xi = \zeta \mu_1 \quad \text{and} \quad \lambda_2 \eta = \zeta \mu_2.$$

Taking any non-zero value for ζ we see that ξ, η are uniquely determined, and then A is uniquely determined. Moreover, different possible values of A (corresponding to different choices of ζ) are scalar multiples of each other and therefore determine the same element g of PGL(2, F). Thus there is only one element g of PGL(2, F) such that $\alpha_i g = \beta_i$ for $1 \leqslant i \leqslant 3$, as required.

EXERCISE 5.14   *The direct product* GL(m, F) $\times$ GL(n, F) *acts on the set* $M_{m \times n}(F)$ *of* $m \times n$ *matrices over the field F by the rule* $(A, (P, Q)) \mapsto P^{-1}AQ$ (*for* $A \in M_{m \times n}(F)$, $P \in$ GL(n, F) *and* $Q \in$ GL(n, F)).
(i)   *Describe the orbits in terms of a single numerical invariant.*
(ii) *Exhibit a set of canonical forms for the orbits of this G-space.*

Solution.   Since P, Q are invertible we know that the rank of $P^{-1}AQ$ must be the same as the rank of A. To show that rank is an invariant of the orbits we need to show conversely that if A, B are $m \times n$ matrices over F that have the same rank then there exist P, Q such that $B = P^{-1}AQ$. Row operations on a matrix X can be achieved by pre-multiplication of X by certain invertible matrices, namely, the well-known elementary matrices. Similarly, column operations can be achieved by post-multiplication by elementary matrices. And X can be reduced to the form

$$\begin{pmatrix} I_r & 0 \\ 0 & 0 \end{pmatrix}$$

by suitable row and column operations (and here r must be the rank of X). Applying this to A and B in turn, we have that if they have the same rank r then there exist matrices P_1, Q_1 such that

$$P_1 A Q_1 = \begin{pmatrix} I_r & 0 \\ 0 & 0 \end{pmatrix}$$

and matrices P_2, Q_2 such that

$$P_2 B Q_2 = \begin{pmatrix} I_r & 0 \\ 0 & 0 \end{pmatrix}.$$

If $P := P_1^{-1}P_2$ and $Q := Q_1 Q_2^{-1}$ then $P^{-1}AQ = B$. Thus the orbits consist of the matrices of given rank r and the matrices

$$\begin{pmatrix} I_r & 0 \\ 0 & 0 \end{pmatrix}$$

for $0 \leqslant r \leqslant \min(m, n)$ may be taken as canonical forms.

EXERCISE 5.18   Let Ω be a transitive G-space and suppose that $H \trianglelefteq G$.
(i)   Show that if Γ is an H-orbit in Ω then so is Γ^g for any $g \in G$.
(ii)   Hence show that G acts transitively on the set of H-orbits; and
(iii)   that if Γ_i, Γ_j are H-orbits then Γ_i and Γ_j are equivalent H-spaces in the sense of Chapter 3, p. 32.

Solution. (i)   Let Γ be an H-orbit in Ω and let g be any element of G. If $\omega \in \Gamma^g$, say $\omega = \gamma^g$, then for any $h \in H$ we have $\omega^h = \gamma^{gh} = (\gamma^{ghg^{-1}})^g$. Since $ghg^{-1} \in H$ and Γ is an H-subspace of Ω it follows that $\omega^h \in \Gamma^g$. Thus Γ^g is an H-subspace of Ω. Next, let ω_1, ω_2 be elements of Γ^g. Choose $\gamma_1, \gamma_2 \in \Gamma$ such that $\gamma_1^g = \omega_1$, $\gamma_2^g = \omega_2$. Since H is transitive on Γ there exists $h \in H$ such that $\gamma_1^h = \gamma_2$. Then $\omega_1^{g^{-1}hg} = \omega_2$ and, since H is a normal subgroup, $g^{-1}hg \in H$. This shows that H is transitive on Γ^g, which therefore is an H-orbit.
    (ii)   The prescription $(\Gamma, g) \mapsto \Gamma^g$ has just been shown to be a map $(\Omega/H) \times G \to \Omega/H$, where Ω/H denotes the set of H-orbits in Ω. Certainly $\Gamma^1 = \Gamma$ and $(\Gamma^{g_1})^{g_2} = \Gamma^{(g_1 g_2)}$ for all $\Gamma \in \Omega/H$ and all $g_1, g_2 \in G$. Therefore G acts

on the set Ω/H of H-orbits. Now let Γ_1, Γ_2 be H-orbits. Choose $\gamma_1 \in \Gamma_1$ and $\gamma_2 \in \Gamma_2$. Since G is transitive on Ω there exists $g \in G$ such that $\gamma_1^g = \gamma_2$. Then we must have $\Gamma_1^g = \Gamma_2$: thus G is transitive on Ω/H.

(iii) Let Γ_i, Γ_j be H-orbits. Choose $g \in G$ such that $\Gamma_i^g = \Gamma_j$ and define $\theta: \Gamma_i \to \Gamma_j$ and $\phi: H \to H$ by

$$\theta: \gamma \mapsto \gamma^g \quad \text{and} \quad \phi: h \mapsto g^{-1}hg.$$

Certainly θ is a bijection on Γ_i to Γ_j and ϕ is an automorphism of H. Also, for any $\gamma \in \Gamma_i$ and any $h \in H$ we have $(\gamma^h)\theta = \gamma^{hg} = (\gamma^g)^{g^{-1}hg} = (\gamma\theta)^{h\phi}$ and so the pair (θ, ϕ) shows that the H-spaces Γ_i, Γ_j are equivalent, as required.

Exercises

5.1.   Let $\phi: \Omega \to \Omega'$ be a G-morphism. Show that if Ω_i is an orbit in Ω then $\Omega_i\phi$ is orbit in Ω'.

5.2.   Suppose that Ω, Π are G-spaces with orbits $\{\Omega_i\}_{i \in I}$, $\{\Pi_j\}_{j \in J}$ respectively. Show that Ω is G-isomorphic with Π if and only if there is a one–one correspondence between I and J (that is, a bijection $\theta: I \to J$) such that Ω_i is G-isomorphic with $\Pi_{i\theta}$.

5.3.   Is the affine group $AGL(n, F)$ transitive on F^n?

5.4.   For which n is the special linear group $SL(n, F)$ transitive on $F^n \backslash \{0\}$?

5.5.   Show that $PGL(n, F)$ is transitive on the set Σ_k of k-flats of $PG(n-1, F)$ for $0 \leqslant k \leqslant n-1$.

The group G is said to be k-fold transitive (or, simply, k-transitive) on Ω if, for any sequences

$$\alpha_1, \ldots, \alpha_k \quad \text{such that } \alpha_i \neq \alpha_j \text{ when } i \neq j,$$

$$\beta_1, \ldots, \beta_k \quad \text{such that } \beta_i \neq \beta_j \text{ when } i \neq j$$

of k elements of Ω, there exists $g \in G$ such that $\alpha_i^g = \beta_i$ for $1 \leqslant i \leqslant k$.

5.6.   Show that S_n is k-fold transitive in its natural action for any $k \leqslant n$, and that A_n is k-fold transitive for any $k \leqslant n-2$.

5.7.   (i) Show that G is doubly transitive on Ω if and only if G has 2 orbits in its action on Ω^2.
    (ii) Show that G is triply transitive on Ω if and only if it has 5 orbits in its action on Ω^3.
    (iii) Show that in general G is k-fold transitive on Ω if and only if G has r_k orbits in Ω^k, where r_k is the number of equivalence relations on the set $\{1, 2, \ldots, k\}$ (so $r_1 = 1, r_2 = 2, r_3 = 5, r_4 = 15, \ldots$).

5.8. Prove that $AGL(n, F)$ is doubly transitive on F^n for any field F. Can it be triply transitive? (Beware of $\mathbb{F}_2$, the field with 2 elements.)

5.9. Prove that for $n \geqslant 2$ the projective group $PGL(n, F)$ is 2-transitive on the points of $PG(n-1, F)$ (that is, $GL(n\ F)$ is 2-transitive in its action on the set of 1-dimensional subspaces of F^n).

5.10. Let $\alpha_1, \alpha_2, \alpha_3$ and $\beta_1, \beta_2, \beta_3$ be triples of distinct points of $PG(1, F)$ (or of $F \cup \{\infty\}$ [see p. 43]). Show that there is exactly one element of $PGL(2, F)$ (or fractional linear transformation over F, respectively) such that $\alpha_i g = \beta_i$ for $1 \leqslant i \leqslant 3$. (One says that $PGL(2, F)$ is sharply triply transitive on the projective line.)

5.11. Recall that circles on the Riemann sphere $\mathbb{C} \cup \{\infty\}$ are of two kinds. There are the ordinary circles in $\mathbb{C}$ (as the Argand plane) and circles of the form $L \cup \{\infty\}$ where L is a straight line in $\mathbb{C}$. Show that the Möbius group M is transitive on the set Σ of circles in $\mathbb{C} \cup \{\infty\}$.

5.12. We define the cross-ratio $[\alpha, \beta, \gamma, \delta]$ of a quadruple of distinct elements $\alpha, \beta, \gamma, \delta$ of $F \cup \{\infty\}$ to be

$$\frac{(\alpha - \beta)(\gamma - \delta)}{(\alpha - \delta)(\gamma - \beta)}$$

with the usual convention that $\xi - \infty = -\infty$, $\infty - \xi = \infty$, $(\infty \times \xi)/(\infty \times \eta) = \xi/\eta$, etc.
   (i) Show that the cross-ratios of quadruples of distinct points take all values in $F \cup \{\infty\}$ except 0, 1, and ∞.
  (ii) Show that there is a fractional linear transformation

$$T : z \longmapsto \frac{az + b}{cz + d}$$

mapping $\alpha, \beta, \gamma, \delta$ to $\alpha', \beta', \gamma', \delta'$ (in that order) if and only if $[\alpha, \beta, \gamma, \delta] = [\alpha', \beta', \gamma', \delta']$. (Thus cross-ratio is an invariant for the fractional linear group acting on quadruples, or for $PGL(2, F)$ acting on quadruples of distinct points from the projective line $PG(1, F)$.)

5.13. Show that if $n \geqslant 3$ then $PGL(n, F)$ has 2 orbits on ordered triples of distinct points in $PG(n-1, F)$.

5.14. The direct product $GL(m, F) \times GL(n, F)$ acts on the set $M_{m \times n}(F)$ of $m \times n$ matrices over the field F by the rule $(A, (P, Q)) \mapsto P^{-1}AQ$ (for $A \in M_{m \times n}(F)$, $P \in GL(m, F)$ and $Q \in GL(n, F)$).
   (i) Describe the orbits in terms of a single numerical invariant.
  (ii) Exhibit a set of canonical forms for the orbits of this G-space.

5.15. The linear group GL(2, F) acts by conjugation on $M_2(F)$. Does
    (i) The characteristic polynomial $c(x)$, or
    (ii) the minimal polynomial $m(x)$, or
    (iii) the pair $(c(x), m(x))$
    constitute a complete set of invariants that will distinguish the orbits?

5.16. Let X be a countably infinite set. Show that Sym(X) has countably many orbits on the power set $\mathscr{P}(X)$, and describe these orbits.

5.17. Let Ω be a transitive G-space, and suppose that $H \leqslant G$ with $|G:H| = k$. Show that if H has t orbits in Ω then $t \leqslant k$.

5.18. Let Ω be a transitive G-space and suppose that $H \trianglelefteq G$.
    (i) Show that if Γ is an H-orbit in Ω then so is Γ^g for any $g \in G$.
    (ii) Hence show that G acts transitively on the set of H-orbits; and
    (iii) that if Γ_i, Γ_j are H-orbits then Γ_i and Γ_j are equivalent H-spaces in the sense of Chapter 3, p. 32.

Chapter 6

The classification of transitive G-spaces

The theorem of Chapter 5, that every G-space is expressible in a unique way as a disjoint union of orbits, reduces many questions about actions of groups to the study of transitive G-spaces. Our next aim is to understand transitive actions, and to produce a catalogue of all transitive G-spaces up to G-isomorphism for a given group G. This is not as hard as it might seem. We already have a good supply: the coset spaces $\cos(G:H)$ that we introduced in Example 4.2 were shown in Example 5.2 to be transitive. What we shall find now is that every transitive G-space is G-isomorphic to a coset space $\cos(G:H)$ for a suitable subgroup H of G. Thus coset spaces provide a sort of 'canonical form' for transitive G-spaces under G-isomorphism.

If G is a group acting on the set Ω, and if $\alpha \in \Omega$, we define the *stabiliser* of α (in some contexts, called the *isotropy subgroup*) by

$$G_\alpha := \text{Stab}(\alpha) := \{g \in G \mid \alpha^g = \alpha\}.$$

Example 6.1. If G acts on itself by conjugation the stabiliser of a is the centraliser $C_G(a)$ (defined, as we recalled in Chapter 1, as the set of all elements of G that commute with a). If G acts on the set of its subgroups by conjugation the stabiliser of a subgroup A is its normaliser (defined by

$$N_G(A) := \{g \in G \mid g^{-1}Ag = A\};$$

this is the greatest subgroup of G in which A is normal).

Example 6.2. Suppose that $H \leqslant G$ and that $\Omega = \cos(G:H)$. Let $\alpha := H \in \Omega$. Then

$$G_\alpha = \{g \in G \mid \alpha^g = \alpha\} = \{g \in G \mid Hg = H\},$$

and so G_α is the subgroup H that we first thought of.

It should be clear that stabilisers are always subgroups: certainly $1 \in G_\alpha$; also, if $g \in G_\alpha$ then $\alpha^g = \alpha$, so $\alpha = \alpha^{g^{-1}}$, which means that $g^{-1} \in G_\alpha$; and if $g, h \in G_\alpha$ then $\alpha^{gh} = \alpha^h = \alpha$, which shows that $gh \in G_\alpha$. Furthermore, $\alpha^x = \alpha^y$ if and only if $xy^{-1} \in G_\alpha$, and this means that x, y are in the same right coset of G_α. We have thus proved

LEMMA 6.1: (i) *The stabiliser G_α is a subgroup of G;*
(ii) *There is a one–one correspondence, given by the mapping $G_\alpha x \mapsto \alpha^x$, between the set of right cosets of G_α and the G-orbit α^G in Ω.*

COROLLARY 6.2: *If G is finite then* $|G| = |G_\alpha| \cdot |\alpha^G|$.

Thus when G is finite the size of any transitive G-space divides $|G|$. One may see this as a variant of Lagrange's Theorem, and in fact it is closer to early (1770 to about 1860) versions of the theorem than the formulation mentioned in Chapter 1 which now bears that name. The main result of the present chapter is a further variation on the same theme:

THEOREM 6.3: *If Ω is a transitive G-space then Ω is G-isomorphic to the coset space* $\cos(G:G_\alpha)$ *for any α in Ω.*

Proof. We are given a transitive G-space Ω and a point α in Ω. If $\omega \in \Omega$ there exists $x \in G$ such that $\alpha^x = \omega$, and we define

$$\omega\theta := G_\alpha x.$$

The lemma tells us that θ is a well-defined mapping on Ω to $\cos(G:G_\alpha)$, and also that θ is bijective. Also, if $g \in G$ then

$$(\omega^g)\theta = (\alpha^{xg})\theta \qquad\qquad \text{[by choice of } x]$$
$$= G_\alpha xg \qquad\qquad \text{[by definition of } \theta]$$
$$= (G_\alpha x)^g \qquad\qquad \text{[definition of action on coset space]}$$
$$= (\omega\theta)^g,$$

which shows that θ respects the G-actions. Thus θ is a G-isomorphism, as required.

In this comparison of a general transitive G-space Ω with a coset space the choice of α has not mattered. The reason is that the group of self-equivalences of Ω, even the group of inner self-equivalences (see Lemma 3.2), is transitive on Ω, and so one point α is as good as any other. Nevertheless, different points do have different stabilisers: so we finish the analysis by asking what effect a change in α has. The answer is as follows.

LEMMA 6.4: (i) *If α, $\beta \in \Omega$ and $\beta = \alpha^x$ then $G_\beta = x^{-1}G_\alpha x$.*
(ii) *If $\theta:\Omega \to \Omega'$ is a G-isomorphism and $\alpha\theta = \beta$ then $G_\beta = G_\alpha$.*

Proof. (i) We have

$$G_\beta = \{g \in G \mid \beta^g = \beta\}$$
$$= \{g \in G \mid \alpha^{xg} = \alpha^x\}$$
$$= \{g \in G \mid \alpha^{xgx^{-1}} = \alpha\}$$
$$= \{g \in G \mid xgx^{-1} \in G_\alpha\} \qquad \text{[by definition of } G_\alpha]$$
$$= \{g \in G \mid g \in x^{-1}G_\alpha x\}$$
$$= x^{-1}G_\alpha x.$$

(ii) Similarly, if $\beta = \alpha\theta$ then

$$G_\beta = \{g \in G \mid \beta^g = \beta\}$$
$$= \{g \in G \mid (\alpha\theta)^g = \alpha\theta\}$$
$$= \{g \in G \mid (\alpha^g)\theta = \alpha\theta\}$$
$$= \{g \in G \mid \alpha^g = \alpha\} \qquad \text{[since } \theta \text{ is one–one]}$$
$$= G_\alpha.$$

PROPOSITION 6.5: *The coset spaces* $\cos(G : H)$ *and* $\cos(G : K)$ *are G-isomorphic if and only if the subgroups H, K are conjugate in G.*

Proof. Suppose that $\theta : \cos(G : H) \to \cos(G : K)$ is a G-isomorphism. Let $\alpha := H \in \cos(G : H)$, $\beta := K \in \cos(G : K)$ and $\gamma := \alpha\theta \in \cos(G : K)$. Then $\gamma = Kx = \beta^x$ for some $x \in G$ and so

$$H = G_\alpha \qquad \text{[see Example 6.2]}$$
$$= G_\gamma \qquad \text{[Part (ii) of Lemma 6.4]}$$
$$= x^{-1}G_\beta x \qquad \text{[Part (i) of Lemma 6.4]}$$
$$= x^{-1}Kx \qquad \text{[Example 6.2]}$$

Thus if $\cos(G : H) \cong \cos(G : K)$ then H, K are conjugate in G.

For the converse suppose that $H = x^{-1}Kx$. The map

$$\theta : \cos(G : H) \to \cos(G : K), \qquad Hy \mapsto Kxy$$

is then a G-isomorphism. One can deduce this from the theorem and the lemma but it is also easy to prove directly. The point of substance in the argument is that the map is well-defined, which requires checking that, in spite of appearances, our prescription depends only on the coset Hy and not on the particular coset representative y. Any other representative for the same coset is of the form hy for some h in H. But, since $xhx^{-1} \in K$, we have that

$$Kx(hy) = K(xhx^{-1})(xy) = Kxy,$$

and this is what we require. A similar routine shows that θ is bijective and a G-morphism:

$$Kt = Kx(x^{-1}t) = (Hx^{-1}t)\theta,$$

so θ is surjective; if $(Hy)\theta = (Hz)\theta$, that is if $Kxy = Kxz$, then $(xy)(xz)^{-1} = xyz^{-1}x^{-1} \in K$, so $yz^{-1} \in H$ and $Hy = Hz$, which shows that θ is injective; and finally, for any g in G we have

$$(Hy)^g\theta = (Hyg)\theta = Kxyg = (Kxy)^g = ((Hy)\theta)^g,$$

which shows that θ is a G-morphism. This completes the proof of the proposition.

For any group G we can now give a complete list of all the transitive G-spaces up to G-isomorphism. If $\mathscr{S} := \{H \,|\, H \leqslant G\}$, then we write $\mathscr{S}$ as the disjoint union of the conjugacy classes $\mathscr{S}_i$ ($i \in I$) of subgroups of G (these are, of course, just the orbits of G acting by conjugation on $\mathscr{S}$). Choose $H_i \in \mathscr{S}_i$, and let $\Gamma_i := \cos(G:H_i)$. If Ω is any transitive G-space then there is precisely one index i such that $\Omega \cong \Gamma_i$.

Specimen solutions to selected exercises

The following specimen solutions, being of the nature of an extension of the text, are placed between the main body of the chapter and the exercises. The wise and conscientious reader will, however, try the exercises unaided before attempting to evaluate what we have offered.

EXERCISE 6.2   *Let $\rho: G \to \mathrm{Sym}(\Omega)$ be the permutation representation associated with the G-space Ω, and let $K := \mathrm{Ker}\,\rho$. Show that $K = \bigcap_{\omega \in \Omega} G_\omega$, and that if G is transitive on Ω then $K = \bigcap_{g \in G} g^{-1}G_\alpha g$ for any α in Ω.*

Solution.   By definition $\mathrm{Ker}\,\rho = \{g \in G \,|\, g\rho = 1\}$, and from the definition of ρ (see p. 31) we have that $g\rho = 1$ if and only if $\omega g = \omega$ for all $\omega \in \Omega$. Thus

$$\mathrm{Ker}\,\rho = \{g \in G \,|\, \omega g = \omega \text{ for all } \omega \in \Omega\} = \bigcap_{\omega \in \Omega} G_\omega.$$

Suppose now that G is transitive on Ω. This means that for any $\omega \in \Omega$ there exists $g \in G$ such that $\alpha g = \omega$ and then $G_\omega = g^{-1}G_\alpha g$. Therefore in this situation

$$\mathrm{Ker}\,\rho = \bigcap_{\omega \in \Omega} G_\omega = \bigcap_{g \in G} g^{-1}G_\alpha g,$$

as required.

EXERCISE 6.5 *Let $\Omega := \cos(G:H)$ and suppose that $K \leqslant G$. Show that two right cosets of H lie in the same K-orbit if and only if they are contained in one and the same double coset HxK in G.*
    Show also that if H, K are finite subgroups of G then

$$|HxK| = |H| \cdot |K|/|x^{-1}Hx \cap K|.$$

Solution.   Suppose that Hx, Hy lie in the same K-orbit in Ω. Then there exists k in K such that $Hxk = Hy$. Therefore $HxK = HxkK = HyK$, so Hx, Hy both lie in the same double coset, namely HxK. Conversely, if Hx, Hy lie in the same double coset then this must be describable both as HxK and as HyK.

Thus there exist $h \in H$, $k \in K$ such that $y = hxk$, and so $Hxk = Hhxk = Hy$, that is, Hx, Hy lie in the same K-orbit.

Let $\alpha := Hx \in \Omega$ and let $\Gamma := \alpha K$, so that Γ is the K-orbit containing Hx. What we have shown in the first part of this answer is that Γ consists of those cosets of H that lie in HxK. Thus $|HxK| = |H| \cdot |\Gamma|$. Now $G_\alpha = x^{-1}Hx$ and so $K_\alpha = \{g \in K \mid \alpha g = \alpha\} = G_\alpha \cap K = x^{-1}Hx \cap K$. Thus $|\Gamma| = |K:K_\alpha| = |K|/|x^{-1}Hx \cap K|$. Therefore $|HxK| = |H| \cdot |K|/|x^{-1}Hx \cap K|$, as required.

EXERCISE 6.6   *Suppose that G acts transitively on Ω and choose α in Ω. Show that there is a one–one correspondence between the family of G-orbits in Ω^2 and the family of G_α-orbits in Ω, given by $\Delta_i \mapsto \Delta_i(\alpha)$ (for G-orbits Δ_i in Ω^2), where $\Delta(\alpha) := \{\beta \in \Omega \mid (\alpha, \beta) \in \Delta\}$. Which is the G-orbit in Ω^2 that corresponds to the G_α-orbit $\{\alpha\}$?*

Solution.   We need to show first that if Δ is a G-orbit in Ω^2 then $\Delta(\alpha)$ is a G_α-orbit in Ω; secondly, that all G_α-orbits arise in this way; and thirdly, that different G-orbits in Ω^2 give rise to different G_α-orbits in Ω.

Let Δ be a G-orbit in Ω^2 and let $\Gamma := \Delta(\alpha)$. We need to show that Γ is a non-empty transitive G_α-subspace of Ω. Suppose that $\gamma \in \Gamma$ and $g \in G_\alpha$. Then $(\alpha, \gamma) \in \Delta$, so $(\alpha g, \gamma g) \in \Delta$, that is $(\alpha, \gamma g) \in \Delta$, whence $\gamma g \in \Gamma$. This shows that Γ is a G_α-subspace of Ω. Since Δ is a G-orbit it is non-empty. Choose (ω_1, ω_2) in Δ. Since G is transitive on Ω, there exists g_1 in G such that $\omega_1 g_1 = \alpha$. Now $(\omega_1, \omega_2)g = (\alpha, \omega_2 g) \in \Delta$, so $\omega_2 g \in \Delta(\alpha) = \Gamma$, and therefore $\Gamma \neq \varnothing$. To see that G_α is transitive on Γ, let γ_1, γ_2 be elements of Γ. By definition of Γ we have $(\alpha, \gamma_1), (\alpha, \gamma_2) \in \Delta$. Since Δ is a G-orbit in Ω^2 there exists g in G such that $(\alpha, \gamma_1)g = (\alpha, \gamma_2)$. This means that $g \in G_\alpha$ and $\gamma_1 g = \gamma_2$. Thus Γ is a G_α-orbit in Ω.

The second point, that all G_α-orbits arise like this, is seen as follows. Let Γ be a G_α-orbit in Ω. Choose β in Γ and let Δ be the G-orbit in Ω^2 containing (α, β). From what we have already proved we know that $\Delta(\alpha)$ is a G_α-orbit in Ω. Since it obviously contains β it must be Γ. Thirdly we need to show that if Δ_i, Δ_j are distinct G-orbits in Ω^2 then $\Delta_i(\alpha)$, $\Delta_j(\alpha)$ must be distinct. But, being orbits, if Δ_i, Δ_j are distinct then they are disjoint and it follows that $\Delta_i(\alpha) \cap \Delta_j(\alpha) = \varnothing$, whence $\Delta_i(\alpha)$, $\Delta_j(\alpha)$ certainly are distinct. Thus $\Delta_i \mapsto \Delta_i(\alpha)$ describes a one–one correspondence between G-orbits in Ω^2 and G_α-orbits in Ω, as required.

Our reasoning shows that the G_α-orbit $\{\alpha\}$ corresponds to the G-orbit containing (α, α) in Ω^2. This is easily identified as the 'diagonal' $\{(\omega, \omega) \mid \omega \in \Omega\}$ in Ω^2.

EXERCISE 6.12   *Let G be a finite group, let $H_1, \ldots, H_m$ be representatives, one from each conjugacy class of subgroups of G, and let $n_i := |G:H_i|$.*
(i)   *Show that the power series generating function for $f_G(n)$ [see Exercise 6.11], namely the series $1 + \sum_1^\infty f_G(n)z^n$, converges in a suitable neighbourhood of 0 in the complex plane.*

(ii)   *Show that if $\phi(z):=1+\sum_1^\infty f_G(n)z^n$ then*

$$\phi(z) = \prod_{i=1}^m \frac{1}{(1-z^{n_i})}$$

within the radius of convergence of the power series.

(iii)   *Deduce that there are roots of unity $\omega_2, \ldots, \omega_k$, and polynomials $p_1, \ldots, p_k$ such that*

$$f_G(n) = p_1(n)+\omega_2^n p_2(n)+\cdots+\omega_k^n p_k(n),$$

and the degree of p_1 is $m-1$, the degree of p_r is at most $m-2$ for $r \geqslant 2$.

(iv)   *Show that*

$$f_G(n) = \frac{n^{m-1}}{A} + O(n^{m-2}), \quad \text{where } A := (m-1)! \prod_1^m n_i.$$

Solution.   We have defined $f_G(n)$ in Exercise 6.11 to be the number of isomorphism classes of G-spaces with n elements. Let Ω_i be the coset space $\cos(G:H_i)$, so that $|\Omega_i| = n_i$. If $a_1, \ldots, a_m$ is a sequence of non-negative integers such that $a_1 n_1 + a_2 n_2 + \cdots + a_m n_m = n$ then $a_1\Omega_1 + a_2\Omega_2 + \cdots + a_m\Omega_m$ (by which we mean the disjoint union of a_1 copies of Ω_1, a_2 copies of Ω_2, etc., so that the symbol $+$ is used in the sense explained formally in Chapter 3, Exercise 10) is a G-space with n elements. By Theorems 5.1 and 6.3, and Proposition 6.5, any G-space with n elements is isomorphic to one of these. By Exercise 5.2 and Proposition 6.5, distinct sequences $a_1, \ldots, a_m$ give non-isomorphic G-spaces. Thus $f_G(n)$ is the number of sequences $a_1, \ldots, a_m$ of non-negative integers such that $a_1 n_1 + \cdots + a_m n_m = n$.

(i)   We certainly must have $0 \leqslant a_i \leqslant n$ for all i, and so $f_G(n) \leqslant (n+1)^m$. Comparison with $\sum(n+1)^m z^n$ shows that $\sum f_G(n)z^n$ converges absolutely if $|z| < 1$. Since the series obviously does not converge for $z = 1$, the radius of convergence is 1.

(ii)   We have that $(1-z^{n_i})^{-1} = \sum_{a=0}^\infty z^{an_i}$. Multiplying these series together for $i := 1, \ldots, m$ and focusing on the z^n term in the product we find that its coefficient is the number of ways of writing n as $a_1 n_1 + a_2 n_2 + \cdots + a_m n_m$ for non-negative integers $a_1, a_2, \ldots, a_m$, that is, it is $f_G(n)$. Thus $\phi(z) = \prod_1^m 1/(1-z^{n_i})$ for $|z| < 1$.

(iii)   Let $g(z):= \prod_1^m (1-z^{m_i})$. Obviously, 1 is a root of $g(z)$ of multiplicity m. Let $\omega_2^{-1}, \ldots, \omega_k^{-1}$ be all its other roots, and let m_r be the multiplicity of ω_r^{-1}, so that

$$g(z) = (1-z)^m \prod_{r=2}^k (1-\omega_r z)^{m_r}.$$

Note that $\omega_2, \ldots, \omega_k$ are roots of unity as the question specifies. Note also that $m_r < m$ for all $r \geqslant 2$. For notational convenience let's set $\omega_1 := 1$ and $m_1 := m$. Now there is a partial fraction expansion

$$\frac{1}{g(z)} = \sum_{r=1}^{k} \sum_{s=1}^{m_r} \frac{c_{rs}}{(1-\omega_r z)^s},$$

for suitable constants c_{rs}. The Binomial Theorem tells us that

$$(1-\omega_r z)^{-s} = \sum_{n=0}^{\infty} \binom{n+s-1}{s-1} (\omega_r z)^n.$$

The coefficient

$$\binom{n+s-1}{s-1}$$

is a polynomial in n of degree $s-1$ and its leading coefficient is $1/(s-1)!$. Thus if we expand each summand in the partial fraction expression for $1/g(z)$ as a power series we find that $1/g(z) = \sum_{n=0}^{\infty} c_n z^n$, where c_n has the form $p_1(n) + p_2(n)\omega_2^n + \cdots + p_k(n)\omega_k^n$. Here p_r is a polynomial of degree at most $m_r - 1$, that is, the degree of p_1 is at most $m-1$ and the degree of p_r is at most $m-2$ if $r \geqslant 2$. From (ii) we know that $f_G(n) = c_n$ and so we have the required expression for $f_G(n)$.

(iv)   Since the numbers ω_r are roots of unity, $|\omega_r^n| = 1$ for all n. Therefore $f_G(n) = Cn^{m-1} + O(n^{m-2})$ where C is the leading coefficient of $p_1(n)$. In the notation introduced above, this leading coefficient is c_{1m} times the coefficient of n^{m-1} in

$$\binom{n+s-1}{s-1},$$

that is, it is $c_{1m}/(s-1)!$. We need to find c_{1m}. Let us write $g(z) = (1-z)^m h(z)$. Then $h(z) = \prod_1^m (1 + z + \cdots + z^{n_i - 1})$ and so $h(1) = \prod_1^m n_i$. It follows that $1/h(z)$ can be expanded as a power series in $1-z$ in a suitably small neighbourhood of $z = 1$, and the constant coefficient in this Taylor expansion will be $\prod_1^m n_i$. On the other hand, this constant coefficient is clearly just c_{1m}. Thus

$$f_G(n) = \frac{n^{m-1}}{A} + O(n^{m-2}), \quad \text{where } A := (m-1)! \prod_1^m n_i,$$

as required.

EXERCISE 6.17   *Suppose that Ω is a transitive G-space and that G has a regular normal subgroup N (that is, $N \trianglelefteq G$ and N acts regularly on Ω).*
(i)   *Show that $G_\alpha \cap N = \{1\}$ and $G_\alpha N = G$ for any $\alpha \in \Omega$.*

(ii)   *Show that N, with G_α acting by conjugation, is G_α-isomorphic to Ω.*
(iii)   *Let A be the group of automorphisms $\{u \mapsto g^{-1}ug \mid g \in G_\alpha\}$ induced by G_α on N. Show that if G acts faithfully on Ω then (Ω, G) is equivalent to (N, X), where X is the set of all transformations $u \mapsto ua + b$ $(a \in A, b \in N)$ of N.*

[NB   since N need not be commutative, the $+$ symbol is perhaps inappropriate here. On the one hand it emphasizes the similarity with affine transformations. On the other hand, we ought really to write $u \mapsto u^a b$.]

Solution. (i)   To say that N acts regularly on Ω means it is transitive and any stabiliser is the trivial group. Choose any point α in Ω. Since $N_\alpha = G_\alpha \cap N$ we have that $G_\alpha \cap N = \{1\}$. Since N is transitive, every coset $G_\alpha x$ contains an element of N and so $G_\alpha N = G$.

   (ii)   The map $\phi : u \mapsto \alpha^u$ from N to Ω is surjective since N is transitive and it is injective since $N_\alpha = \{1\}$. For any $g \in G_\alpha$ and any $u \in N$ we have

$$(g^{-1}ug)\phi = \alpha(g^{-1}ug) = \alpha^{ug} = (\alpha\phi)^g.$$

This shows that ϕ is a G_α-isomorphism from N with G_α acting by conjugation to Ω with G_α acting as a subgroup of G.

   (iii)   Define a map $\theta : G \to X$ as follows. Since $G = G_\alpha N$ and $G_\alpha \cap N = \{1\}$ every element g of G can be uniquely factorised in the form $g = ab$ with $a \in G_\alpha$, $b \in N$. Since G acts faithfully on Ω so does G_α, and therefore G_α acts faithfully on N by conjugation. Thus the group A of automorphisms of N induced by conjugation with elements of G_α is naturally isomorphic with G_α and we can think of a as being an element of A. We define $g\theta$ to be the map $u \mapsto u^a b$ of N. If $g_1, g_2 \in G$ and $g_1 = a_1b_1$, $g_2 = a_2b_2$, then $g_1g_2 = (a_1a_2)(a_2^{-1}b_1a_2b_2)$, from which it follows easily that $(g_1g_2)\theta = g_1\theta g_2\theta$, that is, that θ is an isomorphism. Furthermore, if $y \in N$ and $x \in X$, say $x : u \mapsto u^a b$ for all $u \in N$, then

$$(y^x)\phi = (y^a b)\phi = \alpha^{a^{-1}yab} = \alpha^{yab} = (y\phi)^{ab}.$$

Since $ab = x\theta^{-1}$ this shows that if $\psi := \theta^{-1}$ then (ϕ, ψ) is an equivalence of (N, X) with (Ω, G) as required.

EXERCISE 6.19   *In $PGL(n, F)$ let A be the stabiliser of the hyperplane $\Pi := \{x \in F^n \mid x_n = 0\} \in \Sigma_{n-2}$.*
(i)   *Describe A explicitly (in terms of matrices).*
(ii)   *Now recall that Σ_0, the set of points of $PG(n-1, F)$, is the set of 1-dimensional subspaces of F^n, and let $\Pi_0 := \{P \in \Sigma_0 \mid P \leqslant \Pi\}$. What is the kernel K of the action of A on Π_0?*
(iii)   *Show that $A/K \cong PGL(n-1, F)$ acting in the natural way on Π_0.*
(iv)   *Show that $(\Sigma_0 \setminus \Pi_0, A)$ is equivalent to $(F^{n-1}, AGL(n-1, F))$.*

Solution.   We will use $n \times n$ matrices over F to represent elements of $PGL(n, F)$: thus two matrices represent the same element of the projective group if and only if they are scalar multiples of each other.

(i)   Any matrix of the form

$$\begin{pmatrix} M & 0 \\ b & d \end{pmatrix}$$

with $M \in GL(n-1, F)$, $d \neq 0$ and $b \in F^{n-1}$ certainly stabilises Π and it is easy to see conversely that any matrix that stabilises Π must have this form.

(ii)   An element P of Π_0 is a 1-dimensional subspace of F^n spanned by a vector of the form $(x_1, \ldots, x_{n-1}, 0)$. Let us write this as $(x, 0)$ where $x \in F^{n-1}$. Since

$$(x, 0) \begin{pmatrix} M & 0 \\ b & d \end{pmatrix} = (xM, 0),$$

we see that the element

$$\begin{pmatrix} M & 0 \\ b & d \end{pmatrix}$$

of A maps every such vector to a scalar multiple of itself (which is what is needed if it is to stabilise every 1-dimensional space contained in Π) if and only if M is a scalar matrix.

(iii)   Since every matrix $M \in GL(n-1, F)$ gives rise to elements of A, the calculation in (ii) shows that any map $x \mapsto xM$ is induced by an element of A. Thus $A/K \cong PGL(n-1, F)$ acting in the natural way on Π_0.

(iv)   Each point in $\Sigma_0 \backslash \Pi_0$ has a unique representative vector of the form $(x, 1)$ where $x \in F^{n-1}$. This gives us a map $\phi : \Sigma_0 \backslash \Pi_0 \rightarrow F^{n-1}$. Every element of A has a unique representative matrix of the form

$$\begin{pmatrix} M & 0 \\ b & 1 \end{pmatrix}$$

and we define $\psi : A \rightarrow AGL(n-1, F)$ by

$$\begin{pmatrix} M & 0 \\ b & 1 \end{pmatrix} \mapsto T_{M,b},$$

where $T_{M,b}$ is the transformation $x \mapsto xM + b$ in $AGL(n-1, F)$. It is a matter of routine to check that ψ is an isomorphism of groups and that (ϕ, ψ) is an equivalence of $(\Sigma_0 \backslash \Pi_0, A)$ with $(F^{n-1}, AGL(n-1, F))$ as required.

Exercises

6.1.   Let Ω be a transitive G-space. Show that the following are equivalent:
(i) Ω is G-isomorphic to the right regular G-space;
(ii) $G_\omega = \{1\}$ for all ω in Ω;
(iii) $G_\alpha = \{1\}$ for some α in Ω.

6.2.   Let $\rho: G \to \text{Sym}(\Omega)$ be the permutation representation associated with the G-space Ω, and let $K := \text{Ker } \rho$. Show that $K = \bigcap_{\omega \in \Omega} G_\omega$, and that if G is transitive on Ω then $K = \bigcap_{g \in G} g^{-1} G_\alpha g$ for any α in Ω.

6.3.   Let Ω be a faithful transitive G-space. Show that if G is abelian then Ω is isomorphic to the regular G-space.

6.4.   Suppose that G acts transitively on Ω, and that $A \leqslant G$. Show that the following are equivalent:
   (i) the subgroup A is transitive on Ω;
   (ii) $G_\omega A = G$ for every ω in Ω;
   (iii) $G_\alpha A = G$ for some α in Ω.

6.5.   Let $\Omega := \cos(G:H)$ and suppose that $K \leqslant G$. Show that two right cosets of H lie in the same K-orbit in Ω if and only if they are contained in one and the same double coset HxK in G.

   Show also that if H, K are finite subgroups of G then

$$|HxK| = |H| \cdot |K|/|x^{-1}Hx \cap K|.$$

6.6.   Suppose that G acts transitively on Ω and choose α in Ω. Show that there is a one–one correspondence between the family of G-orbits in Ω^2 and the family of G_α-orbits in Ω, given by

$$\Delta_i \mapsto \Delta_i(\alpha)$$

(for G-orbits Δ_i in Ω^2), where

$$\Delta(\alpha) := \{\beta \in \Omega \,|\, (\alpha, \beta) \in \Delta\}.$$

Which is the G-orbit in Ω^2 that corresponds to the G_α-orbit $\{\alpha\}$? [The G_α-orbits in Ω are known as 'suborbits' of G; the G-orbits in Ω^2 are sometimes known as 'orbitals' of G; but often these terms are used interchangeably. The number of them is the 'rank' of G.]

6.7.   Suppose that G acts transitively on sets Ω_1 and Ω_2. For $\alpha_i \in \Omega_i$ let $G_i := \text{Stab}(\alpha_i)$. Show that
   (i) the number of orbits of G_1 on Ω_2,
   (ii) the number of orbits of G_2 on Ω_1, and
   (iii) the number of orbits of G on $\Omega_1 \times \Omega_2$
      are the same.

   What is this number when G is S_n acting on the sets of r-element and s-element subsets (respectively) of $\{1, 2, \ldots, n\}$?
                     [Oxford Final Honour School 1972, Paper I, Qn. 7.]

6.8. (i) Prove that the following are equivalent for a G-space Ω:
  (a) G is doubly transitive on Ω [see p. 58 for definition];
  (b) G is transitive and the stabiliser G_ω is transitive on $\Omega \setminus \{\omega\}$ for any ω in Ω;
  (c) G is transitive and the stabiliser G_α is transitive on $\Omega \setminus \{\alpha\}$ for some α in Ω.
  (ii) Prove that G is doubly transitive on $\cos(G:H)$ if and only if $G = H \cup HxH$ for any (or for some) x in $G \setminus H$.

6.9. Show that $\cos(G:H)$ and $\cos(G:K)$ are equivalent G-spaces (in the sense of Chapter 3, p. 32) if and only if there is an automorphism θ of G such that $H\theta = K$.

6.10. List the isomorphism classes of transitive G-spaces when G is
  (i) cyclic of prime order;   (ii) S_3;   (iii) A_4.

6.11. If G is a finite group let $f_G(n)$ be the number of (isomorphism classes of) G-spaces with n elements. Show that
  (i) if G is cyclic of prime order p then

$$f_G(n) = 1 + [n/p],$$

  where $[x] := \max\{m \mid m \leqslant x \text{ and } m \in \mathbb{Z}\}$;
  (ii) if G is S_3 then

$$f_G(1) = 1, \quad f_G(2) = 2, \quad f_G(3) = 3,$$
$$f_G(4) = 4, \quad f_G(5) = 5, \quad f_G(6) = 8.$$

  (iii) Calculate $f_G(n)$ for $n \leqslant 12$ when $G = A_4$.

6.12. Let G be a finite group, let $H_1, \ldots, H_m$ be representatives, one from each conjugacy class of subgroups of G, and let $n_i := |G:H_i|$.
  (i) Show that the power series generating function for $f_G(n)$ [see Exercise 6.11], namely the series $1 + \sum_1^\infty f_G(n)z^n$, converges in a suitable neighbourhood of 0 in the complex plane.
  (ii) Show that if $\varphi(z) = 1 + \sum_1^\infty f_G(n)z^n$ then

$$\phi(z) = \prod_{i=1}^m 1/(1 - z^{n_i})$$

  within the radius of convergence of the power series.
  (iii) Deduce that there are roots of unity $\omega_2, \ldots, \omega_k$, and polynomials $p_1, \ldots, p_k$ such that

$$f_G(n) = p_1(n) + \omega_2^n p_2(n) + \cdots + \omega_k^n p_k(n),$$

  and the degree of p_1 is $m-1$, the degree of p_r is at most $m-2$ for $r \geqslant 2$.

(iv) Show that

$$f_G(n) = \frac{n^{m-1}}{A} + O(n^{m-2}), \quad \text{where } A := (m-1)! \prod_1^m n_i.$$

6.13. (i) Show that if n is odd then the dihedral group D_n (of order $2n$) has, up to isomorphism, just one transitive permutation representation of degree n.

(ii) Show that if n is even then D_n has three transitive permutation representations of degree n up to isomorphism and two up to equivalence.

6.14. Show that $\{\Delta, -\Delta\}$ is an orbit of S_n acting on $F[x_1, \ldots, x_n]$, where

$$\Delta := \prod_{i < j} (x_i - x_j),$$

and give an alternative proof of the fact that the set A_n of even permutations is a subgroup of index 2 in S_n (when $n \geqslant 2$).

6.15. Consider again the action of S_n on $F[x_1, \ldots, x_n]$ permuting the variables as in Example 4.3. For $f \in F[x_1, \ldots, x_n]$ let

$$\Phi := f^{S_n} := \{f^\sigma \mid \sigma \in S_n\},$$

$$N := |\Phi|,$$

$$G(f) := \text{Stab}(f) := \{\sigma \in S_n \mid f^\sigma = f\}$$

(i) Show that N divides $n!$. [This was the original 'Lagrange's Theorem'.]

(ii) Give examples of polynomials f for which

(a) $N = 1$; (b) $N = 2$;
(c) $N = n$; (d) $N = \frac{1}{2}n(n-1)$;
(e) $N = n(n-1)$; (f) $N = n!$.

[From about 1845 to perhaps 1870 finite groups were mainly thought of in this way as subgroups of S_n: the symmetric group acted on 'functions' f by 'substitution' of variables; $G(f)$ was the 'group of the function'; N was the number of 'different values' of the function; and $|G(f)|$ the number of 'equal values' of f. The problem of deciding which values of N can occur, and, by implication, what numbers can arise as the orders of subgroups of S_n, was seen as one of the important problems of mathematics for three or four decades of the nineteenth century. It was, for example, the problem posed by the Paris Academy for its Grand Prix of 1860. By about 1880, however, interest in it had waned, and by 1900 it was mentioned in textbooks as a matter of historical interest only.]

6.16. Let *G* be a subgroup of S_n. Show that there is a polynomial *f* in $F[x_1, \ldots, x_n]$ such that, in the notation of Exercise 6.15, $G = G(f)$. [One way to tackle this is to find a 'function' f_0 such that the 'values' f_0^σ are not merely all different for $\sigma \in S_n$ but are actually linearly independent over *F*. Then put $f := \sum_{\sigma \in G} f_0^\sigma$.]

6.17. Suppose that Ω is a transitive *G*-space and that *G* has a regular normal subgroup *N* (that is, $N \lhd G$ and *N* acts regularly on Ω).
   (i) Show that $G_\alpha \cap N = \{1\}$ and $G_\alpha N = G$ for any $\alpha \in \Omega$.
   (ii) Show that *N*, with G_α acting by conjugation, is G_α-isomorphic to Ω.
   (iii) Let *A* be the group of automorphisms $\{u \mapsto g^{-1}ug \mid g \in G_\alpha\}$ induced by G_α on *N*. Show that if *G* acts faithfully on Ω then (Ω, G) is equivalent to (N, X), where *X* is the set of all transformations $u \mapsto ua + b$ ($a \in A$, $b \in N$) of *N*.

6.18. (i) Identify the stabiliser of 0 in the natural action of $\mathrm{AGL}(n, F)$ on F^n.
   (ii) What do the results of Exercise 6.17 tell us in this case?

6.19. In $\mathrm{PGL}(n, F)$ let *A* be the stabiliser of the hyperplane
$$\Pi := \{x \in F^n \mid x_n = 0\} \in \Sigma_{n-2}.$$
   (i) Describe *A* explicitly (in terms of matrices).
   (ii) Now recall that Σ_0, the set of point of $\mathrm{PG}(n-1, F)$, is the set of 1-dimensional subspaces of F^n, and let
$$\Pi_0 := \{P \in \Sigma_0 \mid P \leqslant \Pi\}.$$
   What is the kernel *K* of the action of *A* on Π_0?
   (iii) Show that $A/K \cong \mathrm{PGL}(n-1, F)$ acting in the natural way on Π_0.
   (iv) Show that $(\Sigma_0 \backslash \Pi_0, A)$ is equivalent to $(F^{n-1}, \mathrm{AGL}(n-1, F))$.

6.20. Show that the stabiliser in $O(n)$ of a point of the sphere S^{n-1} (in $\mathbb{R}^n$) is (isomorphic to) $O(n-1)$. What are its orbits in S^{n-1}?

Chapter 7

G-morphisms

We defined *G*-morphisms in Chapter 3: a morphism on the *G*-space Ω to a *G*-space Ω' is a mapping $\theta:\Omega \to \Omega'$ that respects the action of elements of G in the sense that $(\omega^g)\theta = (\omega\theta)^g$ for all $\omega \in \Omega$ and all $g \in G$. Here are some examples of morphisms other than *G*-isomorphisms.

Example 7.1. Let Ω be any *G*-space with orbits Ω_i $(i \in I)$. Let Ω' be I with G acting trivially on it. The map $\theta:\Omega \to \Omega'$ such that $\omega\theta = i$ if $\omega \in \Omega_i$ is a *G*-morphism (because ω and ω^g lie in the same orbit) that collapses each orbit of G to a single point. In many contexts one writes Ω/G for the set of orbits considered as a trivial *G*-space in this way.

Example 7.2. Let Ω be the regular *G*-space and let $\Omega' := \cos(G:H)$ where $H \leqslant G$. The mapping $\theta:x \mapsto Hx$ on Ω to Ω' is a *G*-morphism because

$$(\omega^g)\theta = (\omega g)\theta = H\omega g = (H\omega)^g = (\omega\theta)^g$$

for any ω in Ω and any g in G.

Example 7.3. If G is $\mathrm{GL}(n, F)$ (where F is a field, as usual), Ω is $F^n\backslash\{0\}$, and Ω' is Σ_0, the set of 1-dimensional subspaces of F^n, then the map $x \mapsto \langle x \rangle$ (where $\langle\rangle$ denotes subspace spanned by) is a *G*-morphism.

We shall come back to *G*-morphisms shortly. First we turn to equivalence relations. The equivalence relation ρ on the *G*-space Ω is said to be *G*-invariant (or, simply, invariant) if

$$\alpha \equiv \beta \ (\mathrm{mod} \ \rho) \Rightarrow \alpha^g \equiv \beta^g \ (\mathrm{mod} \ \rho) \ \text{for any } g \text{ in } G.$$

An invariant equivalence relation will be called a *congruence* on Ω. Associated with any equivalence relation ρ is a partition of Ω as the disjoint union of the equivalence classes Γ_i. The condition for ρ to be a congruence says that if α, β are in the same class Γ_i then for any g in G the transforms α^g, β^g will be in one and the same class Γ_j. This means that $\Gamma_i g = \Gamma_j$ (where j depends, of course, on i and on g). Thus elements of G act to permute the equivalence classes bodily amongst themselves. The set $\{\Gamma_i \,|\, i \in I\}$ of ρ-classes with this natural action of G upon it is known as the *quotient space* (or the factor space) Ω/ρ. If we write $\rho(\omega)$ for the equivalence class that contains ω then our action of G on Ω/ρ is given by the rule $\rho(\omega)^g = \rho(\omega^g)$. This shows that the map $\omega \mapsto \rho(\omega)$ is a *G*-morphism

$\Omega \to \Omega/\rho$. It is known as the natural (or canonic) surjection (or projection or epimorphism).

Example 7.1'. Let Ω be any G-space. The relation ρ of equivalence under G defined in Chapter 5 p. 51 by the rule

$$\alpha \equiv \beta \, (\text{mod } G) \Leftrightarrow (\exists g \in G)(\alpha^g = \beta)$$

is a congruence. Here the equivalence classes are the orbits and our projection $\Omega \to \Omega/\rho = \Omega/G$ is essentially the morphism exhibited in Example 7.1.

Example 7.2'. Let Ω be the regular G-space. If $H \leqslant G$ we define ρ by

$$\alpha \equiv \beta \, (\text{mod } \rho) \Leftrightarrow \alpha \beta^{-1} \in H.$$

This is a congruence because if $\alpha \beta^{-1} \in H$ then

$$(\alpha^g)(\beta^g)^{-1} = (\alpha g)(\beta g)^{-1} = \alpha \beta^{-1} \in H$$

Here the ρ-classes are the right cosets of H in G so the quotient space is $\text{cos}(G:H)$ and our map $\Omega \to \Omega/\rho$ is the G-morphism displayed in Example 7.2.

Example 7.3'. Let Ω be $F^n \backslash \{0\}$. We prescribe that vectors x, y are equivalent under ρ if each is a (necessarily non-zero) scalar multiple of the other. Since $GL(n, F)$ acts by linear transformations ρ is a $GL(n, F)$-congruence and, if we identify Ω/ρ with Σ_0 by the natural isomorphism, then the projection $\Omega \to \Omega/\rho$ is the G-morphism described in Example 7.3.

Example 7.4'. Let Ω be a G-space and suppose that $H \lhd G$. Define ρ by the rule

$$\alpha \equiv \beta \, (\text{mod } \rho) \Leftrightarrow (\exists h \in H)(\alpha^h = \beta).$$

The fact that H is normal in G ensures that ρ is G-invariant: if $\alpha^h = \beta$ then

$$\alpha^{g(g^{-1}hg)} = \alpha^{hg} = \beta^g$$

and, as $g^{-1}hg \in H$, we have $\alpha^g \equiv \beta^g \, (\text{mod } \rho)$. In this case Ω/ρ is Ω/H, the set of H-orbits, and we see that G acts to permute amongst themselves the orbits of a normal subgroup (cf. Exercise 5.18, pp. 57, 60).

As our examples suggest, the connection between morphisms and congruences is a close one:

PROPOSITION 7.1: *Let $\theta : \Omega \to \Omega'$ be a G-morphism. Define ρ_θ, a relation on Ω, by the rule*

$$\alpha \equiv \beta \, (\text{mod } \rho_\theta) \Leftrightarrow \alpha\theta = \beta\theta.$$

Then ρ_θ is a congruence on Ω, and the image $\Omega\theta$ is a G-subspace of Ω' that is

G-isomorphic with the quotient Ω/ρ_θ. (In fact θ may be factorised as the composite $\phi\psi$, where ϕ is the projection $\Omega \to \Omega/\rho_\theta$, and ψ is an injective morphism $\Omega/\rho_\theta \to \Omega'$ defined by $\rho_\theta(\alpha) \mapsto \alpha\theta$.)

We leave the proof to the reader (see Exercise 7.2 at the end of this chapter). The idea should not be a new one: although we have had to work with a congruence instead of the kernel of a homomorphism, the situation is essentially the same as that described by the First Isomorphism Theorems of group theory, of ring theory, and of module theory. It is something very general whose proper context is a subject known as 'Universal Algebra'.

The study of morphisms is one of those topics that can be reduced very quickly to the transitive case: one applies the following easy lemma to the orbits.

LEMMA 7.2: *If Ω is a transitive G-space and $\theta:\Omega \to \Omega'$ is a G-morphism then the image $\Omega\theta$ is a transitive G-space.*

The proof is routine: if $\alpha', \beta' \in \Omega\theta$ then $\alpha' = \alpha\theta$ and $\beta' = \beta\theta$ for some $\alpha, \beta \in \Omega$; by transitivity of G on Ω there exists g in G such that $\alpha^g = \beta$; then, since θ is a morphism,

$$\alpha'^g = (\alpha\theta)^g = (\alpha^g)\theta = \beta\theta = \beta'.$$

Thus G is transitive on $\Omega\theta$, as claimed.

From the lemma it follows that any G-morphism $\theta:\Omega \to \Omega'$ maps orbits Ω_i in Ω to orbits in Ω', and of course it acts as a morphism on each orbit. Conversely, given a collection θ_i ($i \in I$) of morphisms, one on each orbit Ω_i of Ω to some orbit Ω_i' of Ω', we obtain a morphism $\theta:\Omega \to \Omega'$ by defining $\omega\theta := \omega\theta_i$, where i is the unique index such that $\omega \in \Omega_i$. It is in this sense that morphisms are determined by morphisms between orbits.

Suppose now that Ω and Ω' are non-empty transitive G-spaces and that $\theta:\Omega \to \Omega'$ is a morphism. Then θ is automatically surjective and so, by the 'First Isomorphism Theorem' (Proposition 7.1 above) there is a congruence ρ on Ω such that $\Omega' = \text{Im } \theta \cong \Omega/\rho$. The ρ-classes are permuted transitively by the action of G, so they must all have the same size, and the effect of θ is to collapse each ρ-class in Ω down to one point in Ω', as portrayed in Fig. 7.1.

To complete the picture we need to know what G-morphisms there are from a given transitive G-space Ω to another. An answer can be given in terms of stabilisers:

LEMMA 7.3: *Let Ω, Ω' be transitive G-spaces and let α, α' be points of Ω, Ω' respectively. There is a G-morphism $\theta:\Omega \to \Omega'$ such that $\theta:\alpha \mapsto \alpha'$ if and only if $G_\alpha \leqslant G_{\alpha'}$. Furthermore, such a morphism, if it exists, is unique.*

Proof. Suppose first that there is a morphism θ such that $\theta:\alpha \mapsto \alpha'$. Then

$$\theta:\alpha^g \mapsto \alpha'^g \qquad \text{for all } g \in G,$$

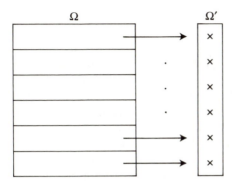

Fig. 7.1. Portrait of a *G*-morphism.

and so

$$\theta: \alpha \mapsto \alpha'^g \qquad \text{for all } g \in G_\alpha.$$

Consequently

$$\alpha'^g = \alpha' \qquad \text{for all } g \in G_\alpha,$$

that is to say, $G_\alpha \leqslant G_{\alpha'}$.

For the converse suppose that $G_\alpha \leqslant G_{\alpha'}$. By transitivity, given $\omega \in \Omega$ there exists $x \in G$ such that $\alpha^x = \omega$, and we define $\omega\theta := \alpha'^x$. On the face of it this may be inadmissible as a definition: but if y is another element of G such that $\alpha^y = \omega$ then $xy^{-1} \in G_\alpha$, so $xy^{-1} \in G_{\alpha'}$, and

$$\alpha'^x = \alpha'^{(xy^{-1})y} = \alpha'^y.$$

Thus our definition of θ is unambiguous. And of course, for any g in G,

$$(\omega^g)\theta = (\alpha^{xg})\theta = \alpha'^{xg} = (\alpha'^x)^g = (\omega\theta)^g,$$

so that θ is a morphism, as required.

The uniqueness is an immediate consequence of the transitivity of G on Ω: for every element of ω is of the form α^g and

$$(\alpha^g)\theta = (\alpha\theta)^g = \alpha'^g.$$

THEOREM 7.4: *Let Ω be a non-empty transitive G-space and α a point of Ω. There is an order-preserving one–one correspondence between congruences on Ω and subgroups H of G that contain the stabiliser G_α.*

Proof. Given a congruence ρ on Ω we define

$$H := H(\rho) := \{g \in G \,|\, \alpha^g \equiv \alpha \ (\text{mod } \rho)\}.$$

Certainly $1 \in H$; if $g \in H$ then $g^{-1} \in H$; and if $g, h \in H$ then $\alpha^{gh} \equiv \alpha^h \equiv \alpha \ (\text{mod } \rho)$

so $gh \in H$. Thus $H \leqslant G$, and it is clear that $G_\alpha \leqslant H$. In fact of course if we put $\alpha' := \rho(\alpha) \in \Omega/\rho$ then H is the stabiliser $G_{\alpha'}$.

Notice that $\rho(\alpha)$, the equivalence class containing α, is an H-orbit in Ω. Certainly it is an H-subspace because for any $g \in H$ we have $\alpha \in \rho(\alpha) \cap \rho(\alpha)^g$, so $\rho(\alpha) = \rho(\alpha)^g$. But H is transitive on $\rho(\alpha)$ because if $\beta \in \rho(\alpha)$ then there exists $g \in G$ such that $\alpha^g = \beta$ and since $\beta \equiv \alpha \pmod{\rho}$ this element g is, by definition, in H. Thus $\rho(\alpha)$ is the H-orbit containing α.

Suppose that ρ, σ are congruences on Ω such that $H(\rho) = H(\sigma)$. Then $\rho(\alpha) = \sigma(\alpha)$ since each of these equivalence classes is the H-orbit containing α. But the equivalence class containing α determines the whole relation if it is a congruence: the ρ-classes in Ω are the transforms $\rho(\alpha)^g$ for $g \in G$; the σ-classes are the sets $\sigma(\alpha)^g$; so, as $\rho(\alpha) = \sigma(\alpha)$, the relations ρ, σ have the same equivalence classes and therefore $\rho = \sigma$.

Now start with a subgroup H such that $G_\alpha \leqslant H$. Let $\Omega' := \cos(G{:}H)$ and let $\alpha' := H \in \Omega'$. Then $G_\alpha \leqslant G_{\alpha'} = H$ (see Example 6.2) and so, by Lemma 7.3, there is a morphism $\Omega \to \Omega'$ such that $\alpha \mapsto \alpha'$. If ρ is the corresponding congruence (defined as in Proposition 7.1), then we have $H(\rho) = G_{\alpha'} = H$, as one sees easily. This shows that every subgroup H arises from some congruence ρ, and completes the proof that our mapping $\rho \mapsto H(\rho)$, from congruences to subgroups containing α, is bijective.

Finally, if $\rho \subseteq \sigma$ (which means that every σ-class is a union of ρ-classes contained in it; or that $\alpha \equiv \beta \pmod{\rho} \Rightarrow \alpha \equiv \beta \pmod{\sigma}$), then from the definitions it follows immediately that $H(\rho) \leqslant H(\sigma)$. Thus our one–one correspondence is order-preserving, and the proof of the theorem is complete.

We finish this chapter and this purely theoretical part of our treatment of group actions with some observations related to our last two results. First we complete the theory that we described in Chapter 6 by treating a question that was left unanswered—and unasked—there. The question is, what are the automorphisms of a transitive G-space Ω? Its relevance to the considerations of Chapter 6 is that an answer should tell us how far from unique an isomorphism of Ω with the appropriate coset space is.

It is an immediate consequence of Lemma 7.3 that, since Ω is a transitive G-space, there is always at most one automorphism of Ω mapping α to β, and that there is such an automorphism if and only if $G_\alpha = G_\beta$. We can give a more precise answer in terms of the normaliser of a stabiliser G_α. Recall that the normaliser of a subgroup H of G is defined by

$$N(H) := N_G(H) := \{g \in G \mid g^{-1}Hg = H\}.$$

It is a subgroup of G, and $H \trianglelefteq N(H)$.

THEOREM 7.5:   *Suppose that Ω is a transitive G-space. Let $A := \mathrm{Aut}(\Omega)$, the group of G-automorphisms of Ω, and let α be any point in Ω. Then*
(i)   *Ω is 'semi-regular' as A-space (that is, every A-orbit in Ω is isomorphic to the regular A-space);*

(ii) *the A-orbits are the ρ-classes, where ρ is the G-congruence on Ω corresponding to N(G$_α$) as in the preceding theorem;*
(iii) $A \cong N(G_α)/G_α$.

Proof. (i) If $ω \in Ω$ then $A_ω = \{1\}$ because the identity is the unique automorphism mapping $ω$ to $ω$. Thus the *A*-orbit containing $ω$ is *A*-isomorphic to $\cos(A:\{1\})$ and this is the regular *A*-space.

(ii) Define the relation $ρ$ on $Ω$ by

$$α \equiv β \pmod ρ \Leftrightarrow (\exists θ \in A)(αθ = β).$$

Certainly $ρ$ is an equivalence relation: it should already be familiar as the relation whose equivalence classes are the *A*-orbits in $Ω$. Also, it is *G*-invariant for, if $α \equiv β \pmod ρ$ then $αθ = β$ for some $θ \in A$, so $(αθ)^g = β^g$; as $θ$ is a *G*-morphism we have $(α^g)θ = β^g$, and this says that $α^g \equiv β^g \pmod ρ$. By the consequence of Lemma 7.3 noted above, if $g \in G$ then $α^g \equiv α \pmod ρ$ if and only if the stabilisers $G_α$ and $G_{α^g}$ are the same. Now $G_{α^g} = g^{-1}G_α g$ (Lemma 6.3), so

$$α^g \equiv α \pmod ρ \Leftrightarrow g \in N(G_α).$$

Thus $ρ$ is the congruence that corresponds to $N(G_α)$ as in Theorem 7.4.

(iii) We have just seen that for $g \in N(G_α)$ there is a unique automorphism $θ_g$ that maps $α$ to $α^g$. Also

$$αθ_g θ_h = α^g θ_h = (αθ_h)^g = α^{hg} = αθ_{hg}$$

and so, since the image of $α$ determines an automorphism uniquely, $θ_g θ_h = θ_{hg}$. Define $t: N(G_α) \to A$ by $t: g \mapsto θ_{g^{-1}}$. Then t is a group homomorphism because

$$(gh)t = θ_{(gh)^{-1}} = θ_{h^{-1}g^{-1}} = θ_{g^{-1}}θ_{h^{-1}} = (gt)(ht).$$

Also t is surjective because (by Lemma 7.3 again) every automorphism is of the form $θ_g$ for some g in $N(G_α)$; and Ker $t = G_α$. Hence $A \cong N(G_α)/G_α$, and the proof is complete.

Our last comments have to do with the notion of primitivity. Let $Ω$ be a transitive *G*-space. If $ρ$ is a congruence on $Ω$ then, as we have seen earlier, the $ρ$-classes form a partition of $Ω$ with the property that its sets are permuted bodily by the action of elements of *G*. Since they are permuted transitively the $ρ$-classes all have the same size, as suggested in Fig. 7.2. We talk of a non-

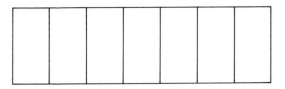

Fig. 7.2. Portrait of a congruence.

trivial congruence if the classes have more than one element (that is, if $\rho \neq =$), and a proper congruence if there is more than one class (that is, if $\rho \neq \Omega^2$). The G-space Ω is said to be *primitive* (or G is said to act primitively, or to be a primitive permutation group on Ω) if it is transitive and non-trivial and there is no non-trivial proper congruence on Ω.

Example 7.5.   If G is doubly transitive (see p. 58) then G is primitive on Ω. For, let ρ be a non-trivial congruence. This means that there exist α, β, in Ω with $\alpha \neq \beta$ and $\alpha \equiv \beta \pmod{\rho}$. But now if γ, δ are any two unequal elements of Ω then there exists g in G such that $\alpha^g = \gamma$ and $\beta^g = \delta$ (this is 2-transitivity), and so $\gamma \equiv \delta \pmod{\rho}$. Thus ρ is the universal relation.

Example 7.6.   Let G be transitive of prime degree p on Ω (recall, the degree is the size of Ω). Then G must be primitive because, as we have seen, the ρ-classes all have the same size if ρ is a congruence, and this size, $|\rho|$, must therefore divide $|\Omega|$. In fact of course $|\Omega| = |\rho| . |\Omega/\rho|$.

The main theorem of this chapter (Theorem 7.4) has the following consequence.

COROLLARY 7.6:   *The G-space Ω is primitive if and only if it is non-trivial and transitive, and a stabiliser G_α is a maximal proper subgroup of G.*

Specimen solutions to selected exercises

The following specimen solutions, being of the nature of an extension of the text, are placed between the main body of the chapter and the exercises. The wise and conscientious reader will, however, try the exercises unaided before attempting to evaluate what we have offered.

EXERCISE 7.1 (i)   *Let Ω, Ω' be G-spaces and $\theta : \Omega \to \Omega'$ a mapping. Show that θ is a G-morphism if and only if, as subset of $\Omega \times \Omega'$, it is a G-subspace.*
(ii)   *Show that the equivalence relation ρ on Ω is a congruence if and only if it is a G-subspace of Ω^2.*

Solution. (i)   The mapping θ is to be identified with the subset $\Theta :=$ $\{(\omega, \omega\theta) | \omega \in \Omega\}$ of $\Omega \times \Omega'$. This is a G-subspace if and only if, for all $g \in G$ we have that $(\omega, \omega') \in \Theta \Rightarrow (\omega, \omega')^g \in \Theta$. Now $(\omega, \omega')^g = (\omega^g, \omega'^g)$ by definition of the action of G on $\Omega \times \Omega'$ and therefore what we require is that $(\omega\theta)^g = (\omega^g)\theta$ for all $\omega \in \Omega$. This is precisely the condition for θ to be a G-morphism.

(ii)   Let ρ be an equivalence relation on Ω. If, as subset of Ω^2, it is G-invariant, then $(\alpha, \beta) \in \rho \Rightarrow (\alpha^g, \beta^g) = (\alpha, \beta)^g \in \rho$, that is, in another notation,

$$\alpha \equiv \beta \pmod{\rho} \Rightarrow \alpha^g \equiv \beta^g \pmod{\rho} \text{ for any } g \text{ in } G,$$

which is precisely the condition that ρ be a congruence. The argument can be reversed to give that if ρ is a congruence then it is *G*-invariant as subset of Ω^2, as required.

EXERCISE 7.10 *Show that if $n \geqslant 5$ then S_n (and also A_n) is primitive on the set of $\frac{1}{2}n(n-1)$ unordered pairs from $\{1, 2, \ldots, n\}$.*

Solution. Let Ω be the set of unordered pairs (of *distinct* elements, of course) from the set $\{1, 2, \ldots, n\}$ with S_n acting in the natural way, and let ρ be a non-trivial congruence on Ω. Since ρ is non-trivial there are distinct pairs $\{a, b\}, \{c, d\}$ such that $\{a, b\} \equiv \{c, d\} \pmod{\rho}$. Suppose first that $\{a, b\}, \{c, d\}$ have one element in common, say (without loss of generality) $a = d$. Now if x, y, z are any three distinct numbers between 1 and n then there is a permutation $f \in S_n$ such that $f: a \mapsto x, b \mapsto y, c \mapsto z$ and so we must have that $\{x, y\} \equiv \{x, z\} \pmod{\rho}$. Therefore $\{1, 2\} \equiv \{1, i\} \equiv \{i, j\} \pmod{\rho}$, from which it follows that every element of Ω is related to $\{1, 2\}$ and so ρ is the universal relation.

If, on the other hand, $\{a, b\}, \{c, d\}$ are disjoint then we can choose e in the range $1 \leqslant e \leqslant n$ and different from each of a, b, c, d, and there is a permutation $g \in S_n$ fixing a, b, c, and mapping d to e. Since ρ is a congruence we have $\{a, b\} \equiv \{c, e\}$ and therefore $\{c, d\} \equiv \{c, e\} \pmod{\rho}$. This brings us to the situation we have already dealt with, and therefore we know that ρ is the universal relation. Thus S_n is primitive on Ω.

The argument needs very little modification to deal with the alternating group A_n: the permutations f, g used in the proof can, with a little care, be taken to be even. Thus A_n is primitive on Ω, as required.

EXERCISE 7.11 *Let Ω be a faithful primitive G-space. Show that every non-trivial normal subgroup of G is transitive on Ω.*

Solution. Let N be a non-trivial normal subgroup of G. Define ρ to be equivalence modulo N (see Chapter 5, p. 51 and Example 7.4), that is, $\alpha \equiv \beta \pmod{\rho}$ if and only if $\alpha^x = \beta$ for some $x \in N$. Thus the ρ-classes are the N-orbits in Ω. If $g \in G$ and $\alpha^x = \beta$ with $x \in N$, then $(\alpha^g)^{g^{-1}xg} = \beta^g$. Since N is normal in G this shows that $\alpha \equiv \beta \pmod{\rho} \Rightarrow \alpha^g \equiv \beta^g \pmod{\rho}$, that is, that ρ is a *G*-congruence on Ω. Since G acts faithfully and N is non-trivial, ρ is a non-trivial relation. Therefore, since Ω is a primitive *G*-space, ρ must be the universal relation, and this means that N is transitive, as required.

Exercises

7.1. (i) Let Ω, Ω' be *G*-spaces and $\theta: \Omega \to \Omega'$ a mapping. Show that θ is a *G*-morphism if and only if, as subset of $\Omega \times \Omega'$, it is a *G*-subspace.

(ii) Show that the equivalence relation ρ on Ω is a congruence if and only if it is a *G*-subspace of Ω^2.

7.2.   Prove Proposition 7.1.

7.3.   Let H, K be subgroups of G with $K \leqslant H$. Show that the prescription $Kx \mapsto Hx$ defines a G-morphism from $\cos(G:K)$ to $\cos(G:H)$.

7.4.   Let G be S_3 and let Ω_n (for $n := 1, 2, 3, 6$) be the (unique up to isomorphism) transitive G-space with n elements. Draw up a table exhibiting the number of different G-morphisms from Ω_m to Ω_n for all m, n.

7.5.   As mentioned in Theorem 7.5. the G-space Ω is said to be *semi-regular* if every orbit is regular (equivalently, if $G_\omega = \{1\}$ for all $\omega \in \Omega$).
   (i) Suppose that Ω is semi-regular and let Σ be a subset containing exactly 1 element from each orbit. Show that for any map $\phi : \Sigma \to \Omega'$ (where Ω' is any G-space) there is a unique G-morphism $\theta : \Omega \to \Omega'$ that extends ϕ in the sense that $\phi = \theta \upharpoonright \Sigma$.
   (ii) Conversely, suppose now that Ω is a G-space with a subset Σ such that every map $\Sigma \to \Omega'$ (for every G-space Ω') extends uniquely to a G-morphism $\Omega \to \Omega'$. Prove that Ω is semi-regular and that Σ contains just 1 element from each orbit of Ω.   [These results have a Universal Algebra interpretation: semi-regular G-spaces are the 'free objects' in the category of G-spaces.]

7.6.   Let Ω be a faithful G-space. Show that $\mathrm{Aut}(\Omega)$ is transitive on Ω if and only if Ω is a semi-regular G-space.

7.7.   For which groups G is the regular G-space primitive?

7.8.   Show that G is primitive on the non-trivial transitive G-space Ω if and only if for all $\Gamma \subseteq \Omega$, if $\Gamma \neq \Omega$ and $|\Gamma| > 1$ then there exists $g \in G$ such that $\Gamma \neq \Gamma g$ and $\Gamma \cap \Gamma g \neq \varnothing$.

7.9.   Tabulate all primitive permutation groups (up to equivalence) of degree n for $n \leqslant 5$.

7.10. Show that if $n \geqslant 5$ then S_n (and also A_n) is primitive on the set of $\frac{1}{2}n(n-1)$ unordered pairs from $\{1, 2, \ldots, n\}$.

7.11. Let Ω be a faithful primitive G-space. Show that every non-trivial normal subgroup of G is transitive on Ω.

Chapter 8

Group actions in group theory

Most of the first seven chapters of this book have been devoted to an exposition of the pure theory of group actions. We turn now to applications of the theory. Later chapters are devoted to applications in combinatorics and geometry. This one is concerned with applications in group theory itself, and we begin by collecting together some elementary results that have already been established.

THEOREM 8.1: *If G is a group, if $H \leqslant G$ and if $g \in G$ then the map $\rho_g: Hx \mapsto Hxg$ is a permutation of $\cos(G:H)$, the set of cosets of H in G. Furthermore the map $g \mapsto \rho_g$ is a homomorphism $\rho: G \to \operatorname{Sym}(\Omega)$ where $\Omega := \cos(G:H)$. The kernel K of ρ is the intersection of all the conjugates $x^{-1}Hx$ of H. In particular, if $|G:H| = n$ then K is a normal subgroup of G such that $K \leqslant H$ and $|G:K|$ divides $n!$.*

THEOREM 8.2: *Let G be a group and for $a \in G$ define*

$$C(a) := C_G(a) := \{g \in G \mid ga = ag\}.$$

Then $C(a)$, the centraliser of a in G, is a subgroup of G and its cosets $C(a)x$ are in one–one correspondence with the conjugates $x^{-1}ax$ of a. In particular, if G is finite then the number of conjugates of a divides $|G|$.

This is an instance of Lemma 6.1. Another is

THEOREM 8.3: *Let G be a group, $H \leqslant G$, and define*

$$N(H) := N_G(H) := \{g \in G \mid g^{-1}Hg = H\}.$$

Then $N(H) \leqslant G$ (and $H \lhd N(H)$), and the conjugate subgroups $x^{-1}Hx$ are in one–one correspondence with the cosets $N(H)x$. In particular, if G is finite then the number of conjugates of any subgroup divides $|G|$.

Notice that every conjugate $x^{-1}Hx$ is a subgroup of G and is isomorphic to H. Indeed, the map $h \mapsto x^{-1}hx$ is an isomorphism $H \to x^{-1}Hx$.

We begin the further investigation of (finite) groups with a fundamental arithmetical result that was known to Galois by about 1830 but which was not published until 1845, when Cauchy gave a long and complicated proof of it.

THEOREM 8.4 [CAUCHY'S THEOREM]: *If G is a finite group and p is a prime divisor of $|G|$ then G contains an element of order p.*

First proof.   We use induction on $|G|$. There is nothing to prove if $|G| = 1$, so suppose that $|G| > 1$ and let p be a prime divisor of $|G|$. As inductive hypothesis we suppose that any smaller group whose order is divisible by p does contain an element of order p.

If G has a proper subgroup H such that p divides $|H|$, then, by hypothesis, there is an element a of order p in H, and this is, of course, an element of order p in G. So we shall suppose that for any proper subgroup H of G, p does not divide $|H|$.

Recall that the centre $Z(G)$ is defined to be the subgroup of all those elements that commute with every element of G; equivalently, it is the set of all elements whose conjugacy classes are of size 1. Let a be any element of G not in $Z(G)$ (if any such exist). Then $C(a)$ is a proper subgroup of G so, by our assumption, p does not divide $|C(a)|$, and therefore p does divide $|G:C(a)|$ which is the size of the conjugacy class containing a. Thus those conjugacy classes that are not of size 1 contain a multiple of p elements. Since G is the disjoint union of its conjugacy classes (the orbits of G acting on itself by conjugation) it follows that the number of classes of size 1, that is $|Z(G)|$, is a multiple of p. We are supposing that p does not divide the orders of proper subgroups of G, so it follows that $Z(G) = G$, that is, that G is abelian.

For abelian groups, however, we know that Cauchy's Theorem is true. This can be proved easily in many ways. We indicated one in Exercise 1.10, and here, for completeness, is another. Take any non-identity element a of G and let $m := \mathrm{ord}(a)$. If $p \mid m$, if say $m = p \cdot m_0$, then a^{m_0} is an element of order p in G. If not, we let $A := \langle a \rangle$ and $\bar{G} := G/A$. Then of course p divides $|\bar{G}|$. Since $|\bar{G}| < |G|$ there is an element $\bar{b} \in \bar{G}$ that has order p. Now $\bar{b} = Ab$ for some $b \in G$, and $b^p \in A$, $b \notin A$. This means that $\mathrm{ord}(b)$ is divisible by p and so again b^k is an element of order p in G, where $k := \mathrm{ord}(b)/p$. Thus G has an element of order p, as required.

The proof that we have just completed is a nineteenth century argument that we have exhibited as a good example of the techniques that one uses for answering questions about finite groups. It is somewhat shorter and easier than Cauchy's original proof, but is still the kind of reasoning to which one is led if one starts from the conjecture that a statement like that of Cauchy's Theorem may be true and seeks to test it. There is another proof, a shorter one, of the kind that takes a combination of experience with good fortune to discover.

Second proof [James H. McKay, *Amer. Math. Monthly* **66**, (1959) p. 119]. We are assuming that G is a finite group and that $|G|$ is divisible by the prime number p. Let

$$\Omega := \{(g_1, g_2, \ldots, g_p) \mid g_i \in G, \ g_1 g_2 \ldots g_p = 1\}.$$

The cyclic group Z_p acts on $G \times G \times \ldots \times G$ (p factors) in an obvious way, permuting the factors cyclically, and we find that Ω is an invariant subset because if $g_1 g_2 \ldots g_p = 1$ then also

$$g_2 \ldots g_p g_1 = g_1^{-1}(g_1 g_2 \ldots g_p)g_1 = g_1^{-1} g_1 = 1.$$

Suppose that Z_p has s fixed points (orbits of size 1) in Ω and r regular orbits (p-cycles). Then $s + pr = |\Omega|$. Now $|\Omega| = |G|^{p-1}$ because we can choose $g_1, \ldots, g_{p-1}$ arbitrarily and then g_p is uniquely determined as $(g_1 \ldots g_{p-1})^{-1}$. Since p divides $|G|$ it also divides $|\Omega|$ and so $p | s$. Now $s > 0$ since $(1, 1, \ldots, 1)$ is a fixed point of Z_p in Ω, and so $s \geqslant p$. There is therefore an element of Ω fixed by Z_p other than $(1, 1, \ldots, 1)$, and this must be a sequence $(a, a, \ldots, a)$ such that $a^p = 1$, $a \neq 1$. Thus G contains an element of order p.

Our next theorem is another famous and extremely useful result of nineteenth century finite group theory, that may be seen as a vast generalisation of Cauchy's Theorem. It deals with what are known as p-subgroups of finite groups: for a prime number p, a finite group X is said to be a *p-group* if its order is a power of p. By Lagrange's Theorem and Cauchy's Theorem this is equivalent with the requirement that $\mathrm{ord}(x)$ be a power of p for all $x \in X$ (which we take as definition of p-group in the infinite case). We shall use repeatedly the fact that if X is a p-group acting on the finite set Ω then the size of every X-orbit is a power of p (this is a corollary of Lagrange's Theorem; see Corollary 6.2). In particular, the size of every non-trivial X-orbit is divisible by p and consequently, if k is the number of fixed points (i.e. orbits of size 1) of X in Ω, then $k \equiv |\Omega| \pmod{p}$.

THEOREM 8.5 [SYLOW'S THEOREMS]: *Let G be a finite group of order $p^r m$ where p is prime, $r \geqslant 1$ and $p \nmid m$. Then*
  (i)   *there is at least one subgroup P of order p^r in G;*
  (ii) *the subgroups of order p^r form one conjugacy class in G;*
  (iii) *if X is any p-subgroup of G then $X \leqslant x^{-1} P x$ for some $x \in G$;*
  (iv) *if n is the number of subgroups of order p^r then $n | m$ and $n \equiv 1 \pmod{p}$.*

Such subgroups of order p^r are known as *Sylow p-subgroups* in honour of a Norwegian mathematician who discovered these results and published them in *Mathematische Annalen* 1872. We have included statement (ii) for emphasis even though it is an immediate consequence of (iii), which tells us that any maximal p-subgroup (that is, a p-subgroup of G not contained in any larger p-subgroup) is a conjugate of P. We will give two proofs here because of the very different insights that they offer. Our first is Sylow's original proof translated freely into our mathematical language (and out of French).

First proof.   Let P be a maximal p-subgroup of G. That is, P has order p^s for some integer s, and there is no subgroup of order $p^{s'}$ with $s' > s$ that contains P. We shall prove (i) by showing that $s = r$.

Let $H := N(P)$, the normaliser of P, and let $m_0 := |H:P|$. Since $P \lhd H$ we can form the quotient group H/P and of course m_0 is its order. If m_0 were a multiple of p then, by Cauchy's Theorem, H/P would contain a subgroup $\bar{A}$ of order p. But $\bar{A} = A/P$ for some subgroup A of H containing P. Then $|A| = p^{s+1}$ in contradiction with the maximality of P. Thus m_0 is not divisible by p.

Observe next that P consists of all those elements of H whose orders are powers of p, and that therefore if Q is a p-subgroup of H then $Q \leqslant P$ (compare Exercise 1.8). For suppose that $a \in H$ and ord(a) is a power of p. Let $\bar{a} := Pa \in H/P$ and let $\bar{A} := \langle \bar{a} \rangle$. Then of course ord$(\bar{a})$ divides ord(a), so $|\bar{A}|$ is a power of p. But $\bar{A} = A/P$ for some subgroup A of H containing P. As $|A| = |P| \cdot |\bar{A}|$, again A is a p-subgroup and, by maximality of P, we must have $A = P$. Hence $\bar{A} = \{1\}$ and so $a \in P$.

Now let $\Omega := \cos(G:H)$ and let $m_1 := |\Omega| = |G:H|$. Let $\alpha := H \in \Omega$, so that $H = \text{Stab}(\alpha)$. We consider the action of P on Ω. Certainly, as $P \leqslant H$, there is one P-orbit, namely $\{\alpha\}$, of length 1. Let $\{\beta\}$ be any P-orbit of length 1. Then $\beta = \alpha^x$ for some $x \in G$ and

$$P \leqslant \text{Stab}(\beta) = x^{-1}\text{Stab}(\alpha)x = x^{-1}Hx.$$

Consequently $xPx^{-1} \leqslant H$ and, as xPx^{-1} is a p-subgroup of G it follows from the previous paragraph that $xPx^{-1} \leqslant P$. Hence $xPx^{-1} = P$, which (as H was defined to be $N(P)$) means that $x \in H$, and so $\beta = \alpha^x = \alpha$. Thus $\{\alpha\}$ is the unique P-orbit of length 1 in Ω. It follows (see the remark preceding the statement of the theorem) that $m_1 \equiv 1 \pmod{p}$. But now

$$|G| = |H|m_1 = |P|m_0m_1 = p^s m_0 m_1,$$

and as p does not divide m_0 or m_1 we must have $s = r$ and $m = m_0 m_1$.

Sylow proves (ii) and (iii) as follows. Let X be any p-subgroup of G, and consider the action of X on Ω. Since $|\Omega| = m_1$, which is not divisible by p, there must be at least one X-orbit $\{\beta\}$ of length 1. Then, as above,

$$X \leqslant \text{Stab}(\beta) = x^{-1}Hx$$

for some $x \in G$. This implies that xXx^{-1} is a p-subgroup of H and so we know that $xXx^{-1} \leqslant P$. Consequently, $X \leqslant x^{-1}Px$, as required.

Finally, (iv) has, in effect already been proved. For, we know now that any subgroup of order p^r is conjugate to P, and so n is the number of conjugates of P. But the conjugates of P are in one–one correspondence with the cosets of $N(P)$, that is, of H in G, and so $n = m_1$.

Second proof [H. Wielandt, *Archiv der Mathematik* 1959].   Let

$$\Omega := \{E \mid E \subseteq G \text{ and } |E| = p^r\},$$

and consider Ω as a G-space with action by right-translation

$$E^g := Eg = \{xg \mid x \in E\}.$$

Now

$$|\Omega| = \binom{p^r m}{p^r} = \frac{p^r m(p^r m - 1) \ldots (p^r m - p^r + 1)}{p^r(p^r - 1) \ldots 1}$$

and this binomial coefficient is not divisible by p (because if $0 \leqslant k < p^r$ then $p^r m - k$ and $p^r - k$ are divisible by exactly the same power of p). We choose a G-orbit Ω_1 such that $|\Omega_1|$ is not divisible by p, choose $E \in \Omega_1$, and let $P := \text{Stab}(E)$, the stabiliser in G of E.

Now on the one hand $|\Omega_1| = |G:P|$ and so, since p does not divide $|\Omega_1|$, we have that p^r divides $|P|$. On the other hand, P acts on E by right multiplication, and so if $\alpha \in E$ then $P_\alpha = \{1\}$: thus every P-orbit in E is isomorphic to the regular P-space and so $|P|$ divides $|E|$, which is p^r. Thus $|P| = p^r$, and we have proved the existence of Sylow p-subgroups.

Now let X be any p-subgroup of G. Since $|\Omega_1|$ is not a multiple of p there must be at least one X-orbit of length 1 in Ω_1. This means that $X \leqslant \text{Stab}(E')$ for some $E' \in \Omega_1$. But $E' = Ex$ for some $x \in G$, and we know that $\text{Stab}(E') = x^{-1}\text{Stab}(E)x$. So we have that $X \leqslant x^{-1}Px$, which proves clause (iii) (and therefore also (ii)) of Sylow's Theorem.

Finally, to prove (iv), let $\{\Omega_i\}_{i \in I}$ be the family of all G-orbits in Ω. If $E_0 \in \Omega_i$ and $g \in E_0$ then $1 \in E_0 g^{-1} \in \Omega_i$, so Ω_i contains at least one set E such that $1 \in E$. We choose $E_i \in \Omega_i$ such that $1 \in E_i$ and let $P_i := \text{Stab}(E_i)$. Notice that if $g \in P_i$ then $g = 1g \in E_i^g = E_i$, and so $P_i \subseteq E_i$. In fact of course, if $x \in E_i$ then $xg \in E_i$ for all $g \in P_i$, that is, $xP_i \subseteq E_i$, and so E_i is the union of a certain number of left cosets of P_i. Therefore $|P_i|$ divides $|E_i|$, which is p^r, and so if P_i is a Sylow p-subgroup then $|\Omega_i| = |G:P_i| = m$, while if P_i is not a Sylow p-subgroup then p must divide $|G:P_i|$, which is $|\Omega_i|$. Also, since $P_i \in E_i$ we see that P_i is a Sylow p-subgroup if and only if $E_i = P_i$. Thus there are precisely n G-orbits Ω_i of size m, one for each Sylow p-subgroup, and the rest of the G-orbits have sizes divisible by p. Consequently for some integer k we have

$$|\Omega| = \sum_i |\Omega_i| = nm + kp$$

that is,

$$\binom{p^r m}{p^r} \equiv nm \pmod{p} \tag{*}$$

Now in fact

$$\binom{p^r m}{p^r} \equiv m \pmod{p}$$

This purely numerical result can be proved by the following device. The congruence (*) has been proved to hold for every group of order $p^r m$ but the only parameter in it that depends on the particular group in question is n. The cyclic group $Z_{p^r m}$ has precisely one subgroup of order p^r, and so in this case (*) is the congruence

$$\binom{p^r m}{p^r} \equiv m \ (\text{mod } p)$$

that we want.

Finally, we put these congruences together to get that $m \equiv mn$ (mod p) and, as p does not divide m, it follows that $n \equiv 1$ (mod p), as required.

The usefulness of Sylow's theorems is enhanced by the fact that we know a great deal about the structure of a finite p-group. Our next two theorems give two of the simplest facts.

THEOREM 8.6 [SYLOW, 1872]:   *Let P be a finite p-group and let Q be a non-trivial normal subgroup of P. Then $Q \cap Z(P) \neq \{1\}$.*

Proof.   Since $Q \lhd P$, P acts on Q by conjugation. Let $k := |Q \cap Z(P)|$, so that k is the number of P-orbits of size 1. We know, since P is a p-group, that

$$k \equiv |Q| \equiv 0 \ (\text{mod } p).$$

But of course $k > 0$ since $1 \in Q \cap Z(P)$. Therefore $k \geqslant p$ and this means that $Q \cap Z(P)$ is non-trivial.

COROLLARY 8.7 [SYLOW, 1872]:   *Suppose that $|P| = p^r$, and let*

$$\{1\} = P_0 < P_1 < P_2 < \ldots < P_{s-1} < P_s = P$$

be a maximal chain of normal subgroups of P. Then $s = r$ and $|P_t| = p^t$ for all t.

For, to say that this chain is maximal means that there is no subgroup X of P such that $X \lhd P$ and $P_t < X < P_{t+1}$. Now P_{t+1}/P_t is a non-trivial normal subgroup of P/P_t. By the theorem it contains a non-trivial subgroup in the centre of P/P_t; thus there exists X such that $P_t < X \leqslant P_{t+1}$ and X/P_t is of order p and central, hence normal, in P/P_t. Then $X \lhd P$ and so $X = P_{t+1}$. This shows that $|P_{t+1}:P_t| = p$ for all t; hence that $|P_t| = p^t$, and that $s = r$.

THEOREM 8.8:   *Let P be a finite p-group and Q a proper subgroup of P. Then Q is properly contained in $N(Q)$.*

Proof.   Let $\Omega := \cos(P:Q)$ and consider the action of Q on Ω. There is one orbit $\{\alpha\}$ of length 1, where $\alpha := Q \in \cos(P:Q)$, and so if k is the number of Q-orbits of length 1, then $k \geqslant 1$. But we know that, since Q is a p-group, $k \equiv |\Omega|$ (mod p). On the other hand $|\Omega| = |P:Q| \equiv 0$ (mod p). Therefore k is a multiple of p and so $k \geqslant p$. Choose $\beta \in \Omega \setminus \{\alpha\}$ such that $\{\beta\}$ is a Q-orbit. Then $Q = P_\beta$, the stabiliser of β in P. But $\beta = \alpha g$ for some $g \in P$, so

$$Q = P_\beta = g^{-1}P_\alpha g = g^{-1}Qg.$$

Thus we have $g \in N(Q)$ and $g \notin Q$, as required.

We finish this chapter with a study of the alternating groups and the first step in the search for simple groups. Recall that the group G is said to be *simple* if $\{1\}, G$ are its only normal subgroups.

THEOREM 8.9: *The alternating group A_5 is simple.*

Proof. Suppose that $H \lhd A_5$ and $H \neq \{1\}$: our assignment is to prove that $H = A_5$. Let Ω be the natural A_5-space $\{1, 2, 3, 4, 5\}$. We know (Chapter 5, Exercise 5.18; see also Chapter 7, Example 7.4) that all the H-orbits must have the same size m, and so, if there are k of them then $km = 5$. Now $m > 1$ since $H \neq \{1\}$ and so $k = 1, m = 5$, and H is transitive (see also Chapter 7, Exercises 7.10, 7.11). It follows that 5 divides $|H|$ and, as $|A_5| = 60$, that 5 does not divide $|A_5/H|$. Consequently H contains all the elements of order 5 in A_5, which are the 5-cycles $(\alpha_1 \alpha_2 \alpha_3 \alpha_4 \alpha_5)$. But it is easy to see that every element of A_5 can be expressed as a product of two 5-cycles. For example

$$(\alpha_1 \alpha_2 \alpha_3) = (\alpha_1 \alpha_3 \alpha_5 \alpha_4 \alpha_2)(\alpha_3 \alpha_2 \alpha_4 \alpha_5 \alpha_1)$$

and

$$(\alpha_1 \alpha_2)(\alpha_3 \alpha_4) = (\alpha_1 \alpha_3 \alpha_5 \alpha_2 \alpha_4)(\alpha_3 \alpha_2 \alpha_5 \alpha_4 \alpha_1).$$

Therefore $H = A_5$ and this proves that A_5 is simple.

THEOREM 8.10: *The alternating group A_n is simple if $n \geq 5$.*

Proof. We use induction on n. The case $n = 5$ has already been settled, so suppose that $n \geq 6$ and that A_{n-1} is already known to be simple. Let H be a non-trivial normal subgroup of A_n. Then H is transitive on the natural A_n-space $\Omega := \{1, 2, \ldots, n\}$ (see Chapter 7, Exercises 7.10, 7.11, for example). Now $H \cap \text{Stab}(n) \lhd \text{Stab}(n)$ and $\text{Stab}(n) \cong A_{n-1}$ which, by inductive hypothesis, is simple. Therefore $H \cap \text{Stab}(n) = \{1\}$ or $H \cap \text{Stab}(n) = \text{Stab}(n)$.

Suppose, for the moment, that $H \cap \text{Stab}(n) = \{1\}$. Then H is a regular normal subgroup of A_n and so (Chapter 6, Exercise 6.17) $\text{Stab}(n)$ acts on H by conjugation in the same way as on Ω. In other words $\text{Stab}(n)$, which is A_{n-1}, acts on $H \setminus \{1\}$ by conjugation in the same way as on $\{1, 2, \ldots, n-1\}$, that is H is a group with the property that any even permutation of $H \setminus \{1\}$ is an automorphism of H. The only groups with this property are Z_1, Z_2, Z_3, and $Z_2 \times Z_2$. For suppose that $n \geq 4$, and consider an automorphism θ of H that acts as a 3-cycle. The set $\{h \in H \,|\, h\theta = h\}$ then consists of exactly $n-3$ elements of H (remember that H has order n). Also it is a subgroup of H, as one checks very easily. By Lagrange's Theorem, therefore, $n-3$ divides n, and so n must be 4 or 6. But $n \neq 6$ because a group of order 6 has elements of order 2 and of order 3, and so its automorphism group cannot even be transitive on the set of non-identity elements.

Thus $H \cap \text{Stab}(n) \neq \{1\}$, so $H \geq \text{Stab}(n)$. As H is transitive it contains at least one element from each coset of $\text{Stab}(n)$ in A_n, and this means that

Stab(n). $H = A_n$. But as Stab(n) $\leqslant H$ we have Stab(n). $H = H$. Therefore $H = A_n$, as required.

Recognition of the importance of simplicity is one of the contributions that Galois made to mathematics. In a famous letter to his friend Auguste Chevalier, written during the night before his fatal duel in 1832, he defines what he means by a simple group and he goes on to assert that the smallest simple group that is not cyclic of prime order has order 5.4.3, that is, 60. We cannot now be certain how Galois knew this for, although the statement can be verified quite easily using Sylow's theorems (see Exercise 8.17) it is hard to see how one could have proceeded in Galois' day, forty years before Sylow's paper was published. (There is, however, some evidence to suggest that Galois might at one time have known or guessed a significant part of the Sylow theorems.) Finding the simple groups of small orders is the first small step in the search for all the finite simple groups, a project that dominated the research effort in group theory for a hundred years. Here is an example of how one can identify a simple group, knowing only rather little about it to begin with.

Let G be a simple group of order 60. By Cauchy's Theorem G contains an element a of order 5. Let $P := \langle a \rangle$ (so that P is a Sylow 5-subgroup of G), and let $\Omega := \{g^{-1}Pg \mid g \in G\}$. We know from Clause (iv) of Sylow's Theorem that $|\Omega|$ divides 12 and is 1 (mod 5). Also, $|\Omega| \neq 1$ since P is not a normal subgroup (after all, G is simple). Hence $|\Omega| = 6$. Our group G acts on Ω by conjugation and the corresponding permutation representation $\rho: G \to S_6$ must be a monomorphism (i.e. injective) since G is simple. Let $H := G\rho \leqslant S_6$. Then of course $H \cong G$, so H is simple. It follows that $H \leqslant A_6$ since $H \cap A_6 \trianglelefteq H$.

Now let $\Gamma := \cos(A_6 : H)$. We have

$$|\Gamma| = \tfrac{1}{2} 6!/60 = 6,$$

and so the action of A_6 on Γ gives us a permutation representation $\sigma: A_6 \to S_6$. Again, as A_6 is simple, σ has to be a monomorphism; and then as $A_6\sigma$ is simple it must be contained in A_6, so $A_6\sigma = A_6$. Thus σ is an automorphism of A_6. As it maps H to a stabiliser in A_6, that is to A_5, we have that $G \cong H \cong A_5$. Thus we have proved

THEOREM 8.11:   *If G is a simple group of order 60 then $G \cong A_5$.*

The argument that we have just completed can be varied and extended in several ways to give further information. It depended originally on the fact that G acts faithfully on the set Ω that has 6 members. Now the group PSL(2, 5) acts faithfully on PG(1, 5) and this has 6 members (see Example 4.5). Using exactly the same argument as before we can therefore produce the conclusion that PSL(2, 5) $\cong A_5$. The same reasoning also shows that PGL(2, 5) $\cong S_5$, and indeed that PGL(2, 5) acting on PG(1, 5) is equivalent to S_5 acting by conjugation on the set of its six cyclic subgroups of order 5.

Finally, let us start from this latter group, namely S_5 acting by conjugation on the set Ω of its six cyclic subgroups of order 5. This gives us a permutation representation $S_5 \cong X \leqslant S_6$. If we let $\Gamma := \cos(S_6 : X)$ then we find that $|\Gamma| = 6$, and so the action of S_6 on Γ produces a permutation representation $\sigma : S_6 \to S_6$. It is easy to see that σ is a monomorphism, and so it is an automorphism of S_6. This means, of course, that Γ is equivalent to the natural S_6-space $\{1, 2, 3, 4, 5, 6\}$. Nevertheless, Γ is not S_6-isomorphic to the natural S_6-space: for, the subgroup X of S_6 stabilises a point in Γ, but it acts on $\{1, 2, 3, 4, 5, 6\}$ in the same way as S_5 on Ω and is therefore transitive on the natural S_6-space.

We have thus produced a remarkable S_6-space Γ that is equivalent to the natural S_6-space but not S_6-isomorphic with it. The automorphism σ is also remarkable for it is an 'outer' automorphism: as it maps the conjugates of X (which, remember, are the stabilisers of points of Γ) to stabilisers S_5 in the natural S_6-space, and as the former are not conjugate to the latter (because, for example, X is transitive on the natural space $\{1, 2, 3, 4, 5, 6\}$, whereas S_5 is not; alternatively, because if they were conjugate then the two S_6-spaces would be isomorphic—see Proposition 6.4) it follows that σ cannot be induced by conjugation by any element of S_6. This is what is meant by an outer automorphism. It is a useful fact, just a little too deep to be proved here, that, for $n \neq 6$, every automorphism of S_n is inner. What we have shown is that S_6 is genuinely exceptional, a fact which led J. A. Todd, a famous Cambridge geometer, to write an article entitled 'On the "odd" number 6'.

Specimen solutions to selected exercises

The following specimen solutions, being of the nature of an extension of the text, are placed between the main body of the chapter and the exercises. The wise and conscientious reader will, however, try the exercises unaided before attempting to evaluate what we have offered.

EXERCISE 8.2 (i)   *Identify the Sylow 2-subgroups and the Sylow 3-subgroups in* S_3. *How many of each are there?*
(ii)   *Exhibit the Sylow 2-subgroups of* S_4.
(iii)   *Exhibit the Sylow 2-subgroups of* A_5. *How many are there?*

Solution. (i)   Since $|S_3| = 6$, the Sylow 2-subgroups have order 2 and the Sylow 3-subgroups have order 3. Therefore they are cyclic, generated by elements of order 2 and order 3 respectively. Thus the Sylow 2-subgroups of S_3 are $\langle (1\ 2) \rangle, \langle (1\ 3) \rangle, \langle (2\ 3) \rangle$, and the Sylow 3-subgroup is $\langle (1\ 2\ 3) \rangle$. There are 3 Sylow 2-subgroups and there is just one Sylow 3-subgroup.

(ii)   Since $|S_4| = 24$ the Sylow 2-subgroups have order 8. If T is any

2-subgroup in S_4 then, since V_4 is a normal 2-subgroup of S_4, the product TV_4 is again a 2-group. Since Sylow 2-subgroups are maximal 2-groups it follows that V_4 is contained in every Sylow 2-subgroup. [Quite generally, a normal p-subgroup of a group G is contained in every Sylow p-subgroup of G.] Take T to be a cyclic subgroup of order 4. Then certainly $T \not\leqslant V_4$ and so TV_4 is a 2-group whose order is strictly greater than 4: thus it must (by Lagrange's Theorem) have order 8 and be a Sylow subgroup. A cyclic subgroup of order 4 in S_4 must be generated by a 4-cycle. There are six 4-cycles, but they occur in inverse pairs and so there are 3 cyclic subgroups of order 4. Thus we find 3 Sylow 2-subgroups, namely

$$\langle V_4, (1\ 2\ 3\ 4)\rangle = \{1, (1\ 2)(3\ 4), (1\ 3)(2\ 4), (1\ 4)(2\ 3), (1\ 2\ 3\ 4),$$
$$(1\ 3), (1\ 4\ 3\ 2), (2\ 4)\};$$

$$\langle V_4, (1\ 2\ 4\ 3)\rangle = \{1, (1\ 2)(3\ 4), (1\ 3)(2\ 4), (1\ 4)(2\ 3), (1\ 2\ 4\ 3),$$
$$(1\ 4), (2\ 3), (1\ 3\ 4\ 2)\};$$

$$\langle V_4, (1\ 3\ 2\ 4)\rangle = \{1, (1\ 2)(3\ 4), (1\ 3)(2\ 4), (1\ 4)(2\ 3), (1\ 3\ 2\ 4), (1\ 4\ 2\ 3),$$
$$(1\ 2), (3\ 4)\}.$$

It is easy to see that there are no others.

(iii)   Since $|A_5| = 60$ the Sylow 2-subgroups have order 4. Now a group of order 4 acting on any set, in this case the set $\{1, 2, 3, 4, 5\}$ of which A_5 is the group of all even permutations, has orbits of lengths 1, 2, or 4. Since 5 is odd, a Sylow 2-subgroup must have at least one orbit of length 1, that is, at least one fixed point. Thus a Sylow 2-subgroup must be contained in one of the natural subgroups isomorphic to A_4. Since in A_4 there is a unique subgroup V_4 of order 4, the Sylow 2-subgroups of A_5 are the subgroups like

$$\{1, (1\ 2)(3\ 4), (1\ 3)(2\ 4), (1\ 4)(2\ 3)\}$$

and there are just 5 of them.

EXERCISE 8.3   *The groups G, H are groups of permutations of the finite sets Γ, Δ respectively. The* wreath *product G wr H is defined to be the group of permutations t of the set $\Gamma \times \Delta$ which are of the form*

$$t:(\gamma, \delta) \longmapsto (\gamma g_\delta, \delta h),$$

where $g_\delta \in G$ and $h \in H$. Thus t is determined by an element h of H and a function $\Delta \to G$. Show that the permutations t with $h = 1$ form a normal subgroup B (known as the 'base' of the wreath product) isomorphic to the direct product of n copies of G, where n is the number of elements of Δ. Show also that the factor group of G wr H by B is isomorphic to H.

Show that if K is a group of permutations of a set Θ and if $(\Gamma \times \Delta) \times \Theta$

is identified with $\Gamma \times (\Delta \times \Theta)$ *in the obvious way, then* $(G \text{ wr } H) \text{ wr } K = G \text{ wr } (H \text{ wr } K)$.

Solution. Let $W := G \text{ wr } H$. We are told that $B := \{t \in W \mid t : (\gamma, \delta) \mapsto (\gamma g_\delta, \delta)\}$. Certainly the identity permutation lies in B (take $g_\delta := 1$ for all $\delta \in \Delta$); also, if $t \in B$ then $t^{-1} \in B$ because $t^{-1} : (\gamma, \delta) \mapsto (\gamma g_\delta^{-1}, \delta)$ for all $(\gamma, \delta) \in \Gamma \times \Delta$; and if $s, t \in B$ then also $st \in B$ because if $s : (\gamma, \delta) \mapsto (\gamma f_\delta, \delta)$ and t is as before then $st : (\gamma, \delta) \mapsto (\gamma f_\delta g_\delta, \delta)$. Thus B is a subgroup of W. If $\Delta = \{\delta_1, \ldots, \delta_n\}$ and $t \in B$ we identify t with the sequence $(g_1, \ldots, g_n)$ of elements of G, where $g_i := g_{\delta_i}$. This provides an isomorphism $B \to G^n$. To show that B is normal in W and to identify W/B we consider the map from W to H defined by $t \mapsto h$. Certainly this is well-defined and a surjective homomorphism with kernel B. Therefore $B \lhd W$ and, by the First Isomorphism Theorem, $W/B \cong H$.

If $t \in (G \text{ wr } H) \text{ wr } K$ then t is given by a formula of the form $t : ((\gamma, \delta), \theta) \mapsto ((\gamma, \delta)s_\theta, \theta k)$, where $s : \Theta \to G \text{ wr } H$ and $k \in K$. Now s_θ is of the form $s_\theta : (\gamma, \delta) \mapsto (\gamma g_{\delta,\theta}, \delta h_\theta)$, and so

$$t : ((\gamma, \delta), \theta) \mapsto ((\gamma g_{\delta,\theta}, \delta h_\theta), \theta k).$$

On the other hand, permutations t' of $G \text{ wr } (H \text{ wr } K)$ are those of the form $t' : (\gamma, (\delta, \theta)) \mapsto (\gamma g_{(\delta,\theta)}, (\delta, \theta)s')$, where $s' \in H \text{ wr } K$. Therefore they are those of the form

$$t' : (\gamma, (\delta, \theta)) \mapsto (\gamma g_{(\delta,\theta)}, (\delta h_\theta, \theta k))$$

where, as the notation is intended to suggest, $g_{\delta,\theta} \in G$ for all $(\delta, \theta) \in \Delta \times \Theta$, $h_\theta \in H$ for all $\theta \in \Theta$, and $k \in K$. Thus, identifying $(\Gamma \times \Delta) \times \Theta$ with $\Gamma \times (\Delta \times \Theta)$ in the natural way, we see that $(G \text{ wr } H) \text{ wr } K = G \text{ wr } (H \text{ wr } K)$, as required.

EXERCISE 8.4 *Let $p^{r(a)}$ be the highest power of the prime number p that divides $(p^a)!$.*
(i) *Show that $r(a) = p^{a-1} + r(a-1)$.*
(ii) *Use induction and a suitable wreath product (as defined in Exercise 8.3) to show that the symmetric group of degree p^a has a subgroup W_a of order $p^{r(a)}$.*

Solution. (i) The factorial $(p^a)!$ is, by definition, the product of the numbers $1, 2, \ldots, p^a$. Contributions to $r(a)$ come only from the members of the subsequence $p, 2p, \ldots, p^a$. There are p^{a-1} such terms. Taking one factor p out of each we have a contribution p^{a-1} to $r(a)$, and we are left with the contribution coming from the product of the numbers $1, 2, \ldots, p^{a-1}$, which, by definition, is $r(a-1)$. Thus $r(a) = p^{a-1} + r(a-1)$, as required.

(ii) If $a = 0$ there is nothing to prove. When $a = 1$ we find that $r(a) = 1$ and the symmetric group of degree p (acting on the set $\Gamma := \{0, 1, \ldots, p-1\}$) has a subgroup P of order p generated by the p-cycle $(0 \; 1 \ldots p-1)$. Suppose therefore as inductive hypothesis that the symmetric group of degree p^{a-1} has a subgroup W_{a-1} of order $p^{r(a-1)}$. A set Ω of size p^a may be identified with the product $\Gamma \times \Delta$, where Γ is a set of size p and Δ is a set of size p^{a-1}. Let $W := P$ wr W_{a-1}. According to Exercise 8.3, W has a normal subgroup B isomorphic to $P^{p^{a-1}}$ with quotient group isomorphic to W_{a-1}. Therefore $|W| = p^{p^{a-1}} \cdot p^{r(a-1)} = p^{r(a)}$, by (i). Taking $W_a := W$ we have our required subgroup of $\mathrm{Sym}(\Omega)$ of order $p^{r(a)}$.

EXERCISE 8.11   *Let G be a finite group, $H \lhd G$, let P be a Sylow p-subgroup of H, and let Ω be the set of all Sylow p-subgroups of H. Show that both H and G act transitively on Ω by conjugation, and deduce that $G = N_G(P) \cdot H$.*

Solution. We know from Sylow's theorems that H acts transitively by conjugation on Ω. If $g \in G$ then $g^{-1} P g$ is contained in H (since H is normal in G) and is a subgroup of the same order as P. Thus G also acts on Ω. Since its subgroup H is transitive, certainly G is transitive on Ω.

The stabiliser of P in this action of G on Ω is its normaliser $N_G(P)$. Since H is transitive on Ω it contains at least one element from each right coset of the stabiliser [see Lemma 6.1 (ii) and Exercise 6.4] and so $G = \bigcup_{h \in H} N_G(P) h = N_G(P) \cdot H$, as required.

EXERCISE 8.13   *Let G be a finite group that has no subgroup of index 2. Let T be a Sylow 2-subgroup of G, let H be a subgroup of index 2 in T, and let t be an element of order 2 in G.*
(i)   *Show that t acts as an even permutation of the G-space $\cos(G:H)$ and hence that t must fix some points of this space.*
(ii)   *Hence show that t must be conjugate to some element of H.*
(iii)   *Let G now be a simple group of order 60. Show that all elements of order 2 must be conjugate in G, that there must be 15 of them, and that they must fall into 5 Sylow subgroups of order 4. Hence show that $G \cong A_5$.*

Solution. (i)   Let $\Omega := \cos(G:H)$. If there were elements of G that induced odd permutations on Ω then the set of all those elements that induce even permutations would be a subgroup of index 2. Since G has no subgroup of index 2 we conclude that all elements of G induce even permutations on Ω, and in particular t does. Since t is of order 2 the permutation it induces on Ω has cycles of lengths 1 and 2 only. Now if $|G| = 2^a m$ where m is odd, then $|G:T| = m$ and $|\Omega| = |G:H| = |G:T||T:H| = 2m$. Since t is an even permutation it cannot be a product of m disjoint 2-cycles in its action on Ω, and therefore it must fix some points.

(ii)   Let α be the coset H thought of as an element of Ω and let β be a fixed point of t in Ω, so that $t \in G_\beta$. But if $\alpha g = \beta$ then $G_\beta = g^{-1} G_\alpha g = g^{-1} H g$. Thus $g t g^{-1} \in H$, as required.

(iii)   Let G now be a simple group of order 60. Let T be a Sylow 2-subgroup, so that $|T| = 4$. Let t_0 be an element of order 2 in T, and let H be the cyclic subgroup generated by t_0. From what has just been proved we know that any element t of order 2 is conjugate in G to an element of H, and therefore to t_0. Thus all elements of order 2 are conjugate in G.

Let $C := C_G(t_0) := \{g \in G \mid g^{-1}t_0 g = t_0\}$, the centraliser of t_0 in G. We know that the number of conjugates of t_0, that is, the number of elements of order 2 in G, is the index $|G:C|$. Being of order 4, T must be abelian and so $T \leqslant C$. Therefore $|G:C|$ divides $|G:T|$, which is 15. Now $|G:C|$ cannot be 1, otherwise $\langle t_0 \rangle$ would be a normal subgroup of G, whereas we are assuming that G is simple. Nor can $|G:C|$ be 3 else we would have a non-trivial homomorphism $G \to S_3$, again contradicting simplicity. If $|G:C|$ were 5 then we would have a non-trivial homomorphism $G \to S_5$, the image would have to be A_5, so G would be isomorphic to A_5, which is impossible since A_5 has 15 elements of order 2, whereas (if $|G:C| = 5$) our group G has only 5. It follows that $|G:C| = 15$ and therefore that G has 15 elements of order 2.

We now know that $|C| = 4$ and therefore that $T = C$. It follows that if T_1 is another Sylow 2-subgroup of G then $T \cap T_1 = \{1\}$, for otherwise $T \cap T_1$ would contain an element of order 2 whose centraliser contains both T and T_1 and therefore is of order strictly bigger than 4, and this is not possible. Let $N := N_G(T)$, the normaliser of T. Then $|G:N|$ divides 15, and the same argument that we applied to C shows that $|G:N|$ cannot be 1 or 3. If $|G:N| = 15$ then there are 15 Sylow 2-subgroups in G, each containing at least one element of order 2, and no two having any element other than the identity in common. There are several ways to see that this cannot be the case: for example, if it were then there would be 45 elements of orders 2 and 4 in G, hence fewer than 15 elements of order 5, and therefore fewer than 4 Sylow 5-subgroups, whereas the simplicity of G combined with Sylow's theorem implies that there are 6; alternatively, we can observe that each Sylow 2-subgroup would have exactly one element of order 2 and therefore would be cyclic of order 4, contradicting the result of Exercise 4.3. It follows that $|G:N| = 5$. Now the action of G on the set of cosets of N by right multiplication, or equivalently its action on the set of conjugates of T by conjugation, gives rise to a homomorphism $G \to S_5$. Since G is simple the kernel must be trivial, so G is isomorphic to the image. The image is simple, therefore must be contained in A_5 (else it would have a normal subgroup of index 2), but has order 60, therefore coincides with A_5. Thus $G \cong A_5$, as required.

EXERCISE 8.17 *Prove the assertion made by Galois in 1832, that there are no simple groups of composite order less than 60.*

Solution. From Sylow's results (Theorem 8.6 or its corollary) we know (1) that a group of order p^a, where p is a prime number and $a \geqslant 2$, cannot be simple.

Suppose now that $|G| = p^a m$ where p is prime, p does not divide m, and p^a does not divide $m!/2$. We will show (2) that G cannot be simple. For, let P be a Sylow p-subgroup, and let $\rho : G \to S_m$ be the permutation representation obtained from the action of G on the coset space $\cos(G:P)$. If G were simple then ρ would have to have trivial kernel, so G would be isomorphic to a subgroup (which we identify with G) of S_m. If G were not contained in A_m then $G \cap A_m$ would be a normal subgroup of index 2 in G, which contradicts the simplicity of G. Thus $G \leqslant A_m$ and so by Lagrange's Theorem, $|G|$ divides $m!/2$. This contradicts our assumption and so G cannot be simple (compare Exercise 4.6).

Using facts (1) and (2) we exclude all composite numbers below 60 as orders of simple groups except for 30 and 56. Consider now a group G of order $2m$ where m is odd and $m > 1$. By Cauchy's theorem there is an element t of order 2 in G. In the action of G on itself by right multiplication (the regular permutation representation of G) the element t is a product of m disjoint 2-cycles. Since m is odd, t is represented by an odd permutation. Therefore the set M of elements represented by even permutations forms a normal subgroup of index 2 (compare Exercise 4.3) and so G cannot be simple. Thus (3) a group of twice-odd order > 2 cannot be simple.

Fact (3) excludes 30 as possible order of a simple group and leaves us just with 56. Let G be a group of order 56. If G has just one Sylow 7-subgroup then this is normal and G is not simple. Otherwise G must have eight Sylow 7-subgroups by Sylow's theorems. Since the Sylow 7-subgroups are cyclic of order 7, any two of them have only the identity element in common. Therefore between them they contain 48 elements of order 7 in G. There must be a Sylow 2-subgroup T, and this contains another 8 elements, all different from the elements of order 7. Thus we have now accounted for 56 elements of G, that is all of them. It follows that T must be the only Sylow 2-subgroup of G and therefore it is normal in G. Thus G cannot be simple.

Exercises

8.1.   (i) Let G be a finite group with just 2 conjugacy classes. Show that $G \cong Z_2$.

(ii) Let G be a finite group with just 3 conjugacy classes. Show that $G \cong Z_3$ or $G \cong S_3$.

(iii) What can you say about finite groups with 4 or 5 conjugacy classes?

8.2.   (i) Identify the Sylow 2-subgroups and the Sylow 3-subgroups in S_3. How many of each are there?

(ii) Exhibit the Sylow 2-subgroups of S_4.

(iii) Exhibit the Sylow 2-subgroups of A_5. How many are there?

8.3.   The groups G, H are groups of permutations of the finite sets Γ, Δ respectively. The *wreath product* G wr H [often written $G \wr H$] is defined to be the group of permutations t of the set $\Gamma \times \Delta$ which are of the form $t: (\gamma, \delta) \mapsto (\gamma g_\delta, \delta h)$, where $g_\delta \in G$ and $h \in H$. Thus t is determined by an element h of H and a function $\Delta \to G$.

Show that the permutations t with $h = 1$ form a normal subgroup B (known as the 'base' of the wreath product) isomorphic to the direct product of n copies of G, where n is the number of elements of Δ. Show also that the factor group of G wr H by B is isomorphic to H.

Show that if K is a group of permutations of a set Θ and if $(\Gamma \times \Delta) \times \Theta$ is identified with $\Gamma \times (\Delta \times \Theta)$ in the obvious way then

$$(G \text{ wr } H) \text{ wr } K = G \text{ wr } (H \text{ wr } K).$$

[Oxford Final Honour School, 1976, Paper I]

8.4.   Let $p^{r(a)}$ be the highest power of the prime number p that divides $(p^a)!$.
   (i) Show that $r(a) = p^{a-1} + r(a-1)$.
   (ii) Use induction and a suitable wreath product (as defined in Exercise 8.3) to show that the symmetric group of degree p^a has a subgroup W_a of order $p^{r(a)}$.   [Essentially Cauchy, 1845, although he did not have wreath products defined in general.]

8.5.   Now let n be any positive integer and let p^r be the highest power of the prime number p that divides $n!$.
   (i) Show that if the 'p-adic form' of n is given by

$$n = n_k p^k + n_{k-1} p^{k-1} + \cdots + n_1 p + n_0$$

where $0 \leqslant n_a < p$ (for $0 \leqslant a \leqslant k$), then

$$r = n_k . r(k) + n_{k-1} . r(k-1) + \cdots + n_1$$

where $r(a)$ is as defined in Exercise 8.4.
   (ii) Show that the symmetric group S_n has a subgroup W of order p^r that can be expressed as a direct product

$$\underbrace{W_k \times W_k \times \cdots \times W_k}_{n_k \text{ factors}} \times \underbrace{W_{k-1} \times \ldots \times W_{k-1}}_{n_{k-1} \text{ factors}} \times \cdots \times \underbrace{W_1 \times \ldots \times W_1}_{n_1 \text{ factors}}$$

where W_a is as defined in Exercise 8.4.   [Cauchy, 1845, slightly modified: thus S_n has a 'Sylow p-subgroup'. Furthermore, a Sylow p-subgroup of S_n can be expressed as a direct product of wreath products.]

8.6.   Let G, H be subgroups of the finite group S and suppose that no element of $G \setminus \{1\}$ is conjugate to any element of H. Show that $|G|.|H|$ divides $|S|$.   [Cauchy, 1845]

8.7.   Deduce from Exercises 8.5, 8.6 that if $G \leqslant S_n$ and if p divides $|G|$ then G contains an element of order p.   [This is Cauchy's proof (1845) of the theorem that now bears his name. The same line of reasoning proves much more, namely the existence of Sylow subgroups, but that does not appear to have been noticed until some 35 years later.]

8.8. Let G be a subgroup of the finite group S that has a Sylow p-subgroup W. By considering the sizes of double cosets WsG for $s \in S$ show that there exists $x \in S$ such that $x^{-1}Wx \cap G$ is a Sylow p-subgroup of G.

Deduce the fact that every finite group G possesses Sylow p-subgroups. [This argument, that if we can find a group which contains G and which has a Sylow p-subgroup then also G must have a Sylow p-subgroup, can be applied in other ways. There are other groups than the symmetric groups into which any finite group may be embedded, some of which can be shown very easily to have Sylow subgroups.]

8.9. (i) Let U denote the set of lower unitriangular matrices in $GL(n, p)$, that is,

$$U := \{(a_{ij}) \mid a_{ij} \in \mathbb{Z}_p, \, a_{ii} = 1, \, a_{ij} = 0 \text{ if } i < j\}.$$

Show that U is a Sylow p-subgroup $GL(n, p)$.
(ii) Show that the symmetric group S_n is isomorphic to a subgroup of $GL(n, p)$ and hence that any group of order n is isomorphic to a subgroup of $GL(n, p)$.
(iii) Deduce the existence of Sylow subgroups in finite groups.

8.10. Give a proof of the existence of Sylow subgroups in a finite group G by using induction on $|G|$ and modifying appropriately our first proof of Cauchy's theorem. [Frobenius 1887]

8.11. Let G be a finite group, $H \trianglelefteq G$, let P be a Sylow p-subgroup of H, and let Ω be the set of all Sylow p-subgroups of H. Show that both H and G act transitively on Ω by conjugation, and deduce that $G = N_G(P) \cdot H$. [The 'Frattini argument']

8.12. Let P be a non-abelian group of order p^3 where p is prime, and let Z denote its centre. Show that $Z \cong Z_p$ and that $P/Z \cong Z_p \times Z_p$.

8.13. Let G be a finite group that has no subgroup of index 2. Let T be a Sylow 2-subgroup of G, let H be a subgroup of index 2 in T, and let t be an element of order 2 in G.
(i) Show that t acts as an even permutation of the G-space $\cos(G:H)$ and hence that t must fix some points of this space.
(ii) Hence show that t must be conjugate to some element of H. ['Thompson's Transfer Lemma'. Compare Chapter 4, Exercise 4.3.]
(iii) Let G now be a simple group of order 60. Show that all elements of order 2 must be conjugate in G, that there must be 15 of them, and that they must fall into 5 Sylow subgroups of order 4. Hence show that $G \cong A_5$.

8.14. Let G be a group of order pq where p, q are prime numbers and $p > q$.
  (i) Show that G has a unique subgroup P of order p.   [There is *no* need for Sylow's theorems here.]
  (ii) Show that either G is cyclic or every element of $G \backslash P$ has order q.
  (iii) Show that if G is not abelian then q divides $p-1$.

8.15. Let G be a group of order $p^2 q$, where p and q are distinct prime numbers.
  (i) Show that if G has distinct Sylow p-subgroups P_1 and P_2 then $P_1 \cap P_2 \lhd G$.
  (ii) Show that if G has more than 1 Sylow p-subgroup, and if $P_1 \cap P_2 = \{1\}$ for any two distinct Sylow p-subgroups P_1, P_2 then there are $q(p^2 - 1)$ elements of p-power order in G, and hence G has just one Sylow q-subgroup.
  (iii) Deduce that G cannot be simple.

8.16. Show that any group of order 36 has a normal subgroup of order 3 or 9.

8.17. Prove the assertion made by Galois in 1832, that there are no simple groups of composite order less than 60.

8.18. Let G be a group of order 60 and let P be a subgroup of order 5. Show that either $G \cong A_5$ or $P \lhd G$.

8.19. Let G be a group of order 120 whose Sylow 5-subgroups are not normal. Show that either $G \cong S_5$ or G has a normal subgroup Z of order 2 such that $G/Z \cong A_5$.

8.20. Prove that if p is prime and $p \geqslant 5$ then the group $\mathrm{PSL}(2, p)$ is simple. [Galois 1832]

Chapter 9

Actions count

In this chapter we discuss a famous counting theorem that is the foundation for a large area of combinatorics known as enumeration theory. Throughout the chapter Ω will be a finite set and G a finite group acting on Ω. For $g \in G$ let

$$\text{fix}_\Omega(g) := \{\omega \in \Omega \,|\, \omega^g = \omega\}.$$

Thus $\text{fix}_\Omega(g)$ consists of those elements of Ω that are in cycles of length 1 in the permutation induced by g; with the conventional rule that cycles of length 1 are not recorded, $\text{fix}_\Omega(g)$ consists of those points of Ω that do not appear in the cycle notation for g. We write $\text{Orb}_\Omega(G)$ for the set of G-orbits in Ω. With this notation the theorem is:

THEOREM 9.1: $\qquad |\text{Orb}_\Omega(G)| = \dfrac{1}{|G|} \sum_{g \in G} |\text{fix}_\Omega(g)|.$

Proof. First suppose that G is transitive on Ω. Let

$$S := \{\omega, g)\,|\,\omega^g = \omega\} \subseteq \Omega \times G.$$

We compare two different expressions for $|S|$. Certainly, on the one hand,

$$|S| = \sum_{g \in G} |\text{fix}_\Omega(g)|.$$

On the other hand,

$$|S| = \sum_{\omega \in \Omega} |G_\omega|,$$

where G_ω is the stabiliser of ω defined in Chapter 6. Now Lagrange's Theorem (Lemma 6.1) tells us (since Ω is, at the moment, a transitive G-space) that $|G_\omega| = |G|/|\Omega|$. Therefore

$$|S| = \sum_{\omega \in \Omega} |G_\omega| = |\Omega| \cdot |G|/|\Omega| = |G|,$$

and so

$$1 = |\text{Orb}_\Omega(G)| = \frac{1}{|G|} \sum_{g \in G} |\text{fix}_\Omega(g)|$$

in this case.

Suppose now that the G-orbits in Ω are $\Omega_1, \ldots, \Omega_t$. Then

$$\text{fix}_\Omega(g) = \text{fix}_{\Omega_1}(g) \cup \ldots \cup \text{fix}_{\Omega_t}(g),$$

and so

$$|\text{fix}_\Omega(g)| = \sum_{i=1}^{t} |\text{fix}_{\Omega_i}(g)|.$$

Therefore

$$\frac{1}{|G|} \sum_{g \in G} |\text{fix}_\Omega(g)| = \frac{1}{|G|} \sum_{g \in G} \sum_{i=1}^{t} |\text{fix}_{\Omega_i}(g)|$$

$$= \sum_{i=1}^{t} \frac{1}{|G|} \sum_{g \in G} |\text{fix}_{\Omega_i}(g)|$$

$$= \sum_{i=1}^{t} 1 = t = |\text{Orb}_\Omega(G)|,$$

and this proves the assertion.

This result has been referred to by many authors since 1960 as 'Burnside's Theorem' (in a preliminary edition (1970) of this text it was even assigned to the wrong Burnside—Oh dear!) but this attribution is mistaken. True, it does appear in the book on group theory published by William Burnside in 1897 (Second Edition 1911, reprinted 1955) but so it should: in the latter part of the nineteenth century it was common knowledge, well understood and widely appreciated by all mathematicians with any training in group theory. It goes back to the early days of the subject, to a paper by Cauchy (1845), and, in its present form, to Frobenius (1887). (See ΠMN, 'A lemma that is not Burnside's', *The Mathematical Scientist*, Vol. 4, 1979, pp. 133–41, for further details.)

Here is an example to illustrate how the theorem may be used.

PROBLEM 9.1: *How many essentially different ways are there of colouring a cube red, white, blue, so that each face is painted with one colour?*

Solution. First we need to agree precisely what we mean by 'essentially different'. Well, surely we want to consider two colourings of the cube as being indistinguishable if one can be obtained from the other by a rotation of the cube. And it seems reasonable to consider colourings as essentially different if not. Since there are 3 colours, any to be used at will on each of the 6 faces of the cube, there are 3^6 formally different colourings of the cube: and we hope that you will agree *amice lector* that two of these are essentially the same precisely when they are in the same orbit of the rotation group G of the cube acting on the set Ω of all these 3^6 different colourings. What we are seeking therefore is the number of orbits of G on Ω.

To find this number we use Theorem 9.1. To use the theorem we need to know, for each rotation g of the cube, how many colourings of the cube it leaves fixed. Consider, for example, a rotation of π about an axis through the mid-points of a pair of opposite edges. This inter-changes the top face and the east face, the bottom face and the west face, the north face and the south face of the cube in Fig. 9.1. Therefore, if the rotated cube is to be indistinguishable from the original cube then top face and east face must have the same colour, bottom face and west face must have the same colour, north face and south face must have the same colour. Consequently, there are 3^3 members of Ω that are fixed by this particular rotation. Similar arguments can be made for each rotation of the cube. The results appear in Table 9.1.

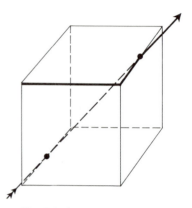

Fig. 9.1. A rotation of a cube.

Table 9.1. Application of Theorem 9.1 to coloured cubes.

Description of rotation	Number of such rotations	Number of colourings fixed by one such rotation	Contribution to $\sum |\text{fix}_\Omega(g)|$
Identity	1	3^6	729
Axis through diagonally opposite vertices; angle $2\pi/3$ or $4\pi/3$	8	3^2	72
Axis through midpoints of pair of opposite edges; angle π	6	3^3	162
Axis through centres of pair of opposite faces; angle π	3	3^4	243
Axis through centres of pair of opposite faces; angle $\pi/2$ or $3\pi/2$	6	3^3	162
Totals	24		1368

Therefore

$$|\text{Orb}_\Omega(G)| = \frac{1}{|G|} \sum_{g \in G} |\text{fix}_\Omega(g)| = \frac{1368}{24} = 57,$$

and there are 57 varieties of cubes coloured red, white, blue.

Here we have included also those cubes where perhaps not all three colours appear. If that had been the problem—to find the number of essentially different cubes that are red, white *and* blue—we would have subtracted the number of monochrome cubes and the number of colourings that use just two of the paints. These can be calculated in the same way. You should find that 27 colourings do not use all the colours, so there are 30 essentially different cubes coloured red, white *and* blue.

Specimen solutions to selected exercises

The following specimen solutions, being of the nature of an extension of the text, are placed between the main body of the chapter and the exercises. The wise and conscientious reader will, however, try the exercises unaided before attempting to evaluate what we have offered.

EXERCISE 9.1   *Suppose that G is transitive on the finite set* Ω. *Show that if* $|\Omega| > 1$ *then there exist elements of G that have no fixed points in* Ω.

Solution.   There is nothing lost if we assume that G acts faithfully on Ω. Then G must be finite and we can apply Theorem 9.1 (Not Burnside's Lemma): $|G| = \sum_{g \in G} |\text{fix}_\Omega(g)|$. Now all terms in the sum are non-negative integers and $|\text{fix}_\Omega(1)| = |\Omega| > 1$. If we had $|\text{fix}_\Omega(g)| > 0$ for all $g \in G$ then the sum would exceed $|G|$. Thus there must exist $g \in G$ with $\text{fix}_\Omega(g) = \varnothing$.

Aliter.   We may assume (as above) that G is finite. Let $n := |\Omega|$ and let $m := |G|/n$, the order of the stabilisers in G of elements of Ω. The stabilisers G_ω all have the identity element in common and therefore

$$\left| \bigcup_{\omega \in \Omega} G_\omega \right| \leqslant n(m-1) + 1 = nm - (n-1) = |G| - (n-1).$$

Any element not in the union of the stabilisers is fixed-point-free and therefore there are at least $n-1$ such elements in G.

EXERCISE 9.3   *Show that if G is m-fold transitive on* Ω *(see p. 58) then*

$$\frac{1}{|G|} \sum_{g \in G} |\text{fix}_\Omega(g)|^k = r_k \quad \text{for } 0 \leqslant k \leqslant m,$$

where $r_0 = 1, r_1 = 1, r_2 = 2, r_3 = 5, r_4 = 15$, *and in general* r_k *is the number of different equivalence relations on the set* $\{1, 2, \ldots, k\}$.

Solution.   We are assuming that G is finite and that $|\Omega| \geqslant m$. For an equivalence relation ρ on the set $\{1, \ldots, k\}$ define a subset $\Gamma_\rho \subseteq \Omega^k$ by

$$\Gamma_\rho := \{(\omega_1, \omega_2, \ldots, \omega_k) \mid \omega_i = \omega_j \Leftrightarrow i \equiv j \ (\text{mod } \rho)\}.$$

Certainly, if $(\omega_1, \ldots, \omega_k) \in \Gamma_\rho$ and $g \in G$ then $(\omega_1, \ldots, \omega_k)^g \in \Gamma_\rho$. Suppose that $(\alpha_1, \alpha_2, \ldots, \alpha_k), (\beta_1, \beta_2, \ldots, \beta_k) \in \Gamma_\rho$, and let $i_1, \ldots, i_r$ be representatives of the distinct ρ-classes in $\{1, \ldots, k\}$. Then $\alpha_{i_1}, \ldots, \alpha_{i_s}$ and $\beta_{i_1}, \ldots, \beta_{i_s}$ are sequences of distinct elements of Ω. Since G is m-fold transitive and $s \leqslant m$ it is certainly s-fold transitive, and so there exists $g \in G$ such that $\alpha_{i_j} g = \beta_{i_j}$ for $j := 1, \ldots, s$. It follows easily that $(\alpha_1, \ldots, \alpha_k)g = (\beta_1, \ldots, \beta_k)$. Thus G is transitive on Γ_ρ, that is, Γ_ρ is a G-orbit in Ω^k. If $\rho \neq \rho'$ then certainly $\Gamma_\rho \cap \Gamma_{\rho'} = \varnothing$. Since $\bigcup \Gamma_\rho = \Omega^k$, we have found the orbit decomposition of the G-space Ω^k, and we see that the number of G-orbits in Ω^k is r_k, where r_k is the number of equivalence relations on $\{1, \ldots, k\}$. The fact that $r_0 = r_1 = 1$ should be clear. The two G-orbits in Ω^2 are $\{(\omega, \omega) \mid \omega \in \Omega\}$ (the diagonal in Ω^2) and $\{(\omega_1, \omega_2) \mid \omega_1 \neq \omega_2\}$. The five G-orbits in Ω^3 are represented by triples of the forms $(\alpha, \alpha, \alpha), (\alpha, \alpha, \beta), (\alpha, \beta, \alpha), (\beta, \alpha, \alpha),$ and (α, β, γ). And so on.

Now $(\omega_1, \ldots, \omega_k)^g = (\omega_1^g, \ldots, \omega_k^g)$ and so for any $g \in G$ we have that $\mathrm{fix}_{\Omega^k}(g) = (\mathrm{fix}_\Omega(g))^k$. Applying Theorem 9.1 (Not Burnside's Lemma) we see that

$$\frac{1}{|G|} \sum_{g \in G} |\mathrm{fix}_\Omega(g)|^k = \frac{1}{|G|} \sum_{g \in G} |\mathrm{fix}_{\Omega^k}(g)| = r_k,$$

as required.

EXERCISE 9.9   *A manufacturer turns out necklaces consisting of 18 coloured glass beads on a circular nylon string whose ends are joined with an invisible weld. There are 6 white beads, 6 pink beads and 6 coral beads in each necklace. How many essentially different necklaces can she produce?*

Solution.   Two necklaces can be compared by arranging them on a flat surface with the string arranged as a circle and the beads equally spaced round it. They are then deemed to be essentially the same if the one can be rotated through some angle $k.2\pi/18$ or can be lifted up, turned over and replaced, so that it matches the other. The group G of possible operations here is the dihedral group of order 36, and the number of essentially different necklaces that can be produced is the number of its orbits in the set of all conceivable necklaces. We calculate this number using Not Burnside's Lemma.

Rotations of a necklace form a cyclic group of order 18 containing elements of orders 1, 2, 3, 6, 9, 18. An operation of turning a necklace over, which we shall call a 'reflection', is of one of two kinds: either two diametrically opposite beads are fixed or the line of reflection meets the string midway between two neighbouring beads. We shall denote a rotation of order r by f_r, reflections of the first kind by g_2, and reflections of the second kind by h_2. Thinking of these symmetry operations as permutations of the beads, a necklace is fixed by the operation if and only if each cycle of the permutation consists of beads of the same colour. Thus rotations f_9, f_{18} cannot fix any necklaces. The identity rotation f_1 fixes all the necklaces, of which there are $18!/(6!)^3$. A necklace is fixed by the rotation f_2 if and only if three of its 2-cycles consist of white beads, three consist of pink beads and three consist of coral beads. Thus the number of necklaces fixed by f_2 is $9!/(3!)^3$. Very similar calculations give the remaining entries in Table 9.2.

Table 9.2. Application of Theorem 9.1 to coloured necklaces.

$x \in G$	$\#(x \in G)$	$|\text{fix}(x)|$	Contribution
f_1	1	$18!/(6!)^3$	17 153 136
f_2	1	$9!/(3!)^3$	1 680
f_3	2	$6!/(2!)^3$	180
f_6	2	$3!$	12
f_9	6	0	0
f_{18}	6	0	0
g_2	9	$9!/(3!)^3$	15 120
h_2	9	$9!/(3!)^3$	15 120
Totals:	36		17 185 248

Thus by Not Burnside's Lemma the number of essentially distinct necklaces is $17\ 185\ 248 \div 36$, which is 477 368.

Exercises

9.1. Suppose that G is transitive on the finite set Ω. Show that if $|\Omega| > 1$ then there exist elements of G that have no fixed points in Ω.   [Jordan 1872].

9.2. Let G be a finite group and Ω_1, Ω_2 finite G-spaces. Show that

$$\text{fix}_{\Omega_1 \times \Omega_2}(g) = \text{fix}_{\Omega_1}(g) \times \text{fix}_{\Omega_2}(g)$$

for any $g \in G$, and deduce that

$$|\text{Orb}_{\Omega_1 \times \Omega_2}(G)| = \frac{1}{|G|} \sum_{g \in G} \phi_1(g)\phi_2(g)$$

where $\phi_i(g) := |\text{fix}_{\Omega_i}(g)|$.

9.3. Show that if G is m-fold transitive on Ω (see p. 58) then

$$\frac{1}{|G|} \sum_{g \in G} |\text{fix}_\Omega(g)|^k = r_k \quad \text{for } 0 \leqslant k \leqslant m,$$

where $r_0 = 1$, $r_1 = 1$, $r_2 = 2$, $r_3 = 5$, $r_4 = 15$, and in general r_k is the number of different equivalence relations on the set $\{1, 2, \ldots, k\}$.

9.4.   Let $n := |\Omega|$ and let

$$h_s := |\{g \in G \,|\, |\text{supp}(g)| = s\}|$$

$$= |\{g \in G \,|\, g \text{ moves } s \text{ points of } \Omega\}|$$

(see Chapter 1, Exercise 16). Show that if G is m-fold transitive then

$$\sum_{r=0}^{n} r(r-1)\dots(r-k+1)h_{n-r} = |G| \qquad \text{for } 0 \leqslant k \leqslant m.$$

[Cauchy 1845]

9.5.   Let S be a finite group and A, B subgroups of S. Show that

$$|\{(a, x, b)\,|\, a \in A, x \in S, b \in B, a^{-1}xb = x\}|$$

$$= |A| . |B| . |S:A, B|,$$

where $|S:A, B|$ denotes the number of double cosets AsB in S. [Frobenius (1887). He proved this by direct calculation with cosets and conjugates in S, and deduced the counting theorem from it. But it can easily be deduced in turn from the counting theorem by taking $\Omega := S$, $G := A \times B$, with action given by $\omega^{(a,b)} := a^{-1}\omega b$.]

9.6.   Take $S := \text{Sym}(\Omega)$, $A := \text{Stab}_{\text{Sym}(\Omega)}(\alpha)$, and $B := G \leqslant \text{Sym}(\Omega)$. Exhibit a one–one correspondence between double cosets AsB in S and G-orbits in Ω (compare Chapter 6, Exercise 5), and deduce the counting theorem from the result of Exercise 9.5 above.   [Frobenius, 1887. This appears to be how Frobenius came across the counting theorem. He realised the importance of the result better than Cauchy did, and gave as a second proof the direct line of argument that we have explained in the text.]

9.7.   The number r of orbits of a stabiliser G_α is known as the *rank* of the group G acting transitively on the set Ω (see Exercise 6.6). Show that the rank is given by

$$r = \frac{1}{|G|} \sum_{g \in G} |\text{fix}_\Omega(g)|^2.$$

9.8.   Show that if X is a finite group then

$$\sum_{x \in X} |C(x)| = m . |X|$$

where m is the number of conjugacy classes of X and $C(x)$ is the centraliser of x. Deduce that the probability that two random elements of X commute is $|X|/m$.

9.9.   A manufacturer turns out necklaces consisting of 18 coloured glass beads on a circular nylon string (whose ends have been joined with a weld, not a knot). There are 6 white beads, 6 pink beads, and 6 coral beads in each necklace. How many essentially different necklaces can she produce?

9.10. How many essentially different ways are there of colouring a regular dodecahedron if each of its twelve pentagonal faces is to be black or white?

9.11. A desk calendar is made by printing tables for each month on a regular dodecahedron, one month on each of the twelve faces. How many essentially different ways of doing this are there?

Chapter 10

Geometry: an introduction

The discussion in this chapter is preliminary to the formal developments which follow. Geometry is in the first instance a description of physical space, and this description becomes algebraic with the introduction of coordinates. Fundamental to our intuition of physical space is the notion of a displacement of a rigid object. Intuitively, a rigid object is one which retains its size and shape when moved about. The property of rigidity can be explained in terms of separation or distance: a physical object is rigid if the separation of any two of its particles is unchanged when the object is displaced. These ideas lead us to the concept of a transformation of the whole of physical space (rather than of an object occupying a piece of physical space) which preserves separation of points. Such a transformation is called an *isometry*.

If in plane geometry we use rectangular cartesian coordinates, a point P is represented by an ordered pair (x, y) of (real) numbers, and the distance between $P(x_1, y_1)$ and $Q(x_2, y_2)$ is given by

$$d(P, Q) = +\{(x_1 - x_2)^2 + (y_1 - y_2)^2\}^{1/2},$$

where $+\{\ \}^{1/2}$ denotes the positive square root. Likewise, in 3-dimensional geometry, a point is represented by an ordered triple of real numbers, and the distance between $P(x_1, y_1, z_1)$ and $Q(x_2, y_2, z_2)$ is given by

$$d(P, Q) = +\{(x_1 - x_2)^2 + (y_1 - y_2)^2 + (z_1 - z_2)^2\}^{1/2}.$$

Put in this form, the isometries of the plane (or 3-space) are identified with maps $\mathbb{R}^2 \to \mathbb{R}^2$ (or $\mathbb{R}^3 \to \mathbb{R}^3$) which preserve the distance d.

It is easy to make the generalisation from $\mathbb{R}^2$ and $\mathbb{R}^3$ to $\mathbb{R}^n$ (for any finite n), defining for $a = (\alpha_1, \ldots, \alpha_n)$, $b = (\beta_1, \ldots, \beta_n)$ the distance $d(a, b)$ by

$$d(a, b) := +\left\{ \sum_{i=1}^{n} (\alpha_i - \beta_i)^2 \right\}^{1/2}.$$

The reader is probably familiar with $\mathbb{R}^n$ as a topological space with metric given by d, and we recall (without proof) the triangle inequality

$$d(a, b) + d(b, c) \geqslant d(a, c).$$

We now make the formal definition: *An isometry of $\mathbb{R}^n$ is a bijective map $\sigma : \mathbb{R}^n \to \mathbb{R}^n$ such that $d(a\sigma, b\sigma) = d(a, b)$ for all $a, b \in \mathbb{R}^n$.*

Clearly, the product of two isometries is an isometry, and the identity map is an isometry. Also any isometry σ has an inverse map σ^{-1}, and since $d(a\sigma^{-1}, b\sigma^{-1}) = d(a\sigma^{-1}\sigma, b\sigma^{-1}\sigma) = d(a, b)$, σ^{-1} is an isometry. The isometries of $\mathbb{R}^n$ therefore form a group, which we shall denote by $I(n)$. We now seek a more explicit algebraic description of how an isometry acts on an element of $\mathbb{R}^n$.

We denote by 0 the n-tuple $(\lambda_1, \ldots, \lambda_n)$ with all $\lambda_i = 0$, and by e_j the n-tuple with $\lambda_j = 1$, $\lambda_i = 0$ for $i \neq j$. We show first that an isometry σ which fixes $0, e_1, \ldots, e_j, \ldots, e_n$ is the identity. Suppose that for any $a = (\alpha_1, \ldots, \alpha_n) \in \mathbb{R}^n$ we have $a\sigma = a^* = (\alpha_1^*, \ldots, \alpha_n^*)$. Then $d(a, 0) = d(a^*, 0)$ gives

$$\sum_{i=1}^{n} \alpha_i^2 = \sum_{i=1}^{n} \alpha_i^{*2}.$$

Further, $d(a, e_j) = d(a^*, e_j)$ gives

$$\sum_{i=1}^{n} \alpha_i^2 - 2\alpha_j + 1 = \sum_{i=1}^{n} \alpha_i^{*2} - 2\alpha_j^* + 1,$$

whence $\alpha_j = \alpha_j^*$ $(j = 1, \ldots, n)$, and $a = a^*$. Hence $\sigma = 1$.

Now let σ be an isometry which fixes 0, and suppose that $e_i\sigma = e_i^* = (\tau_{i1}, \ldots, \tau_{in})$, $(i = 1, \ldots, n)$. Then $d(0, e_i) = d(0, e_i^*)$ gives

$$\sum_{k=1}^{n} \tau_{ik}^2 = 1, \quad (i = 1, \ldots, n). \tag{1}$$

Also, for $i \neq j$, $d(e_i, e_j) = d(e_i^*, e_j^*)$ gives

$$\sum_{k=1}^{n} (\tau_{ik} - \tau_{jk})^2 = 2,$$

whence, using (1), we obtain

$$\sum_{k=1}^{n} \tau_{ik}\tau_{jk} = 0, \quad (i, j = 1, \ldots, n; i \neq j). \tag{2}$$

The relations (1) and (2) are precisely the conditions that the $n \times n$ matrix $T = [\tau_{ij}]$ satisfies the matrix equation $TT' = I$ (where T' denotes the transpose of T). We recall from Chapter 2 that a matrix T such that $T^{-1} = T'$ is called *orthogonal*. Consider now the bijective map $\rho: \mathbb{R}^n \to \mathbb{R}^n$, given by $u\rho = u^* = uT$ (where we are treating u, u^* as $1 \times n$ (row-) matrices, and uT denotes the matrix product). We have $0\rho = 0$, $e_i\rho = e_i^*$. Moreover, since the matrix product $(a-b)(a-b)'$ is the 1×1 matrix $[d(a, b)]$, we have

$$[d(a\rho, b\rho)] = ((a-b)T)((a-b)T)' = (a-b)TT'(a-b)'$$

$$= (a-b)(a-b)' = [d(a, b)],$$

so ρ is an isometry. Hence $\sigma\rho^{-1}$ is an isometry which fixes $0, e_1, \ldots, e_n$. It follows that $\sigma\rho^{-1} = 1$, and $\sigma = \rho$.

Finally, let σ be any isometry, and let $0\sigma = a$. Then $\tau: u \mapsto u + a$ (addition of row-matrices) is an isometry, with inverse $\tau^{-1}: u \mapsto u - a$, and $\sigma\tau^{-1}$ is an isometry fixing 0. Hence $u(\sigma\tau^{-1}) = uT$, where T is an orthogonal matrix, and $u\sigma = (uT)\tau = uT + a$. The isometries are thus precisely the group of transformations given in matrix form by

$$u \mapsto uT + a, \quad T \text{ orthogonal,}$$

which was described in Chapter 2.

We now have a well-grounded framework which (for $n = 2, 3$) is what we need for the handling of classical Euclidean geometry. In talking of $\mathbb{R}^n$ in this context, we shall use geometrical terminology; thus, an element of $\mathbb{R}^n$ will be called a *point*, and any subset of $\mathbb{R}^n$ will be called a *figure*. Two figures S, S^* will be *equivalent* or *congruent* if and only if $S^* = S\sigma$ for some isometry σ of $\mathbb{R}^n$. Congruence so defined is clearly an equivalence relation, because the isometries are a group.

Classical geometry also deals with relationships between figures other than congruence; for instance, *similarity*, two figures being similar if they have the same shape but not necessarily the same size, so that one is congruent to an enlargement of the other. It is not difficult to see how to formulate similarity in algebraic terms; S, S^* will be similar figures if and only if $S^* = S\sigma$ for some transformation $\sigma: \mathbb{R}^n \to \mathbb{R}^n$ of the form

$$\sigma: u \mapsto u(\lambda T) + a, \quad \lambda \in \mathbb{R}, \ \lambda \neq 0, \ T \text{ orthogonal.}$$

These transformations form a group, which contains the isometries as a subgroup. The similarity transformations do not preserve distance, but they preserve ratios of distances, that is to say

$$d(a\sigma, b\sigma)/d(p\sigma, q\sigma) = d(a, b)/d(p, q).$$

Certain theorems of Euclidean plane geometry are statements which do not involve the notion of distance, but only intersection and parallelism of lines. As examples we quote the 'little Desargues theorem', which says that if AA', BB', CC' are parallel, and AB, BC are parallel to $A'B'$, $B'C'$ respectively, then CA is parallel to $C'A'$; and the 'big Desargues theorem', which asserts the same conclusion when parallelism of AA', BB', CC' is replaced by concurrence in the hypotheses (see Fig. 10.1).

Now, a line in $\mathbb{R}^2$ is a figure

$$l = \{(x, y): \alpha x + \beta y + \gamma = 0, \ \alpha, \beta, \gamma \in \mathbb{R}, \ \alpha \neq 0 \text{ or } \beta \neq 0\},$$

and two lines either coincide, or intersect in a single point, or have empty intersection (in which case they are parallel). If $\sigma: \mathbb{R}^2 \to \mathbb{R}^2$ is a bijection which maps lines onto lines and l, m are two lines, then $(l \cap m)\sigma = l\sigma \cap m\sigma$. Such a 'line-preserving' map is called a *collineation* of $\mathbb{R}^2$, and necessarily maps

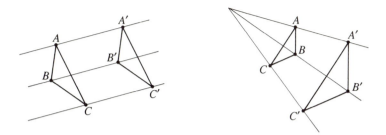

Fig. 10.1. The two versions of Desargues Theorem.

parallel lines to parallel lines. The isometries and the similarity transformations are collineations; so indeed are all transformations of the form

$$\sigma : u \mapsto uT + a, \quad T \text{ invertible.}$$

These transformations form a group, which is called the *affine group*. Figures
S, S^* such that $S^* = S\sigma$ for some affine transformation σ will be called *affine-
equivalent*.

   The definition of affine transformations extends immediately to $\mathbb{R}^n$, and
indeed to F^n, where F is any field: the affine group of dimension n over a field F
is the set of transformations $\sigma : F^n \to F^n$ of the form $\sigma : u \mapsto uT + a$, where T is an
invertible $n \times n$ matrix.

   In terms of the metric topology of $\mathbb{R}^n$, the affine transformations are
continuous. Thus, although an affine transformation does not (in general)
preserve the metric, it is always a homeomorphism of $\mathbb{R}^n$ with the metric
topology. If S, S^* are affine-equivalent figures, then S, S^* are homeomorphic
(with the topologies induced on them as subspaces of the metric space $\mathbb{R}^n$).

   By way of illustration, we examine the equivalence classes of some figures in
$\mathbb{R}^2$, under the Euclidean, similarity, and affine groups. We consider loci of the
second degree, i.e. non-empty point sets

$$\Gamma = \{(x, y) : ax^2 + 2hxy + by^2 + 2gx + 2fy + c = 0, a \neq 0 \text{ or } h \neq 0 \text{ or } b \neq 0\}.$$

Such loci have the traditional name of *conics*. We confine our attention to
central conics, which are those for which $ab \neq h^2$. To examine the effect of an
affine (or Euclidean, or similarity) transformation on a conic Γ, it is
convenient to express the equation of Γ in matrix form, as

$$uAu' + 2uk' + [c] = 0,$$

where

$$u = [x, y], \quad A = \begin{bmatrix} a & h \\ h & b \end{bmatrix}, \quad k = [g, f],$$

and $[c]$ is the 1×1 matrix. If now we make an affine transformation

$\sigma: u \mapsto u^* = [x^*, y^*]$, we can write $u = u^*T + w$, and obtain the matrix equation

$$(u^*T + w)A(T'u^{*\prime} + w') + 2(u^*T + w)k' + [c] = 0,$$

which on simplifying becomes

$$u^*A^*u^{*\prime} + 2u^*k^{*\prime} + [c^*] = 0$$

where $A^* = TAT'$, $k^* = kT' + wAT'$, $[c^*] = [c] + 2wk' + wAw'$. This is again a second-degree equation, which asserts that (x^*, y^*) lies on a conic

$$\Gamma^* = \{(x, y): a^*x^2 + 2h^*xy + b^*y^2 + 2g^*x + 2f^*y + c^* = 0,$$

$$a^* \neq 0 \text{ or } h^* \neq 0 \text{ or } b^* \neq 0\}.$$

Since $\det A^* = (\det T)^2 . \det A$ (and $\det T \neq 0$), we have $ab \neq h^2$ if and only if $a^*b^* \neq h^{*2}$; in other words, central conics are transformed into central conics. For such a conic, we can choose $w = -kA^{-1}$, giving $k^* = 0$. The point represented by the vector $-kA^{-1}$ is called the *centre* of the conic Γ. Using this value of w, we obtain

$$[c^*] = [C] - kA^{-1}k',$$

which on evaluation gives

$$c^* = \Delta/(ab - h^2), \text{ where } \Delta = abc + 2fgh - af^2 - bg^2 - ch^2.$$

If we are dealing with Euclidean transformations, the foregoing analysis is valid; the only reservation which we need to make is that the matrix T must be orthogonal. We now appeal to a well-known result about matrices $A^* = TAT'$, where A is symmetric and T is orthogonal, namely that, given A, T may be chosen so that A^* is diagonal, and the diagonal elements of A^* are the eigenvalues of A. In the present case, this means that there is an orthogonal matrix T such that $A^* = \text{diag}(\lambda, \mu)$, where λ, μ are the roots of the equation $t^2 - (a+b)t + ab - h^2 = 0$. We can therefore find a Euclidean transformation which takes the central conic Γ into the conic Γ^* with equation

$$\lambda x^2 + \mu y^2 + \Delta/(ab - h^2) = 0.$$

If $\Delta \neq 0$, this equation can be re-written as

$$\alpha x^2 + \beta y^2 = 1,$$

where $\alpha + \beta = -(a+b)(ab - h^2)/\Delta$, $\alpha\beta = (ab - h^2)^3/\Delta^2$.

Now, the (unordered) pair of numbers $\{\alpha, \beta\}$ characterises the Euclidean equivalence class of Γ; for if Γ is Euclidean-equivalent to a conic Γ^{**} with equation

$$\gamma x^2 + \delta y^2 = 1,$$

then there is a transformation $[x, y] \to [x, y]U + c$ (U orthogonal) which takes Γ^* to Γ^{**}; and it is easily seen that we must have $c = 0$, and

$$\begin{bmatrix} \gamma & 0 \\ 0 & \delta \end{bmatrix} = U \begin{bmatrix} \alpha & 0 \\ 0 & \beta \end{bmatrix} U'$$

which is possible if and only if $\{\gamma, \delta\} = \{\alpha, \beta\}$. The unordered pair $\{\alpha, \beta\}$ determines and is determined by the ordered pair $(\alpha + \beta, \alpha\beta)$. If we now refer back to the discussion of invariants in Chapter 5, we see that we have obtained for central ($ab \neq h^2$) non-degenerate ($\Delta \neq 0$) conics Γ in $\mathbb{R}^2$ under the action of the Euclidean group, a complete set of invariants, namely

$$-(a+b)(ab-h^2)/\Delta, \quad (ab-h^2)^3/\Delta^2.$$

Under the similarity group, any central non-degenerate conic is again equivalent to a conic with equation $\alpha x^2 + \beta y^2 = 1$ ($\alpha, \beta \neq 0$); but now this conic is equivalent to the conic with equation $\gamma x^2 + \delta y^2 = 1$ if and only if *either* $\gamma/\delta = \alpha/\beta$ or $\gamma/\delta = \beta/\alpha$. The unordered pair $\{\alpha/\beta, \beta/\alpha\}$, or equivalently the number $(\alpha/\beta) + (\beta/\alpha)$, or equivalently the number

$$(\alpha + \beta)^2/\alpha\beta = (a+b)^2/(ab-h^2),$$

is a complete set of invariants for such conics under the similarity group.

Under the affine group, the conics with equation $\alpha x^2 + \beta y^2 = 1, \gamma x + \delta y^2 = 1$ are equivalent if and only if $\alpha\beta$ and $\gamma\delta$ have the same sign. (We remind the reader that since we are dealing *ex hypothesi* with non-empty loci, α and β cannot both be < 0; and since we are assuming the central property, $\alpha\beta \neq 0$.) There are thus only two affine equivalence classes of central non-degenerate conics Γ, namely those for which $ab - h^2 > 0$ (the *ellipses*) and those for which $ab - h^2 < 0$ (the *hyperbolae*). This classification is valid for the Euclidean group in the sense that if two such (central, non-degenerate) conics are Euclidean equivalent, they must both be ellipses or both hyperbolae. However, in affine geometry, *any* two ellipses are equivalent; in Euclidean geometry they are only equivalent if they have the same pair of Euclidean invariants.

Finally, we examine the homeomorphism classes of the affine-equivalence classes which we have found. All ellipses are affine-equivalent to the locus with

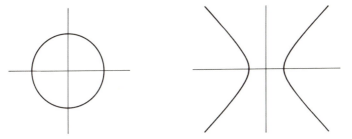

Fig. 10.2. Canonical forms for affine-equivalence classes of ellipses and hyperbolae.

equation $x^2+y^2=1$, i.e. the circle, denoted by S^1. All hyperbolae are affine-equivalent to the locus with equation $x^2-y^2=1$ (Fig. 10.2). This is a disconnected set with two components, namely

$$\{(x, y):x^2-y^2=1, x>0\} \quad \text{and} \quad \{(x, y):x^2-y^2=1, x<0\}.$$

Each of these components is homeomorphic to $\mathbb{R}$ (as is shown by the map $(x, y)\mapsto y$), and the hyperbola is therefore homeomorphic to $\mathbb{R}\cup\mathbb{R}$ (the *disjoint* union of two copies of $\mathbb{R}$).

Exercises

10.1. Discuss the classification of the degenerate central conics in $\mathbb{R}^2$, namely, those given by equations $ax^2+2hxy+by^2+2gx+2fy+c=0$, with $ab\neq h^2$ and $\Delta = abc+2fgh-af^2-bg^2-ch^2=0$.

10.2. Discuss the classification of the non-central conics in $\mathbb{R}^2$, namely, those given by equations $ax^2+2hxy+by^2+2gx+2fy+c=0$ with $ab=h^2$.

10.3. Consider the figures S in $\mathbb{R}^2$ consisting of an ordered pair of lines (l_1, l_2) given by equations $a_1x+b_1y+c_1=0$, $a_2x+b_2y+c_2=0$ (a_1, b_1 not both zero, a_2, b_2 not both zero). Show that

$$(a_1a_2+b_1b_2)^2/(a_1^2+b_1^2)(a_2^2+b_2^2)$$

is an invariant of such line-pairs under the similarity group. Is it a complete set of invariants under (a) the similarity group, (b) the Euclidean group? What are the affine-equivalence classes of ordered line-pairs?

10.4. Discuss the Euclidean, similarity and affine equivalence classes of non-singular central quadrics in $\mathbb{R}^3$, i.e. loci

$$\{(x, y, z):ax^2+by^2+cz^2+2fyz+2gzx$$
$$+2hxy+2ux+2vy+2wz+d=0\}$$

with

$$\begin{vmatrix} a & h & g \\ h & b & f \\ g & f & c \end{vmatrix}\neq 0, \quad \begin{vmatrix} a & h & g & u \\ h & b & f & v \\ g & f & c & w \\ u & v & w & d \end{vmatrix}\neq 0.$$

Show in particular that there are three affine equivalence classes, and find simple representatives of their homeomorphism classes.

10.5. Show that a distance-preserving map of a metric space S to itself, that is a map $\sigma:S\to S$ such that $d(a\sigma, b\sigma)=d(a, b)$ for all $a, b\in S$, is necessarily injective but need not be surjective.

Show that for $S = \mathbb{R}^2$ (with the usual metric), such a map is necessarily surjective.   [Hint: follow carefully the lines of the treatment of isometries given in the text.]

10.6. Some theorems of Euclidean geometry involve only incidence of lines, and ratios of distances measured on the same line; for instance, the theorem that the point of intersection of the diagonals of a parallelogram is the mid-point of each diagonal. Give two other examples of theorems of this kind.

Show that, if a, b, c are collinear points in $\mathbb{R}^2$, and σ is an affine transformation of $\mathbb{R}^2$, then (with the usual distance function d in $\mathbb{R}^2$), $d(a\sigma, b\sigma)/d(a\sigma, c\sigma) = d(a, b)/d(a, c)$, that is, affine transformations of $\mathbb{R}^2$ preserve ratios of Euclidean distances of collinear points.

10.7. Show that the area of the triangle with vertices $P(x_1, y_1)$, $Q(x_2, y_2)$, $R(x_3, y_3)$ in the cartesian plane is $|\Delta|$, where

$$\Delta = \begin{vmatrix} x_1 & y_1 & 1 \\ x_2 & y_2 & 1 \\ x_3 & y_3 & 1 \end{vmatrix}.$$

Show that if σ is an affine transformation of the plane given in matrix form by $\sigma : u \mapsto uA + a$, then the area of the triangle with vertices $P\sigma$, $Q\sigma$, $R\sigma$ is $|\Delta . \det A|$.

Chapter 11

The axiomatisation of geometry

Classical geometry was presented in the *Elements* of Euclid in an axiomatic form, that is, as a body of theorems developed by deductive logic from initial postulates. It was, however, what we should now call a physical theory, for it was seeking to describe the spatial relationships of the physical world. The modern notion of an axiomatic theory, which requires only that the axioms shall be logically consistent, did not develop until much later. Thus the possibility of there being more than one 'valid' geometry was not envisaged, and was accepted only with difficulty by the learned world in the 19th century when 'non-Euclidean' geometries were described.

Algebraic methods have had a decisive influence on geometrical thinking. Although the notion of coordinates can be traced back a long way in the history of the subject, it was not until the 17th century that algebraic notation and technique had developed to the point which enabled coordinate methods to be elevated by Descartes to the status of a 'universal method' applicable to any geometrical situation. Since then, geometry has become progressively more dominated by algebraic theory, and it can not now be thought of as a subject distinct from algebra.

We nowadays require any mathematical theory to be satisfactorily axiomatised. It is essential to know what assumptions are granted before one can say what constitutes a valid proof. The drive towards formal consistency has loosened the ties with the physical intuition which was the original source of geometrical theory and has sharpened the distinction between a formal proof of a theorem on the one hand, and on the other hand the motivation and intuitive thinking which lies behind, and without which the theorem would not have been formulated. For Descartes, the plane was something known, and coordinates and linear equations provided a *description* of points and lines. There are great formal advantages in shifting our stance, and *defining* a plane as $\mathbb{R}^2$, a point as an ordered pair $(x, y) \in \mathbb{R}^2$, a line as a locus $\{(x, y) : ax + by + c = 0\}$, and so forth; but in order to retain an understanding of the theory, we must not lose sight of the situation in physical space which motivates the theory. If the subject of study is, formally, $\mathbb{R}^2$ and its subsets, then figures drawn in a physical plane (sheet of paper, blackboard, etc.) must be regarded as representations or *pictures* of situations in the formal theory. A theorem must be validated by strict deduction from the axioms of the formal theory, and not by intuitive arguments from pictures. It is however common enough in mathematics to present 'proofs' of a somewhat informal nature, provided that they can be made strict, and it is often convenient to give such informal indications of proofs in terms of pictures. What must above all be emphasised is the importance of pictures in understanding any formal theory, not only of Euclidean plane geometry, but also of other theories which extend and generalise classical geometry. In writing a book such as this, there is necessarily a concentration on the formal development of the theory, and also a practical limitation on the number of pictures which can be included. It is essential that the reader should always make pictures (either by drawing or mental visualisation) of the situation which is being discussed.

We now consider how best to axiomatise the geometries which have been mentioned in Chapter 10. It is tempting to do this in terms of a space $\Omega = \mathbb{R}^2$ (or $\mathbb{R}^3$, or more generally $\mathbb{R}^n$) acted on by a certain group G of bijective transformations $\Omega \to \Omega$, which can be specified by matrix equations. A *figure* will then be a subset of Ω, and G acts naturally on the set $\mathscr{P}(\Omega)$ of figures. Two figures will be *equivalent* in the geometry if one is mapped to the other by a transformation in G. A property of figures can be identified with the set $\mathscr{S} \subset \mathscr{P}(\Omega)$ of those figures which have the property, and a property of figures will be called a *geometrical property* (in relation to the geometry under discussion) if whenever a figure belongs to $\mathscr{S}$, all figures equivalent to it belong to $\mathscr{S}$. This means that $\mathscr{S}$ is invariant under the action of G, or equivalently, that $\mathscr{S}$ is a union of orbits of $\mathscr{P}(\Omega)$ under the action of G. This approach, which brings the formal theory firmly under the umbrella of the theory of G-spaces, is the line which we shall follow. It is the approach to geometry which was first clearly enunciated by Felix Klein in 1872, in an address to a mathematical conference at Erlangen, and has thus become known as the *Erlangen* (or *Erlanger*) *Programm*.

In talking of a geometry as a G-space, we shall find it necessary to mention the pair (Ω, G) explicitly. We shall not need to mention the action μ of G on Ω, as G will always be defined as a group of transformations of Ω, and there is no ambiguity about the action. It will also be natural and unambiguous to write $(\omega, g)\mu \, (= \omega^g)$ as ωg, this being the normal way of denoting the image of $\omega \in \Omega$ under a map $g : \Omega \to \Omega$. We shall make free use of the notion of *equivalence* discussed in Chapter 3. Two geometries (Ω, G) and (Ω^*, G^*) will be equivalent if and only if there exist a bijective map $\theta : \Omega \to \Omega^*$ and an isomorphism $\phi : G \to G^*$ such that $(\omega g)\theta = (\omega\theta)(g\phi)$.

It will now be clear that, for Euclidean plane geometry, a formal algebraic description of the structure which we wish to discuss is provided by the action on $\Omega = \mathbb{R}^2$ of the group $G = \mathrm{I}(2)$, and it is tempting to take $(\mathbb{R}^2, \mathrm{I}(2))$ as the *definition* of Euclidean plane geometry. However, before deciding precisely how to formulate the definition, it will be helpful to look again at the situation in cartesian geometry. Here we have a 'known' plane Π, a distance $\delta(p, q)$ defined for pairs of points of Π, and the notion of a group G of isometries of Π (bijective mappings $\Pi \to \Pi$ which preserve distance). These notions are independent of coordinate systems. Now, we can choose a cartesian coordinate system with arbitrary origin and arbitrary rectangular axes, and any such choice of coordinate system sets up a representation of Π by $\mathbb{R}^2$, yielding a bijection $\theta:\Pi \to \mathbb{R}^2$ such that $\delta(p, q) = d(p\theta, q\theta)$, where d is the standard metric in $\mathbb{R}^2$. The coordinate system also defines an isomorphism $\phi:G \to I(2)$, which copies the isometries of Π by isometries of $\mathbb{R}^2$ in a natural way, namely $g\phi = \theta^{-1}g\theta$. The pair (θ, ϕ) defines an equivalence of the geometries (Π, G) and $(\mathbb{R}^2, I(2))$, as we have $(\pi g)\theta = (\pi\theta)(g\phi)$ for all $\pi \in \Pi$. If now we take two different coordinate systems, we obtain pairs (θ_1, ϕ_1),

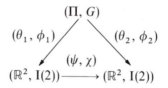

(θ_2, ϕ_2), and if $\psi = \phi_1^{-1}\theta_2$, $\chi = \theta_1^{-1}\phi_2$, the pair (ψ, χ) defines a self-equivalence of $(\mathbb{R}^2, I(2))$. Now, ψ is an isometry of $\mathbb{R}^2$ because

$$d(a, b) = \delta(a\theta_1^{-1}, b\theta_1^{-1}) = d(a\theta_1^{-1}\theta_2, b\theta_1^{-1}\theta_2) = d(a\psi, b\psi)$$

for all $a, b \in \mathbb{R}^2$, and ψ is certainly a bijection. Also, from the fact that (ψ, χ) is an equivalence, we have

$$(x\alpha)\psi = (x\psi)(\alpha\chi) \quad \text{for all } x \in \mathbb{R}^2, \alpha \in I(2)$$

giving $\psi(\alpha\chi) = \alpha\psi$, or $\alpha\chi = \psi^{-1}\alpha\psi$, so that the automorphism χ is simply conjugation by ψ, which we shall denote by κ_ψ. The self-equivalence (ψ, χ) of $(\mathbb{R}^2, I(2))$ is therefore an *inner* self-equivalence of the type described in Lemma 3.2.

We thus have for (Π, G) a whole host of coordinate systems provided by $(\mathbb{R}^2, I(2))$, any two of which are related by an inner self-equivalence of $(\mathbb{R}^2, I(2))$; and given any one coordinate system, all others can be derived from it by applying such inner self-equivalences. No one coordinate system has any special status. Even a modest experience of coordinate methods shows the value of freedom of choice of the coordinate system, and the importance of the judicious exercise of this choice in dealing with a specific problem. If we were to define Euclidean plane geometry as $(\mathbb{R}^2, I(2))$, we should in effect be committing ourselves at the start to one particular coordinate system. To avoid this unhappy situation, we proceed as follows.

We define Euclidean plane geometry as a G-space (Ω, G) equivalent to $(\mathbb{R}^2, I(2))$. This equivalence is represented by a pair of maps (θ, ϕ) where $\theta:\Omega \to \mathbb{R}^2$ is a bijection, and $\phi:G \to I(2)$ is an isomorphism, and $(\omega g)\theta = (\omega\theta)(g\phi)$ for all $\omega \in \Omega, g \in G$. Now, for any $\psi \in I(2)$, the pair of maps (θ^*, ϕ^*), where $\theta^* = \theta\psi, \phi^* = \phi\kappa_\psi$, also defines an equivalence of (Ω, G) with $(\mathbb{R}^2, I(2))$. Any such equivalence will be called a *coordinate system* for (Ω, G) (or a *coordinatisation* of (Ω, G), or a *coordinate representation* of (Ω, G)). From the definition, any two coordinate systems for (Ω, G) are related by an inner self-equivalence of $(\mathbb{R}^2, I(2))$.

Similar principles can clearly be applied to the definition of Euclidean geometry of n dimensions, using $(\mathbb{R}^n, \mathrm{I}(n))$ in the place of $(\mathbb{R}^2, \mathrm{I}(2))$. Likewise, the definition of affine geometry of n dimensions over a field F can be based on $(F^n, \mathrm{AGL}(n, F))$.

The coordinate space $\mathbb{R}^n$ (or F^n) has a natural structure of a vector space, and in many contexts it is convenient to think of the coordinates as vectors rather than as n-tuples. The affine group is easy to define in terms of a vector space $V_n(F)$ as the set of maps $V_n(F) \to V_n(F)$ of the form $u \to u\theta + a$ where $\theta: V_n(F) \to V_n(F)$ is an invertible linear map (that is, an automorphism of $V_n(F)$ as a vector space) and $a \in V_n(F)$. This group will be denoted by $\mathrm{AGL}(V_n(F))$, the automorphism group of $V_n(F)$ being denoted by $\mathrm{GL}(V_n(F))$. It is clear that any choice of basis for $V_n(F)$ defines an equivalence of $(V_n(F), \mathrm{AGL}(V_n(F)))$ with $(F^n, \mathrm{AGL}(n, F))$. For an affine geometry of dimension n over F, we may regard $(V_n(F), AGL(V_n(F)))$ as a partial coordinatisation, in which the origin (the zero vector of $V_n(F)$) has been selected, but the choice of axes (basis of $V_n(F)$) has been left open. We shall call any equivalence of the geometry (Ω, G) with $(V_n(F), \mathrm{AGL}(V_n(F)))$ a *vectorial representation* of the geometry. It is worth noting that, if we apply to a vectorial representation the inner self-equivalences of $(V_n(F), \mathrm{AGL}(V_n(F))$ determined by the translations $\tau: u \mapsto u + a, a \in V_n(F)$, the vectorial representations thus obtained yield all the coordinate representations. We shall generally find it convenient to use vectorial representations but it must of course be borne in mind that although we are representing the space as a vector space, the group of transformations is not one which preserves the vector space structure: in particular, a translation $u \mapsto u + a$, with $a \neq 0$, is *not* an automorphism of $V_n(F)$. (See Exercise 11.6.)

If we seek a vectorial representation of Euclidean geometry, in terms of a vector space $V_n(\mathbb{R})$ acted on by a suitable group of transformations, we meet a further difficulty. It was easy to define the affine group in basis-free terms, but we lack a basis-free definition of distance in general vector space $V_n(\mathbb{R})$. In fact, the definition we have adopted for Euclidean plane geometry reflects the situation in cartesian geometry in which only *rectangular* axes are used, a restriction represented in our formulation by the fact that in any coordinate system the distance between (x_1, y_1) and (x_2, y_2) is given by the standard metric $d((x_1, y_1), (x_2, y_2)) = +\{(x_1 - x_2)^2 + (y_1 - y_2)^2\}^{1/2}$. In vector space terminology, this means that the choice of basis is not entirely free.

We therefore consider how a Euclidean metric may be defined in a general vector space $V_n(\mathbb{R})$. This can be achieved in terms of a *positive-definite symmetric inner product*, that is, a mapping $V \times V \to \mathbb{R}$ in which the image of (x, y) is denoted by $x \cdot y$ and satisfies the conditions:

(i) $(\lambda_1 x_1 + \lambda_2 x_2) \cdot y = \lambda_1 (x_1 \cdot y) + \lambda_2 (x_2 \cdot y),$

$x \cdot (\mu_1 y_1 + \mu_2 y_2) = \mu_1 (x \cdot y_1) + \mu_2$ (bilinearity),

(ii) $x \cdot y = y \cdot x$ (symmetry),

(iii) $x \cdot x \geqslant 0$ for all $x \in V$,   and $x \cdot x = 0$ if and only if $x = 0$,

(positive-definiteness).

A transformation $\theta \in GL(V_n(\mathbb{R}))$ is said to preserve the inner product if $x\theta \cdot y\theta = x \cdot y$ for all $x, y \in V_n(\mathbb{R})$. Such tranformations are called orthogonal (with respect to the given inner product) and they form a subgroup $O(V_n(\mathbb{R}))$, which is called the *orthogonal group*.

An *orthonormal basis* of V_n is a set of elements $e_1, \ldots, e_n \in V_n$ such that $e_i \cdot e_j = \delta_{ij}$. Such a set is necessarily linearly independent. The existence of orthonormal bases requires proof, the standard method being the construction of an orthonormal basis from an arbitrary basis by the Gram–Schmidt orthogonalisation process. Clearly, an orthogonal transformation takes an orthonormal basis to an orthonormal basis.

If for $x \in V_n$ we define the *norm* of x as $\|x\| = +\{x \cdot x\}^{1/2}$, then for $\theta \in O(V_n)$ we have $\|x\theta\| = \|x\|$. Conversely, if $\theta \in GL(V_n)$ is norm-preserving, that is $\|x\theta\| = \|x\|$ for all $x \in V_n$, then

$$x\theta \cdot y\theta = \tfrac{1}{2}\left(\|x\theta + y\theta\|^2 - \|x\theta\|^2 - \|y\theta\|^2\right)$$

$$= \tfrac{1}{2}\left(\|(x+y)\theta\|^2 - \|x\theta\|^2 - \|y\theta\|^2\right)$$

$$= \tfrac{1}{2}\left(\|x+y\|^2 - \|x\|^2 - \|y\|^2\right)$$

$$= x \cdot y,$$

so that $\theta \in O(V_n)$.

Now consider the group $I(V_n)$ of transformations $\phi: V_n \to V_n$ given by $u\phi = u\theta + a$, where $\theta \in O(V_n)$ and $a \in V_n$. We have $x\phi - y\phi = x\theta - y\theta = (x-y)\theta$, so that

$$\|x\phi - y\phi\| = \|(x-y)\theta\| = \|x - y\| \quad \text{for all } x, y \in V_n.$$

If we take an orthonormal basis for V_n, and $x = \sum \alpha_i e_i$, $y = \sum \beta_i e_i$, so that x, y are represented by the n-tuples $a = (\alpha_1, \ldots, \alpha_n)$, $b = (\beta_1, \ldots, \beta_n) \in \mathbb{R}^n$, then

$$\|x - y\| = +\left\{\sum(\alpha_i - \beta_i)^2\right\}^{1/2} = d(a, b).$$

It is now easy to check that the choice of orthonormal basis has set up an equivalence between $(V_n, I(V_n))$ and $(\mathbb{R}^n, I(n))$, in which the subgroup $O(V_n)$ of $I(V_n)$ corresponds to the subgroup $O(n)$ of $I(n)$. Any equivalence of a Euclidean n-dimensional geometry with $(V_n, I(V_n))$ will be called a *vectorial representation*, and any choice of orthonormal basis for V_n then yields a *coordinate representation*.

Infinitely many positive definite inner products can be defined on a vector space $V_n(\mathbb{R})$; for instance, taking any basis $e_1, \ldots, e_n$ of $V_n(\mathbb{R})$, the product defined by $(\sum \alpha_i e_i) \cdot (\sum \beta_i e_i) = \sum \alpha_i \beta_i$ is easily verified to have the required properties; and the definition gives in particular $e_i \cdot e_j = \delta_{ij}$, so that the arbitrarily selected basis is orthonormal with respect to the inner product so defined. Given any one inner product and any $\psi \in \mathrm{GL}(V_n)$, we can define a new inner product (denoted by $*$) by writing

$$x * y = (x\psi^{-1}) \cdot (y\psi^{-1}) \quad \text{for all } x, y \in V_n;$$

this is readily verified to have all the required properties of a positive-definite symmetric inner product. If $e_1, \ldots, e_n$ is an orthonormal basis with respect to the first inner product, and $e_i^* = e_i\psi$, $(i = 1, \ldots, n)$, then $e_i^* * e_j^* = e_i \cdot e_j = \delta_{ij}$, and hence $e_1^*, \ldots, e_n^*$ is an orthonormal basis with respect to the new inner product. Moreover, all positive-definite symmetric inner products can be obtained from the given one in this way. For any second inner product we can find an orthonormal basis $e_1^*, \ldots, e_n^*$, and a (unique) $\psi \in \mathrm{GL}(V_n)$ such that $e_i\psi = e_i^*$, and then the product defined by $x * y = (x\psi^{-1}) \cdot (x\phi^{-1})$ also has $e_1^*, \ldots, e_n^*$ as an orthonormal basis. Now it follows immediately from bilinearity that two inner products with a common orthonormal basis must be identical, hence the second inner product is precisely that given by $x * y = (x\psi^{-1}) \cdot (y\psi^{-1})$.

Now consider the two distinct inner products defined on $V_n(\mathbb{R})$. These yield different orthogonal groups $\mathrm{O}(V_n)$, $\mathrm{O}^*(V_n) \leqslant \mathrm{GL}(V_n)$ and different isometry groups $\mathrm{I}(V_n)$, $\mathrm{I}^*(V_n) \leqslant \mathrm{AGL}(V_n)$. We claim, however, that the geometries $(V_n, \mathrm{I}(V_n))$ and $(V_n, \mathrm{I}^*(V_n))$ are equivalent; in fact, that if the two inner products are related as above by a transformation $\psi \in \mathrm{GL}(V_n)$, then the pair of maps (ψ, κ_ψ) defines such an equivalence. To show this, it is only necessary to show that conjugation by ψ maps $\mathrm{I}(V_n)$ onto $\mathrm{I}^*(V_n)$, and it will be sufficient to show that $\mathrm{I}(V_n)$ is mapped into $\mathrm{I}^*(V_n)$, as the argument is then easily completed by reversing the roles of the two products. We therefore have to establish that $\phi \in \mathrm{I}(V_n)$ implies $\phi^* = \psi^{-1}\phi\psi \in \mathrm{I}^*(V_n)$. Now $u\phi = u\theta + a$, where $\theta \in \mathrm{O}(V_n)$, and

$$u\phi^* = ((u\psi^{-1})\phi)\psi = ((u\psi^{-1})\theta + a)\psi = u(\psi^{-1}\theta\psi) + a\psi,$$

so to show that $\phi^* \in \mathrm{I}^*(V_n)$, we need only to show that $\psi^{-1}\theta\psi = \theta^* \in \mathrm{O}^*(V_n)$, that is to say, that θ^* preserves the second inner product. Now this follows from

$$(x\theta^*) * (y\theta^*) = (x\theta^*\psi^{-1}) \cdot (y\theta^*\psi^{-1})$$
$$= (x\psi^{-1}\theta) \cdot (y\psi^{-1}\theta) = (x\psi^{-1}) \cdot (y\psi^{-1}) = x * y.$$

Before leaving the general discussion of coordinate systems, we draw attention to a point which not infrequently gives rise to confusion. To fix ideas, consider a coordinate representation by $(\mathbb{R}^n, \mathrm{I}(n))$ of Euclidean n-dimensional geometry. If we consider a figure and a transformation in the geometry, these are represented by a figure (or subset) S of $\mathbb{R}^n$, and a transformation $\omega \in \mathrm{I}(n)$. Then $S^* = S\psi$ represents, in the given coordinate system, the image of the given figure under the given transformation. *But* there is also an alternative coordinate system related to the given one by maps (ψ, κ_ψ), and S_ψ represents in the second coordinate system the same figure as that represented by S in the first coordinate system. The application of the transformation ψ to figures in $\mathbb{R}^n$ can arise *either* as a transformation of figures represented in a fixed coordinate system, *or* as a change of coordinate system which does not alter the geometrical entities, but alters their coordinate representation. Similarly, for $\phi \in I(n)$, the change from ϕ to $\phi^* = \psi^{-1}\phi\psi$ arises *either* as a change in a fixed coordinate system from the transformation ϕ to a conjugate transformation, *or* as a change of coordinate system which alters the representation of a given transformation of the group under consideration. This ambivalence is inherent in the notion of a coordinate system, and makes it necessary for the reader to keep clearly in mind whether in a particular context we are discussing different figures and transformations relative to a fixed coordinate system, or the same figures and transformations relative to different coordinate systems. At the same time, these two alternative ways of looking at the algebraic situation can be extremely useful. For instance, if we wish to examine the conjugacy class of a member ϕ of $I(n)$, we can address ourselves to the abstractly identical problem of examining the representations of ϕ in alternative coordinate systems.

The following chapters pursue a more detailed investigation of some of the geometries which have been described, and we shall follow the formally convenient path from the general to the particular. This reverses the course of historical development, but has the advantage that any results obtained in a general situation are valid in a special situation. For example, any theorem of affine geometry over an arbitrary field F holds in affine geometry over $\mathbb{R}$; any theorem of affine geometry over $\mathbb{R}$ is valid in Euclidean geometry (which is also defined in terms of $\mathbb{R}^n$, but uses a smaller group of transformations).

Exercises

11.1. (i) The following subsets of $\Omega = \{A, B, C, D, E, F, G\}$ will be called *lines*:

$$\{A, B, F\}, \{B, C, D\}, \{A, C, E\}, \{A, D, G\},$$
$$\{B, E, G\}, \{C, F, G\}, \{D, E, F\}.$$

Try to draw a picture of this configuration, and show that it can not be achieved in such a way that the points of each line (as defined) are represented by collinear points in the picture.

(ii) Consider the group G of permutations of Ω which map lines to lines. Show that

(a) G is 2-transitive;

(b) the subgroup of G which fixes two points of Ω is isomorphic to $Z_2 \times Z_2$;

(c) $|G| = 168$.

(iii) Consider the set Ω^* of points of $\mathbb{F}_2^3$ other than $(0,0,0)$, and define a line in Ω^* as

$$\{(x, y, z) \in \Omega^* : ax + by + cz = 0,\ a, b, c \in \mathbb{F}_2,\ a, b, c \text{ not all } 0\}.$$

Show that there is a one–one correspondence between Ω and Ω^* in which the lines of Ω correspond to the lines of Ω^*. Hence show that $G \cong GL(3, \mathbb{F}_2)$.

11.2. (i) It is known that certain cubic curves in the complex plane have 9 points of inflexion, and that a line joining any two such points meets the curve again in a third point of inflexion. Through each of the 9 points there are thus 4 lines, each containing a further 2 of the 9 points; and there are in all 12 such lines. Try to draw a picture of the configuration of points and lines. Show that a picture can be drawn on a torus in such a way that the lines are represented by closed curves, any two of which meet in not more than one point.

(ii) Consider the configuration in set-theoretic terms, as a set Ω of 9 points, in which a certain 12 subsets of 3 points are called lines, and let G be the group of permutations of Ω which map lines to lines. Show that

(a) G is 2-transitive;

(b) the subgroup of G which fixes two points of Ω is isomorphic to S_3;

(c) $|G| = 432$.

(iii) In $\mathbb{F}_3^2$ a *line* has the usual (affine) definition:

$$\{(x, y) : ax + by + c = 0,\ a, b, c \in \mathbb{F}_3,\ a, b \text{ not both } 0\}.$$

Show that there is a one–one correspondence between Ω and $\mathbb{F}_3^2$ in which the lines of Ω correspond to the lines of $\mathbb{F}_3^2$. Hence show that $G \cong AGL(2, \mathbb{F}_3)$.

11.3. The five points $P_1, \ldots, P_5$ in (physical) 3-space are such that no four are coplanar (and hence no three are collinear). A plane Π (which contains none of the five points) is met by the line $P_i P_j$ in a point $P_{ij} (= P_{ji})$. Draw a picture of the 10 points P_{ij}, showing their collinearities.

Triangles ABC, $A'B'C'$ in a plane Π are said to be *centrally in perspective* if AA', BB', CC' are concurrent. They are said to be *axially in perspective* if $(BC \cap B'C')$, $(CA \cap C'A')$, $(AB \cap A'B')$ are collinear. A celebrated theorem of Desargues asserts that triangles centrally in perspective are axially in perspective, and conversely. By considering your figure and its mode of construction, derive a demonstration of this theorem. (Ignore the special cases which arise when AA', BB', CC' are parallel, or two corresponding sides of the triangles are parallel.)

How many pairs of triangles in perspective can you find in your picture of the 10 points P_{ij}?

Consider the group of collineations of the set of 10 points, that is, the permutations of the 10 points which map collinear subsets to collinear subsets. Show that

(i) G is transitive;

(ii) $|G_\alpha| = 12$, where G_α is the stabiliser of a point α in the set;

(iii) $|G| = 120$;

(iv) $G \cong S_5$.

11.4. In this exercise, a *triad* will mean an ordered set of three distinct collinear points. The classical Theorem of Pappus states that if (A, B, C) and (A', B', C') are triads on distinct lines and BC', $B'C$ meet in A'', CA', $C'A$ meet in B'', AB', $A'B$ meet in C'' then (A'', B'', C'') is a triad, which we shall call the *Pappus triad* of the triads (A, B, C) and (A', B', C'). Draw a picture of the 9 points and their collinearities, and show that each of the triads (A, B, C), (A', B', C'), (A'', B'', C'') is the Pappus triad of the other two. Such a set of three triads will be called a *Pappus set*. How many sets of three lines can you find in your picture such that the points on them form a Pappus set of triads?

   Consider now the group G of collineations of the Pappus configuration, and show that

(i) G is transitive;

(ii) $|G_A| = 12$;

(iii) $|G| = 108$;

(iv) G has a normal subgroup $H \cong S_3 \times Z_3$.

11.5. (i) It is assumed in the *Elements* of Euclid that given any two points A, B, then the circle with centre A and passing through B meets (in two points) the circle with centre B and passing through A. Show that this assumption is valid in $\mathbb{R}^2$, a circle being defined as a non-empty locus

$$\{(x, y) : x^2 + y^2 + 2gx + 2fy + c = 0, \; g, f, c \in \mathbb{R}\},$$

the centre of the circle being the point $(-g, -f)$.

(ii) Let us now examine the corresponding situation in $\mathbb{Q}^2$, a circle with centre $(-g, -f)$ being defined as a non-empty locus

$$\{(x, y) \in \mathbb{Q}^2 : x^2 + y^2 + 2gx + 2fy + c = 0, \; c \in \mathbb{Q}\}.$$

(a) Show that the circles with equation $x^2 + y^2 = 1$, $x^2 + y^2 - 2x = 0$ have empty intersection, thus contradicting the assumption described in (i), with $A = (0, 0)$, $B = (1, 0)$.

(b) Show that there are no points in $\mathbb{Q}^2$ satisfying the equation $x^2 + y^2 = 3$.

(c) Show that if there is one point of $\mathbb{Q}^2$ which satisfies an equation $x^2 + y^2 = k$ $(k \in \mathbb{Q})$, then there are infinitely many.

(d) Deduce from (c) that every circle in $\mathbb{Q}^2$ has infinitely many points.

(iii) We define now a subfield of $\mathbb{R}$ in the following way. We first define a sequence $\mathbb{Q} = F_0 \subset F_1 \subset F_2 \subset \ldots$ of subfields of $\mathbb{R}$ by the rule that F_{i+1} is the smallest subfield of $\mathbb{R}$ which contains F_i and the square roots of all positive elements of F_i. We then define

$$\mathbb{E} := \bigcup_{i=0}^{\infty} F_i.$$

Show that $\mathbb{E}$ is a subfield of $\mathbb{R}$, and that every element > 0 of $\mathbb{E}$ has a square root in $\mathbb{E}$. Show that each field F_i is countable, and hence that $\mathbb{E}$ is countable.

(iv) Show that in $\mathbb{E}^2$ (with obvious definitions of lines and circles), any of the ruler and compass constructions of classical geometry can be performed. What would be the objections (if any) to taking $\mathbb{E}$ instead of $\mathbb{R}$ as the coordinate field in a formal definition of Euclidean geometry?

11.6. (i) It has been noted that an affine geometry may be defined as $(V_n(F), \mathrm{AGL}(V_n))$, but that the group of transformations $\mathrm{AGL}(V_n)$ does not preserve the vector space structure of V_n. In particular, a translation $\tau_a: V_n \to V_n$ given by $\tau_a: x \mapsto x + a$, is not an automorphism of V_n as a vector space. Show that, by redefining addition and scalar multiplication in V_n by

$$x \ddagger y = (x\tau^{-1} + y\tau^{-1})\tau, \quad \lambda * x = (\lambda(x\tau^{-1}))\tau,$$

V_n can be given a new structure of a vector space V_n^*, such that the mapping $x \mapsto x\tau$ is an isomorphism $V_n \to V_n^*$ of vector spaces, and the mapping $\phi \mapsto \tau^{-1}\phi\tau$ is an isomorphism $\mathrm{GL}(V_n) \to \mathrm{GL}(V_n^*)$ of groups.

(ii) Suppose that we have a positive-definite symmetric inner product defined on $V_n(\mathbb{R})$, and that as in (i) we define a new vector space structure $V_n^*(\mathbb{R})$ by means of a translation τ. Show that we can define a positive-definite symmetric inner product on $V_n^*(\mathbb{R})$ by

$$x * y = (x\tau^{-1}) \cdot (y\tau^{-1})$$

and that $\phi \to \tau^{-1}\phi\tau$ is an isomorphism $\mathrm{O}(V_n) \to \mathrm{O}^*(V_n^*)$ of groups.

11.7. (i) Consider the set Ω of all symmetric bilinear forms defined on a real n-dimensional vector space $V_n(\mathbb{R})$; that is, maps $\pi: V_n \times V_n \to \mathbb{R}$ satisfying the bilinear and symmetry conditions. Show that we can define an action μ of $\mathrm{GL}(V_n)$ on Ω by $(\pi, \theta)\mu = \pi^\theta$, where

$$(x, y)\pi^\theta = (x\theta, y\theta)\pi.$$

Show that the positive-definite forms (the inner products) are an orbit of Ω under the action of $\mathrm{GL}(V_n) (= G)$, and that for such an inner product π the stabiliser G_π is the associated orthogonal group $\mathrm{O}(V_n)$.

(ii) Take a fixed basis $e_1, \ldots, e_n$ for V_n, and for any $\pi \in \Omega$ let $(e_i, e_j)\pi = \lambda_{ij}$. Show that $\lambda_{ij} = \lambda_{ji}$, and that

$$\left(\sum_i \alpha_i e_i, \sum_j \beta_j e_j \right)\mu = \sum_i \sum_j \alpha_i \beta_j \lambda_{ij}.$$

We say that μ is represented by the symmetric $n \times n$ matrix $A = [\lambda_{ij}]$. Show that, if $\theta \in GL(V_n)$ is given by $e_i \theta = \sum_j \tau_{ij} e_j$, so that $T = [\theta_{ij}]$ is an invertible $n \times n$ matrix, then π^θ is represented by the matrix TAT'. Show that, if Ω^* is the set of $n \times n$ symmetric matrices, then an action μ^* of the group $GL(n, \mathbb{R})$ of invertible matrices on Ω^* is defined by

$$(A, T)\mu^* = TAT',$$

and that $(\Omega, GL(V_n), \mu)$ and $(\Omega^*, GL(n, \mathbb{R}), \mu^*)$ are equivalent.

(iii) Investigate the orbits of Ω^* under the action of $GL(n, \mathbb{R})$, and find a complete set of invariants to describe these orbits. Find also canonical representatives of the orbits.

Chapter 12

Affine geometry

In the previous chapter we laid the foundation for the following formal definition:

An affine geometry $\mathscr{A}_n(F)$ of n dimensions over a field F is a G-space (Ω, G) equivalent to $(F^n, \mathrm{AGL}(n, F))$.

Such a geometry has coordinate representations and vectorial representations, and generally if we want to discuss a geometry in any detail, we must conduct our discussion in terms of a representation. We shall for instance wish to define figures in the geometry by the way they are described in a representation. Thus, we may seek to define a *line* in $\mathscr{A}_2(\mathbb{R})$ as a locus given by an equation of the form $ax + by + c = 0$ in a coordinate representation; but if this is to be a valid definition of a class of figures in $\mathscr{A}_2(\mathbb{R})$, we must satisfy ourselves that the definition is independent of the coordinate system, or equivalently, since two coordinate systems are related by a transformation in $\mathrm{AGL}(2, \mathbb{R})$, that any $\psi \in \mathrm{AGL}(2, \mathbb{R})$ transforms a locus of the type described into another locus of the same type. Otherwise expressed, the set of figures described by the definition must be invariant under the transformations in the group $\mathrm{AGL}(2, \mathbb{R})$. In discussing $\mathscr{A}_n(F)$ we shall generally choose to work with a vectorial representation $(V_n(F), \mathrm{AGL}(V_n))$, and then definitions in terms of $V_n(F)$ will need to be checked for invariance under transformations in the group $\mathrm{AGL}(V_n)$.

Our motivation for introducing affine geometry $\mathscr{A}_2(\mathbb{R})$ was that the group $\mathrm{AGL}(2, \mathbb{R})$ is the largest group of matrix transformations of the form $u \mapsto uA + a$ which preserves lines, so that $\mathscr{A}_2(\mathbb{R})$ appears to be the appropriate plane geometry in which to discuss non-metrical incidence properties of lines such as the two theorems of Desargues which were mentioned. A line-preserving transformation $\mathbb{R}^2 \to \mathbb{R}^2$ will be called a *collineation* and every affine transformation of $\mathbb{R}^2$ is a collineation. However, it is by no means clear that the affine group contains *all* the collineations of $\mathbb{R}^2$. That this is in fact the case will be shown in the course of the present chapter, when it will be seen to depend on a somewhat special property of the field $\mathbb{R}$.

Our first task is to define the analogues of lines in spaces of dimension > 2 over an arbitrary field F, and as we shall be using a vectorial representation we start by reviewing the salient features of the group $\mathrm{AGL}(V_n)$. Every $\phi \in \mathrm{AGL}(V_n)$ is of the form $\phi : u \mapsto u\theta + a$, $\theta \in \mathrm{GL}(V_n)$, $a \in V_n$. These transformations include the translations $\tau_a : u \mapsto u + a$, which form an abelian subgroup T of $\mathrm{AGL}(V_n)$, isomorphic to the additive group of V_n. Thus, any

$\phi \in \text{AGL}(V_n)$ is of the form $\phi = \theta\tau$, $\theta \in \text{GL}(V_n)$, $\tau \in T$, and $\text{AGL}(V_n)$ is generated by its subgroups $\text{GL}(V_n)$ and T. Also, $\text{GL}(V_n) \cap T = \{1\}$, hence the expression of ϕ in the form $\theta\tau$ is unique. Since

$$u(\theta_1\tau_a)(\theta_2\tau_b) = (u\theta_1 + a)\theta_2 + b = u\theta_1\theta_2 + a\theta_2 + b,$$

we have $(\theta_1\tau_a)(\theta_2\tau_b) = (\theta_1\theta_2)\tau_c$, where $c = a\theta_2 + b$, and the mapping $\phi = \theta\tau \mapsto \theta$ is a (surjective) homomorphism $\text{AGL}(V_n) \to \text{GL}(V_n)$ with kernel T. Thus T is a normal subgroup of $\text{AGL}(V_n)$, and $\text{GL}(V_n) \cong \text{AGL}(V_n)/T$. The previous calculation shows in particular that if $\phi = \theta\tau$, then

$$\phi^{-1}\tau_a\phi = \theta^{-1}\tau_a\theta = \tau_{a\theta}.$$

We shall define the *determinant* of $\phi = \theta\tau \in \text{AGL}(V_n)$ by $\det \phi = \det \theta$, and the preceding remarks show that

$$\det(\phi_1\phi_2) = \det(\theta_1\theta_2) = \det\theta_1 . \det\theta_2 = \det\phi_1 . \det\phi_2,$$

and indeed that the map $\det : \text{AGL}(V_n) \to F^*$ is a homomorphism of $\text{AGL}(V_n)$ onto the multiplicative group F^* of non-zero elements of F.

We recall that a line in $\mathbb{R}^2$ is determined by two distinct points (x_1, y_1), (x_2, y_2), and is the set of points $(\lambda x_1 + (1-\lambda)x_2, \lambda y_1 + (1-\lambda)y_2)$, $\lambda \in \mathbb{R}$. The definitions which follow generalise this idea.

DEFINITION 1:   *The affine span of* $(k+1)$ *vectors* $u_0, u_1, \ldots, u_k \in V_n(F)$ *is the set of vectors*

$$\{\lambda_0 u_0 + \lambda_1 u_1 + \cdots + \lambda_k u_k : \lambda_i \in F, \lambda_0 + \lambda_1 + \cdots + \lambda_k = 1\}.$$

To show the invariance of this definition, let $v = \sum \lambda_i u_i$ with $\sum \lambda_i = 1$, and let $\phi = \theta\tau_a \in \text{AGL}(V_n)$; then

$$v\phi = \left(\sum \lambda_i u_i\right)\theta + a = \left(\sum \lambda_i(u_i\theta)\right) + a = \sum \lambda_i(u_i\theta + a) = \sum \lambda_i(u_i\phi),$$

so that ϕ takes the affine span of $u_0, u_1, \ldots, u_k$ to the affine span of $u_0\phi, u_1\phi,$ $\ldots, u_k\phi$. Note also that, if ϕ fixes all the u_i, it fixes all points in their affine span.

DEFINITION 2:   *The vectors* $u_0, u_1, \ldots, u_k \in V_n(F)$ *are* affine-independent *if*

$$\lambda_0 u_0 + \lambda_1 u_1 + \cdots + \lambda_k u_k = 0, \quad \lambda_i \in F, \lambda_0 + \lambda_1 + \cdots + \lambda_k = 0,$$

imply $\lambda_0 = \lambda_1 = \cdots = \lambda_k = 0$.

The definition of affine independence is equivalent to the statement that none of the vectors u_i is in the affine span of the others, and its invariance under $\text{AGL}(V_n)$ is therefore assured. Note also that the affine independence of $u_0, u_1,$ $\ldots, u_k$ is equivalent (for any $i = 0, 1, \ldots, k$) to the linear independence of

$u_0 - u_i, \ldots, u_{i-1} - u_i, u_{i+1} - u_i, \ldots, u_k - u_i$, and in particular that *two* vectors are affine-independent if and only if they are distinct.

DEFINITION 3: *An affine k-flat in $V_n(F)$ is the affine span of $(k+1)$ affine-independent vectors.*

Among the affine k-flats of V_n are the k-dimensional vector subspaces of V_n, for if $u_1, \ldots, u_k$ are linearly independent, their linear span is precisely the same as the affine span of the affine-independent set $0, u_1, \ldots, u_k$. We can in fact give an alternative description of the affine k-flats of V_n in terms of the k-dimensional vector subspaces of V_n. Before doing so, we make the following further definition.

DEFINITION 4: *A figure (= subset) S of $V_n(F)$ is parallel to a figure S*, and we write S∥S*, if and only if $S^* = S + a$, for some $a \in V_n(F)$.*

The definition asserts that there is a translation $\tau_a \in T$ such that $S\tau_a = S^*$, a relation which we can express by saying that S^* is a *translate* of S. Parallelism, as here defined, is an equivalence relation, and every figure is parallel to itself. We must check the definition for invariance, and show that if S∥S* then $S\phi ∥ S^*\phi$ for $\phi \in \mathrm{AGL}(V_n)$. Writing $S^* = S\tau_a$ and $\phi = \theta\tau_c$, we have

$$S^*\phi = S\tau_a\theta\tau_c = S\theta\tau_{a\theta}\tau_c = S(\theta\tau_c)\tau_{a\theta} = (S\phi)\tau_{a\theta}.$$

We now show that the affine k-flats of V_n are simply the translates of vector subspaces.

THEOREM 12.1: *The affine k-flats of $V_n(F)$ are the figures $W + a$, where W is a k-dimensional vector subspace of $V_n(F)$, and $a \in V_n(F)$.*

Proof. If A_k is an affine k-flat, it is the affine span of an affine-independent set of vectors $u_0, u_1, \ldots, u_k$. Let W_k be the linear span of the linearly independent set of vectors $u_1 - u_0, \ldots, u_k - u_0$. Then $A_k = W_k + u_0$.

Conversely, if W_k is a k-dimensional subspace of V_n spanned by a linearly independent set of vectors $v_1, \ldots, v_k$, and $A_k = W_k + a$, then A_k is the affine span of the affine-independent set of vectors $a, a + v_1, \ldots, a + v_k$.

Theorem 12.1 asserts that the affine k-flats of V_n are the cosets in the additive group of V_n of the k-dimensional vector subspaces of V_n. Any k-flat is a coset of a unique subspace, which will be called the *direction* of the k-flat. The k-dimensional directions are simply the k-dimensional subspaces of V_n. Two k-flats are parallel if and only if they have the same direction.

It should be noted that our definitions are valid for $k = 0$ and $k = n$. A 0-flat is simply an element of V_n, or *point*; the only n-flat is V_n. A 1-flat will be called a *line*, and a 2-flat a *plane*.

We see quite easily that the affine group is transitive on k-flats, for given any two k-flats $A = W + a$, $A^* = W^* + a^*$ where W, W^* are k-dimensional subspaces of V_n, there is a transformation $\theta \in \mathrm{GL}(V_n)$ such that $W\theta = W^*$, and writing $\phi = \tau_a^{-1}\theta\tau_a$, we have $A\phi = A^*$. This result also follows from the next theorem.

THEOREM 12.2: *If $u_0, u_1, \ldots, u_k$ and $v_0, v_1, \ldots, v_k$ are two affine independent sets of $(k+1)$ points, there is an affine transformation ϕ such that $v_i = u_i\phi$, $i = 0, 1, \ldots, k$; and if $k = n$, then ϕ is unique.*

Proof. Writing $u_i^* = u_i - u_0$, $v_i^* = v_i - v_0$, $(i = 1, \ldots, n)$, each of the sets $\{u_1^*, \ldots, u_k^*\}$ and $\{v_1^*, \ldots, v_k^*\}$ is linearly independent; hence there is a transformation $\theta \in GL(V_n)$ such that $v_i^* = u_i^*\theta$, $i = 1, \ldots, n$. If now $\phi = \tau_{u_0}^{-1}\theta\tau_{v_0}$, we have $v_i = u_i\phi$, $i = 0, 1, \ldots, n$. Clearly θ (and hence ϕ) is not unique if $k < n$. If $k = n$, suppose we have affine transformations ϕ_1, ϕ_2 such that $v_i = u_i\phi_1, v_i = u_i\phi_2, i = 0, 1, \ldots, n$. Then $\phi_1\phi_2^{-1}$ fixes $u_0, u_1, \ldots, u_n$, and hence every point of their affine span, which is the whole of V_n. Hence $\phi_1\phi_2^{-1} = 1$, and $\phi_1 = \phi_2$.

COROLLARY 12.3:   *An affine transformation in n dimensions is completely determined by its action on a set of $(n+1)$ affine-independent points.*

Since any two distinct points are affine-independent, Theorem 12.2 tells us that the affine group is 2-transitive (see Chapter 5). In affine geometry there can therefore be no invariant (such as the Euclidean distance) which distinguishes different pairs of distinct points. We could define a metric invariant under the affine group by: $d(a, b) = 1$ if $a \neq b$, $d(a, b) = 0$ if $a = b$; but this metric gives to the space the discrete topology, which adds nothing to its set-theoretic structure.

Our next objective is a study of the conjugacy classes of the affine group. In any group, conjugate elements have a strong family resemblance. In the case of a G-space (Ω, G), conjugate elements of G resemble each other not only in group-theoretic properties but also in their action on Ω. This resemblance is expressed in formal terms by the fact that, for any $\psi \in G$, the maps (ψ, κ_ψ) define a self-equivalence of (Ω, G). If we consider vectorial representations $(V_n(F), AGL(V_n))$ of $\mathscr{A}_n(F)$, then $\phi_1, \phi_2 \in AGL(V_n)$ are conjugate in $AGL(V_n)$ if and only if they represent the *same* affine transformation in two different vectorial representations. A similar statement applies to coordinate representations, and conjugacy in $AGL(n, F)$. We shall in fact find it convenient to investigate the conjugacy classes by examining the coordinate representations of an arbitrary affine transformation. For a given transformation we shall seek a particularly simple coordinate representation, which will serve as a *canonical representative* of the corresponding conjugacy class of $AGL(n, F)$ (cf. Example 5.11).

It is seen by considering the homomorphism $AGL(V_n) \to GL(V_n)$ defined by $\theta\tau \mapsto \theta$ that, if $\phi_1 = \theta_1\tau_1$ and $\phi_2 = \theta_2\tau_2$ are conjugate in $AGL(V_n)$, then θ_1 and θ_2 are conjugate in $GL(V_n)$. The problem of finding canonical representatives of conjugacy classes of $GL(n, F)$ is (in matrix terms) that of finding for a given (invertible) matrix A a conjugate matrix $T^{-1}AT$ of pleasantly simple form, and is equivalent to the problem of finding for a given $\theta \in GL(V_n)$ a basis of V_n relative to which θ is represented by such a matrix. The reader is probably familiar with this important problem of linear algebra, and may know general results about the rational and Jordan canonical forms of linear transformations. If we were to tackle the general question of conjugacy classes of $AGL(n, F)$, we should have to draw on such results. We shall however confine our detailed examination to affine geometry $\mathscr{A}_2(\mathbb{R})$ and thus to the group $AGL(2, \mathbb{R})$. We therefore preface our discussion by a self-contained investigation of the conjugacy classes of $GL(2, \mathbb{R})$, which will be conducted in terms of the matrix representation of a transformation $\theta \in GL(V_2(\mathbb{R}))$. We do this partly for the sake of completeness, and also because some aspects of the problem which are special to the field $\mathbb{R}$ are not always covered by general treatments of the theory of canonical matrices.

For a given $\theta \in GL(V_2(\mathbb{R}))$, it is natural to start the search for a 'good' basis of $V_2(\mathbb{R})$ by looking for 1-dimensional subspaces fixed by θ. This is the same as finding non-zero vectors x such that $x\theta = \lambda x$, $\lambda \in \mathbb{R}$. If such can be found, λ is necessarily non-zero, since $x\theta \neq 0$ (because $x \neq 0$ and θ is invertible). From $x(\theta - \lambda . 1) = 0$ it follows that $\theta - \lambda . 1 : V_2(\mathbb{R}) \to V_2(\mathbb{R})$ has a non-trivial kernel, hence $\det(\theta - \lambda . 1) = 0$. If θ is represented, relative to some basis e, f of $V_2(\mathbb{R})$, by the matrix

$$A = \begin{bmatrix} \alpha & \beta \\ \gamma & \delta \end{bmatrix},$$

then

$$\det(\theta - \lambda . 1) = \lambda^2 - (\alpha + \delta)\lambda + \alpha\delta - \beta\gamma = \lambda^2 - (\operatorname{tr}\theta)\lambda + (\det\theta)$$

where the coefficients $\operatorname{tr}\theta$ (the *trace* of θ) and $\det\theta$ (the *determinant* of θ) depend only on θ, and not on the particular matrix representation. The polynomial

$$\chi(t) = t^2 - (\operatorname{tr}\theta)t + (\det\theta)$$

is the *characteristic polynomial* of θ, and its zeroes are the *eigenvalues* of θ. We shall need to use *Cayley's Theorem*, which asserts that $\chi(\theta) = 0$, where $\chi(\theta)$ stands for $\theta^2 - (\operatorname{tr}\theta) . \theta + (\det\theta) . 1$. In the limited case under consideration, this theorem is easily proved in terms of a matrix representation, for we have $e\theta = \alpha e + \beta f$, $f\theta = \gamma e + \delta f$, whence $e(\theta - \alpha . 1) = \beta f$, $f(\theta - \delta . 1) = \gamma e$, giving

$$e(\theta - \alpha . 1)(\theta - \delta . 1) = \beta\gamma e, \quad f(\theta - \delta . 1)(\theta - \alpha . 1) = \beta\gamma f,$$

which reduce to $e . \chi(\theta) = 0$, $f . \chi(\theta) = 0$, showing that $\chi(\theta) = 0$.

The following possible cases arise for the solutions of the equation $\chi(t) = 0$:
(i) two distinct real roots λ, μ (necessarily non-zero): $\chi(t) = (t - \lambda)(t - \mu)$;
(ii) one real root $\lambda \neq 0$: $\chi(t) = (t - \lambda)^2$;
(iii) no real roots: the roots are a complex conjugate pair $\{\rho e^{i\alpha}, \rho e^{-i\alpha}\}$ ($\rho > 0$, $\alpha \in (0, \pi)$): $\chi(t) = t^2 - 2(\rho \cos \alpha)t + \rho^2$.

In case (i), we have non-zero vectors e, f such that $e\theta = \lambda e, f\theta = \mu f$. These vectors are linearly independent, since otherwise $f = ve$ ($v \neq 0$), and $\mu f = f\theta = ve\theta = v\lambda e = \lambda f$, giving $\lambda = \mu$. We can therefore take e, f as a basis for $V_2(\mathbb{R})$, and θ is then represented by the matrix

$$\begin{bmatrix} \lambda & 0 \\ 0 & \mu \end{bmatrix}.$$

In case (ii), we have $(\theta - \lambda.1)^2 = 0$, and we consider separately the alternative possibilities: (i) $(\theta - \lambda.1) = 0$, (ii) $(\theta - \lambda.1) \neq 0$. In the first case, $\theta = \lambda.1$, and relative to any basis θ is represented by the matrix

$$\begin{bmatrix} \lambda & 0 \\ 0 & \lambda \end{bmatrix}.$$

In the second case, we take any vector e which is not in the kernel of $\theta - \lambda.1$, so that $f = e(\theta - \lambda.1) = e\theta - \lambda e \neq 0$. Then e, f are linearly independent, since otherwise $f = ve$ ($v \neq 0$), giving $e\theta = (\lambda + v)e$, and $\lambda + v$ would have to be a solution of $\chi(t) = 0$. We can therefore take e, f as a basis for $V_2(\mathbb{R})$. Since $(\theta - \lambda.1)^2 = 0$, we have $f(\theta - \lambda.1) = e(\theta - \lambda.1)^2 = 0$, so that $e\theta = \lambda e + f$, $f\theta = \lambda f$, and relative to the basis e, f, θ is represented by the matrix

$$\begin{bmatrix} \lambda & 1 \\ 0 & \lambda \end{bmatrix}.$$

In case (iii) we have

$$\chi(\theta) = \theta^2 - 2\rho \cos \alpha . \theta + \rho^2 . 1 = (\theta - \rho \cos \alpha . 1)^2 + \rho^2 \sin^2 \alpha . 1 = 0,$$

and writing

$$\psi = (\rho \sin \alpha)^{-1}(\theta - \rho \cos \alpha . 1),$$

we have $\psi^2 = -1$. Taking *any* non-zero vector e, and writing $f = e\psi$, we have $f\psi = -e$. The vectors e, f are linearly independent, since otherwise $f = ve$, giving $-e = f\psi = ve\psi = vf = v^2 e$, and hence $v^2 = -1$, which is impossible for $v \in \mathbb{R}$. The relations $f = e\psi$, $-e = f\psi$ may be written as

$$e\theta = \rho \cos \alpha . e + \rho \sin \alpha . f, \quad f\theta = -\rho \sin \alpha . e + \rho \cos \alpha . f,$$

so, taking e, f as a basis for $V_2(\mathbb{R})$, θ is represented by the matrix $\rho . R_\alpha$, where

$$R_\alpha = \begin{bmatrix} \cos \alpha & \sin \alpha \\ -\sin \alpha & \cos \alpha \end{bmatrix} = \cos \alpha . I + \sin \alpha . J, \text{ with } J = \begin{bmatrix} 0 & 1 \\ -1 & 0 \end{bmatrix}.$$

The designation of this 2×2 matrix by J will be adopted as a standard notation.

We have now shown that any $\theta \in \mathrm{GL}(V_2(\mathbb{R}))$ may be represented by one of the following matrices:

$$\begin{bmatrix} \lambda & 0 \\ 0 & \mu \end{bmatrix} \quad (\lambda \neq \mu, \ \lambda, \mu \neq 0);$$

$$\begin{bmatrix} \lambda & 0 \\ 0 & \lambda \end{bmatrix} \quad (\lambda \neq 0);$$

$$\begin{bmatrix} \lambda & 1 \\ 0 & \lambda \end{bmatrix} \quad (\lambda \neq 0);$$

$$\rho . R_\alpha \quad (\rho > 0, \ \alpha \in (0, \pi)).$$

If θ is representable by

$$\begin{bmatrix} \lambda & 0 \\ 0 & \mu \end{bmatrix},$$

then (simply by interchanging the basis elements of $V_2(\mathbb{R})$) it is also representable by

$$\begin{bmatrix} \mu & 0 \\ 0 & \lambda \end{bmatrix}.$$

Subject to this reservation, a given $\theta \in \mathrm{GL}(V_n(\mathbb{R}))$ cannot be represented by more than one matrix in the set, as the characteristic polynomials of these matrices are distinct, except for pairs

$$\begin{bmatrix} \lambda & 0 \\ 0 & \lambda \end{bmatrix}, \quad \begin{bmatrix} \lambda & 1 \\ 0 & \lambda \end{bmatrix};$$

and for such a pair, the first fixes every one-dimensional subspace of $\mathbb{R}^2$, while the second only fixes one such subspace, so they certainly cannot both represent the same transformation θ. Provided that we include in the set only one of each pair

$$\begin{bmatrix} \lambda & 0 \\ 0 & \mu \end{bmatrix}, \quad \begin{bmatrix} \mu & 0 \\ 0 & \lambda \end{bmatrix},$$

we thus have a set of canonical representatives of the conjugacy classes of the multiplicative group of invertible 2×2 matrices with elements in $\mathbb{R}$, and each such matrix characterises a conjugacy class of $\mathrm{GL}(2, \mathbb{R})$, or equivalently of $\mathrm{GL}(V_2(\mathbb{R}))$.

We now embark on a similar description, by means of canonical representatives in $\mathrm{AGL}(2, \mathbb{R})$, of the conjugacy classes of the affine group. If such a transformation has a fixed point, we may take a vectorial representation in which the fixed point is the zero vector of $V_2(\mathbb{R})$, and the transformation is then represented by a transformation in $\mathrm{GL}(V_2(\mathbb{R}))$, and by suitable choice of basis is represented by a transformation $u \mapsto uA$ in $GL(2, \mathbb{R})$, where A is a matrix of the canonical set which we have described. To complete our description, we need to examine affine transformations with no fixed point. Again, we work with a vectorial representation.

The fixed points u of a transformation $\phi = \theta\tau_a \in AGL(V_2(\mathbb{R}))$ are given by $u = u\phi = u\theta + a$, and thus are the solutions of the equation $u(1 - \theta) = a$. If $1 - \theta$ is invertible, this equation has a solution $u = a(1 - \theta)^{-1}$; hence if ϕ has no fixed points, $1 - \theta$ must be singular, which means that θ has 1 as an eigenvalue. We can therefore choose a basis of $V_2(\mathbb{R})$ so that θ is represented by a matrix of one (and only one) of the following forms:

$$\begin{bmatrix} 1 & 0 \\ 0 & 1 \end{bmatrix}; \quad \begin{bmatrix} 1 & 0 \\ 0 & \lambda \end{bmatrix} \quad (\lambda \neq 0, 1); \quad \begin{bmatrix} 1 & 1 \\ 0 & 1 \end{bmatrix}.$$

In the first case, $\theta = 1$ and $\phi = \tau_a$, which has no fixed points for $a \neq 0$. We can take a as the first basis vector of $V_2(\mathbb{R})$, and then ϕ is represented in $AGL(2, \mathbb{R})$ by the matrix transformation

$$[x, y] \mapsto [x^*, y^*] = [x, y] + [1, 0].$$

In the second case, the transformation is represented in $AGL(2, \mathbb{R})$ by the matrix transformation

$$[x, y] \mapsto [x^*, y^*] = [x, y]\begin{bmatrix} 1 & 0 \\ 0 & \lambda \end{bmatrix} + [\xi, \eta],$$

and this has a fixed point if and only if $\xi = 0$, in which case all points $(x, \eta/(1 - \lambda))$ are fixed. We must therefore have $\xi \neq 0$. Now, from $y^* = \lambda y + \eta$ we have $y^* - (\eta/(1 - \lambda)) = \lambda(y - (\eta/(1 - \lambda)))$, so by choosing $(0, \eta/(1 - \lambda))$ as the origin of a new coordinate system (or equivalently taking a new vectorial representation with cf as the new zero vector, where $c = \eta/(1 - \lambda)$), we obtain a representation of the same transformation by a matrix equation of the same form, but with $\eta = 0$. Finally, by replacing the basis e, f in this new vectorial representation by $\xi e, \xi f$, we obtain a matrix representation

$$[x, y] \mapsto [x^*, y^*] = [x, y]\begin{bmatrix} 1 & 0 \\ 0 & \lambda \end{bmatrix} + [1, 0].$$

We apply a similar procedure in the third case, obtaining first a matrix representation

$$[x, y] \mapsto [x^*, y^*] = [x, y]\begin{bmatrix} 1 & 1 \\ 0 & 1 \end{bmatrix} + [\xi, \eta],$$

and observing that this has a fixed point if and only if $\xi = 0$, in which case every point $(-\eta, y)$ is fixed. We thus have $\xi \neq 0$, and by moving the origin to $(-\eta, 0)$ and adjusting the basis by the factor ξ, we obtain a matrix representation

$$[x, y] \mapsto [x^*, y^*] = [x, y] \begin{bmatrix} 1 & 1 \\ 0 & 1 \end{bmatrix} + [1, 0].$$

We now list (in equational form) representatives of all the types of affine transformations in $\mathscr{A}_2(\mathbb{R})$, distinguishing the *fixed-point-free* (with no fixed point), the *central* (with one fixed point), and the *axial* (with a line of fixed points). Each transformation listed characterises a conjugacy class of the affine group (with the reservation that in C.1 below it is the *unordered* pair $\{\lambda, \mu\}$ which characterises the conjugacy class).

Fixed-point-free
  F.1 $x^* = x+1,\ y^* = y.$
  F.2 $x^* = x+1,\ y^* = \lambda y,\ (\lambda \neq 0,1).$
  F.3 $x^* = x+1,\ y^* = x+y.$

Central (with centre $(0, 0)$).
  C.1 $x^* = \lambda x,\ y^* = \mu y,\ (\lambda,\ \mu \neq 0,\ 1,\ \lambda \neq \mu).$
  C.1 $x^* = \rho \cos \alpha . x - \rho \sin \alpha . y,$
       $y^* = \rho \sin \alpha . x + \rho \cos \alpha . y,\quad (\rho > 0,\ \alpha \in (0, \pi)).$
  C.3 $x^* = \lambda x,\ y^* = \lambda y,\ (\lambda \neq 0,1).$
  C.4 $x^* = \lambda x,\ y^* = x+\lambda y,\ (\lambda \neq 0,1).$

Axial
  A.1 $x^* = x,\ y^* = \lambda y,\ (\lambda \neq 0,1).$ Axis $y = 0.$
  A.2 $x^* = x,\ y^* = x+y.$ Axis $x = 0.$

Identity
  I $x^* = x,\ y^* = y.$

Certain of these transformations have been given special names.
  C.3 is a *central dilatation*, and if $\lambda = -1$ is a *central reflexion*.
  A.1 is an *axial dilatation*, and if $\lambda = -1$ is an *axial reflexion*.
  A.2 is a *shear*, and deserves special mention.
In both A.1 and A.2 the line containing a point and its image is parallel to the line with equation $x = 0$, and this for A.2 is the axis of the transformation.

Axial transformations are particularly easy to discuss in terms of pictures. We shall show that an axial transformation is uniquely determined if we are given the axis, and the image P^* of one point P not on the axis. The fact that the image X^* of an arbitrary point X can then be found by a geometrical construction is of significance. (See Exercises 12.14 and 12.15.)

If we are given the axis of an axial transformation and the image P^* of a point P not on the axis, we know that for any point X not on the axis, the line joining X and its image X^* is parallel to PP^* (Fig. 12.1). If PP^* is not parallel to the axis, we have an axial dilatation. If X is not on PP^* and the line PX meets the axis in Y, then the line PY is transformed into the line P^*Y (since the transformation fixes Y), so that the image X^* of X must lie on P^*Y, and is the intersection of P^*Y with the line through X parallel to PP^*. If PX does not meet (and is therefore parallel to) the axis, the line P^*Y in this construction must be replaced by the line through P^* parallel to the axis. Finally, if Q lies on PP^*, we can use X, X^* to find the image Q^* of Q by the construction indicated in Fig. 12.1.

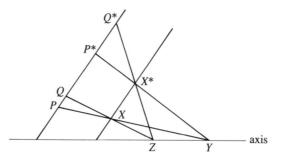

Fig. 12.1. The geometry of an axial dilatation.

If PP^* is parallel to the axis, the transformation is a shear, and a similar construction enables us to find the image of any point X not on PP^*, and then the image of any point Q on PP^*. (Fig. 12.2).

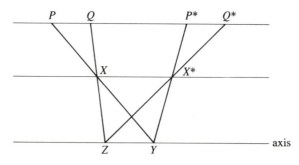

Fig. 12.2. The geometry of a shear.

The axial transformations generate the whole of the affine group; indeed, the axial dilatations are sufficient. The reader may wish to try the exercise of proving this directly. We shall later give a proof that the axial reflexions generate all affine transformations of determinant ± 1. The transformations listed above which do not fall into this class are

$$\text{F.2 } (\lambda \neq -1); \quad \text{C.1 } (\lambda\mu \neq \pm 1); \quad \text{C.2 } (\rho \neq 1); \quad \text{C.3 } (\lambda \neq -1);$$
$$\text{C.4 } (\lambda \neq -1); \quad \text{A.1 } (\lambda \neq -1).$$

It is relatively easy to see that all of these can be obtained by combining a transformation of determinant ± 1 with *one* or *two* axial dilatations.

Transformations of type C.2 combine a central dilatation $x^* = \rho x, y^* = \rho y$ with a transformation $x^* = \cos \alpha . x - \sin \alpha . y, y^* = -\sin \alpha . x + \cos \alpha . y$. The latter transformation has matrix R_α, and is a special orthogonal transformation of $\mathbb{R}^2$. It is natural to draw pictures of $\mathbb{R}^2$ using rectangular axes (e.g. on ordinary graph paper), and such a picture will be called a *standard picture*. In a standard picture, the transformation with matrix R_α appears as a rotation about the origin through an angle α. We have $R_\alpha R_\beta = R_{\alpha + \beta}$; R_α is defined for any real number α and $R_\alpha = R_\beta$ if and only if $\alpha \equiv \beta \pmod{2\pi}$. The mapping $\alpha \mapsto R_\alpha$ is a homomorphism of the additive group of $\mathbb{R}$ onto the multiplicative group of special orthogonal 2×2 matrices, with kernel the additive subgroup of $\mathbb{R}$ consisting of integral multiples of 2π, which we denote by $2\pi\mathbb{Z}$. In listing the affine transformations, we used R_α, where α was a real number, and for any such matrix we may take α (uniquely) in the range $[0, 2\pi)$. The set of such real numbers is not however closed under addition, and we shall therefore in future talk of matrices R_α, where $\alpha \in \mathbb{R}/2\pi\mathbb{Z}$. The (multiplicative) group of special orthogonal 2×2 matrices is isomorphic to the (additive) group $\mathbb{R}/2\pi\mathbb{Z}$, and we note that any non-trivial finite subgroup of $\mathbb{R}/2\pi\mathbb{Z}$ is cyclic, being generated by the element with the smallest representative > 0 in $[0, 2\pi)$.

We now develop some of the basic properties of $\mathscr{A}_2(\mathbb{R})$, using a vectorial representation. It follows from the definitions made early in this chapter that, if a, b, c are collinear points with $a \neq b$, then the line in question is the affine span of a, b, and c can be expressed uniquely as $c = (1 - \lambda)a + \lambda b$, $\lambda \in \mathbb{R}$. The discussion of Definition 1 also shows that if ϕ is any affine transformation, then $c\phi = (1 - \lambda) . a\phi + \lambda . b\phi$. The number λ is therefore an affine invariant of the ordered triad (a, b, c). We shall call it the *affine ratio* of the triad, and write it as $\lambda = (\delta ac/\delta ab)$; note that we have not defined δab, but only the whole symbol $(\delta ac/\delta ab)$. It will, however, be seen that in a standard picture, the number λ is in fact the ratio of the distance between a and c to the distance between a and b (regard being had to sign). We shall be able to prove results about the affine ratio which are suggested by the notation; for example, we already know that if $\lambda = 0$, then $c = a$, and if $\lambda = 1$, then $c = b$, so we have $(\delta aa/\delta ab) = 0$, $(\delta ab/\delta ab) = 1$.

If $a \neq b$, we denote by (a, b) the line which contains them (Fig. 12.3). If c, d lie on a line parallel to (a, b) and not coinciding with it, then $a \neq c$, and there is a unique line through d parallel to (a, c). This line is not parallel to (a, b) as otherwise (a, c) would be parallel to (a, b) and would therefore coincide with it, giving $c \in (a, b)$ contrary to hypothesis. Hence the line meets (a, b) in a point e. We now define

$$(\delta cd/\delta ab) := (\delta ae/\delta ab).$$

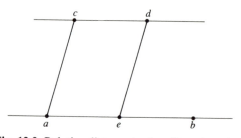

Fig. 12.3. Relative distance in the affine plane (I).

Finally, for any two points $f, g \in (a, b)$, we define

$$(\delta fg/\delta ab) := (\delta ag/\delta ab) - (\delta af/\delta ab).$$

If $x, y, z \in (a, b)$, it follows from this definition that

$$(\delta xy/\delta ab) + (\delta yz/\delta ab) = (\delta xz/\delta ab),$$

whence, by putting $z = x$, we have

$$(\delta xy/\delta ab) = -(\delta yx/\delta ab).$$

It is now easy to show, using the first definition, that these results also hold for points x, y, z which lie on a line parallel to (a, b).

The number $(\delta xy/\delta ab)$ has now been defined for any $a \neq b$, and any points x, y which lie on a line parallel to (a, b). This number may be thought of as a 'relative distance', measuring the separation of x, y in terms of the separation of a, b. Note however that this 'relative distance' may be positive or negative, and must be seen in terms of *directed* or *oriented* separation.

It is a matter of straightforward calculation to show that if a, b, x, y are collinear, with $a \neq b$ and $x \neq y$, then

$$(\delta xy/\delta ab) . (\delta ab/\delta xy) = 1,$$

for if $x = (1 - \lambda)a + \lambda b$, $y = (1 - \mu)a + \mu b$, we have

$$a = -\frac{\mu}{\lambda - \mu}x + \frac{\lambda}{\lambda - \mu}y, \quad b = \frac{1 - \mu}{\lambda - \mu}x + \frac{\lambda - 1}{\lambda - \mu}y,$$

giving $(\delta xy/\delta ab) = \lambda - \mu$, $(\delta ab/\delta xy) = 1/(\lambda - \mu)$. This result is now easily extended to the case in which (x, y) is parallel to, but not coincident with, the line (a, b).

Any point-pair $\{a, b\}$ has a *midpoint* $c = \frac{1}{2}a + \frac{1}{2}b$. If $a \neq b$, then c lies on (a, b) and $(\delta ac/\delta cb) = 1$. We can prove 'euclidean-like' theorems about midpoints, for example:

(i) If a, b, c, d is a proper parallelogram, that is to say, (a, b) and (c, d) are parallel and not coincident, (b, c) and (d, a) are parallel and not coincident, then the midpoints of $\{a, c\}$ and $\{b, d\}$ coincide.

(ii) If a, b, c are not collinear, and d, e, f are the midpoints of $\{b, c\}$, $\{c, a\}$, $\{a, b\}$ respectively, then the lines (a, d), (b, e), (c, f) have a common point g, and $(\delta ag/\delta gd) = (\delta bg/\delta ge) = (\delta cg/\delta gf) = 2$. (The proofs of these results are left as an exercise for the reader.)

If a, b, c are not collinear, $b' \in (a, b)$, $c' \in (a, c)$, and $(b', c')\|(b, c)$, then the ordered triad (a, b', b) is affine equivalent to the ordered triad (a, c', c); for there is a shear with axis through a and parallel to (b, c) which takes b to c, and

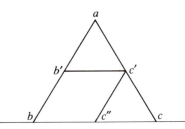

Fig. 12.4. Relative distance in the affine plane (II).

hence also takes b' to c', leaving a fixed (see Fig. 12.4). We therefore have

$$(\delta ab'/\delta ab) = (\delta ac'/\delta ac).$$

Also, if $c'' \in (b, c)$ with $(c', c'')\|(a, b)$, we have

$$(\delta b'c'/\delta bc) = (\delta bc''/\delta bc) = 1 - (\delta c''c/\delta bc) = 1 - (\delta cc''/\delta cb).$$

The previous result shows that $(\delta cc''/\delta cb) = (\delta cc'/\delta ca)$, hence

$$(\delta b'c'/\delta bc) = 1 - (\delta cc'/\delta ca) = (\delta c'a/\delta ca) = (\delta ac'/\delta ac).$$

We thus have

$$(\delta ab'/\delta ab) = (\delta ac'/\delta ac) = (\delta b'c'/\delta bc) = \rho \text{ (say)}.$$

We know from Theorem 12.2 that there is a unique affine transformation taking one affine-independent ordered triad of points in the affine plane to another such triad. Three points are affine-independent if and only if they are non-collinear; a non-collinear ordered triad of points will be called a *triangle*. Given two triangles (a, b, c) and (a^*, b^*, c^*), there is an associated real number given by det ϕ, where ϕ is the unique affine transformation such that $(a, b, c)\phi = (a^*, b^*, c^*)$. We shall write

$$(\Delta a^*b^*c^*/\Delta abc) := \det \phi.$$

The symbol $(\Delta a^*b^*c^*/\Delta abc)$ has the good multiplicative property

$$(\Delta a^{**}b^{**}c^{**}/\Delta a^*b^*c^*) \cdot (\Delta a^*b^*c^*/\Delta abc) = (\Delta a^{**}b^{**}c^{**}/\Delta abc),$$

which follows from the fact that det $\phi\psi = \det \phi \cdot \det \psi$. Also, the symbol is an affine invariant of pairs of triangles, for if $(a, b, c)\psi = (p, q, r)$ and $(a^*, b^*, c^*)\psi = (p^*, q^*, r^*)$, then $(p^*, q^*, r^*) = (p, q, r)\psi^{-1}\phi\psi$, and since $\det(\psi^{-1}\phi\psi) = \det \phi$, we have

$$(\Delta p^*q^*r^*/\Delta pqr) = (\Delta a^*b^*c^*/\Delta abc).$$

Note that $(\Delta acb/\Delta abc) = -1$, for we interchange b, c leaving a fixed by an axial reflexion, which has determinant -1. Any permutation is a product of transpositions, and if a', b', c' is a permutation of a, b, c, we have $(\Delta a' b' c'/\Delta abc) = +1$ or -1, according as the permutation is even or odd.

We may think of the symbol $(\Delta pqr/\Delta abc)$ as representing a 'relative area' of the two triangles, but note that it may be positive or negative, and must be thought of in terms of *oriented* area. In a standard picture, the number $(\Delta pqr/\Delta abc)$ is numerically the ratio of the areas of the two triangles drawn on the graph paper, with a positive sign if the triangles in the picture are similarly oriented (that is, if the cyclic orders p, q, r and a, b, c of their vertices are both clockwise or both anticlockwise), and with a negative sign if the triangles in the picture are oppositely oriented.

The following results depend on finding the affine transformation ϕ which takes one triangle to another, and the details of the proofs are left to the reader.

(i) $(\Delta a'bc/\Delta abc) = 1$ if and only if (a, a') and (b, c) are (distinct) parallel lines. (The required transformation is a shear.)

(ii) $(\Delta a'bc/\Delta abc) = -1$ if and only if the mid-point of $\{a, a'\}$ lies on the line (b, c). (The required transformation is an axial reflexion.)

(iii) In the situation represented by Fig. 12.4 $(\Delta ab'c'/\Delta abc) = \rho^2$. (The required transformation is a central dilatation.)

An affine transformation ϕ with $\det \phi = 1$ is called *equi-affine*. Such transformations form a subgroup (the *equi-affine group*) of the affine group. We prove

THEOREM 12.4: *An equi-affine transformation is the product of an even number of axial reflexions.*

Proof. Let $\det \phi = 1$, let (a, b, c) be any triangle, and let $(a, b, c)\phi = (a^*, b^*, c^*)$. We show that we can map a, b, c to a^*, b^*, c^* respectively by a sequence $\psi_1, \ldots, \psi_r$ of axial reflexions. Then, since there is a unique affine transformation which performs this mapping, we have $\phi = \psi_1\psi_2 \ldots \psi_r$. Also, since $\det \psi_i = -1$, r is necessarily even.

We now proceed in steps. If $a \neq a^*$, there will be an axial reflexion taking a to a^*, and b, c to b', c', say. If b' does not lie on (a^*, b^*), there is an axial reflexion (with axis through a^* and the mid-point of $\{b', b^*\}$) which fixes a^* and takes b' to b^*. If b' does lie on (a^*, b^*), we apply an axial reflexion with axis through a^* but not through b^*; this maps b' to a point b'' not on (a^*, b^*), and we can then proceed as before, finding an axial reflexion which fixes a^* and maps b'' to b^*. Thus, after *two* or *three* axial reflexions, we have mapped a to a^*

and b to b^*; if we have only used two, we apply a further axial reflexion with axis (a^*, b^*). We have now mapped c to $c^\dagger$, say, and we have $(\Delta a^* b^* c^\dagger / \Delta a^* b^* c^*) = -1$; there is therefore an affine transformation of determinant -1 which fixes a^* and b^* and takes $c^\dagger$ to c^*, and this is necessarily an axial reflexion with (a^*, b^*) as axis. We have therefore expressed ϕ as a product of four axial reflexions.

If $a = a^*$, we can map b to b^* (leaving a^* fixed) using *one* or *two* axial reflexions and complete the operation with a total of *two* or *four* axial reflexions. In fact, the number of four reflexions is not best possible; it can be shown that any equi-affine transformation is a product of two axial reflexions. The proof however requires a rather tedious consideration of different cases.

COROLLARY 12.5: *If ϕ is an affine transformation with* det $\phi = -1$, *then ϕ is a product of an odd number of axial reflexions.*

Proof. If ψ is any axial reflexion, $\det(\phi\psi) = 1$, hence $\phi\psi = \psi_1\psi_2 \ldots \psi_{2s}$, and $\phi = \psi_1\psi_2 \ldots \psi_{2s}\psi$.

We have shown that the axial reflexions generate a subgroup of the affine group consisting of all affine transformations ϕ with $|\det \phi| = 1$. This group contains the equi-affine group as a normal subgroup of index 2.

Finally, we give the proof promised at the beginning of this chapter, that every collineation of $\mathscr{A}_2(\mathbb{R})$ is an affine transformation. To prepare the ground, let χ be a collineation, (a, b, c) any triangle, and let ϕ be the (unique) affine transformation such that $(a\phi, b\phi, c\phi) = (a\chi, b\chi, c\chi)$. (Note that $a\chi, b\chi, c\chi$ are necessarily non-collinear, so that $(a\chi, b\chi, c\chi)$ is a triangle.) Then $\phi\chi^{-1}$ is a collineation, which fixes a, b, c. If we can show that such a collineation is the identity, we shall have shown that $\chi = \phi$, and shall have proved the theorem.

We shall work in a coordinate representation, taking a, b, c as $(0, 0)$, $(1, 0)$, $(0, 1)$.

LEMMA 12.6: *If χ is a collineation fixing $(0, 0)$, $(1, 0)$, $(0, 1)$, then $(x, y)\chi = (x\alpha, y\alpha)$, where $\alpha: \mathbb{R} \to \mathbb{R}$ is a field automorphism.*

Proof. The x-axis, consisting of points $(x, 0)$, is invariant under χ, and we define a mapping $\alpha: \mathbb{R} \to \mathbb{R}$ by $(x, 0)\chi = (x\alpha, 0)$. Since χ maps the x-axis one–one onto itself, α is bijective. We show that α has the homomorphism properties $(a+b)\alpha = a\alpha + b\alpha$, $(ab)\alpha = (a\alpha)(b\alpha)$; we shall then have shown that α is an automorphism of $\mathbb{R}$.

Consider Fig. 12.5, in which $0, A, B, C$ are the points $(0, 0)$, $(a, 0)$, $(b, 0)$, $(c, 0)$ respectively, J is the point $(0, 1)$, JK and AK are parallel to the x-axis and y-axis respectively, and KC is parallel to JB.

Easy calculations show that K is $(a, 1)$, JB has equation $x + by = b$, KC has

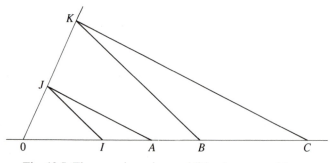

Fig. 12.5. The mapping α is an additive homomorphism.

equation $(x-a)+b(y-1)=0$, and C is $(a+b, 0)$, giving $c=a+b$. If now we apply the collineation χ to this figure, we get a like figure in which A, B, C have moved to A^*, B^*, C^* with x-coordinates $a\alpha, b\alpha, c\alpha$ respectively, with $c\alpha = a\alpha + b\alpha$. We thus have $(a+b)\alpha = a\alpha + b\alpha$ for all $a, b \in \mathbb{R}$.

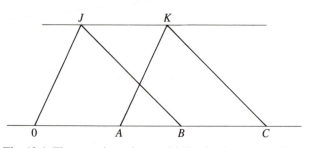

Fig. 12.6. The mapping α is a multiplicative homomorphism.

To show that $(ab)\alpha = (a\alpha)(b\alpha)$, we consider Fig. 12.6, in which $0, I, A, B, C$ are the points $(0, 0), (1, 0), (a, 0), (b, 0), (c, 0)$ respectively. J is $(0, 1)$, BK and KC are parallel to IJ and JA respectively. Easy calculations show that $c = ab$, and so by an argument very similar to that explained above, we obtain the required result.

To conclude the proof, we appeal to the situations shown in Fig. 12.7.

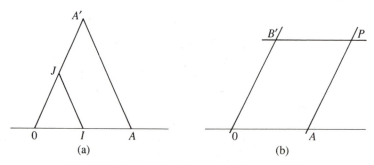

(a)   (b)

Fig. 12.7. Completing the proof of Lemma 12.6.

In Fig. 12.7(a), $0, I, A, J$ are $(0, 0), (1, 0), (a, 0), (0, 1)$ respectively, and AA' is parallel to IJ; we find that A' is $(0, a)$. Since I and J are fixed by χ, and χ takes A to $(a\alpha, 0)$, it must take A' to $(0, a\alpha)$, so we have $(0, a)\chi = (0, a\alpha)$.

In Fig. 12.7(b), $0, A, B'$ are $(0, 0), (a, 0), (0, b)$ respectively, and $B'P$, AP are parallel to the x-axis and y-axis. We find that P is (a, b), and since χ takes A to $(a\alpha, 0)$ and B' to $(0, b\alpha)$, it takes P to $(a\alpha, b\alpha)$, so that we have $(a, b)\chi = (a\alpha, b\alpha)$, and the proof of the lemma is complete.

The arguments used in the proof of Lemma 12.6 have not appealed to any special properties of the field $\mathbb{R}$. They are valid for $\mathscr{A}_2(F)$ over an arbitrary field F, and indeed are easily extended to $\mathscr{A}_n(F)$ for any $n \geq 2$. The field $\mathbb{R}$ has the special property that its only automorphism is the identity. When we have shown this, we shall know that the collineation χ of Lemma 12.6 must be the identity, and our theorem will be established.

LEMMA 12.7:   *The only automorphism of $\mathbb{R}$ is the identity.*

Proof. Let $\alpha: \mathbb{R} \to \mathbb{R}$ be an automorphism. Then $0\alpha = 0$, $1\alpha = 1$, and $(-x)\alpha = -(x\alpha)$. Also $2\alpha = (1+1)\alpha = 1\alpha + 1\alpha = 1 + 1 = 2$, and by induction $n\alpha = n$ for any integer $n \geq 0$. Also, $(-n)\alpha = -(n\alpha) = -n$. Hence α restricted to the integers is the identity.

Now consider any rational number r/s, with $s > 0$. The s-fold sum $(r/s) + \cdots + (r/s)$ is equal to r, so $s.((r/s)\alpha) = r\alpha = r$, and $(r/s)\alpha = r/s$. Hence α restricted to the rational field $\mathbb{Q}$ is the identity.

Now comes the clever bit: the field $\mathbb{R}$ is ordered, and we show that α preserves the order, that is to say: $a > b \Rightarrow a\alpha > b\alpha$. This is the same as saying that $a - b > 0 \Rightarrow a\alpha - b\alpha = (a-b)\alpha > 0$, so it is sufficient to show that $c > 0 \Rightarrow c\alpha > 0$. Now in the real field $c > 0$ if and only if c is a non-zero square, so if $c > 0$ we have $c = d^2$ for some $d \neq 0$. Hence $c\alpha = (d\alpha)^2$ and $d\alpha \neq 0$, so $c\alpha > 0$; that is, $c > 0 \Rightarrow c\alpha > 0$.

If now $a \in \mathbb{R}$ and $q \in \mathbb{R}$, we have $a > q \Rightarrow a\alpha > q\alpha = q$ and $q > a \Rightarrow q = q\alpha > a\alpha$. But if $a\alpha \neq a$, there is a rational number q between a and $a\alpha$, and we have either $a < q < a\alpha$ or $a\alpha < q < a$, contradicting the result just stated. Hence $a\alpha = a$ for all $a \in \mathbb{R}$, and $\alpha = 1$.

We have now completed the proof of

THEOREM 12.8:   *The group of collineations of the real affine plane is the affine group.*

Exercises

12.1. In the course of this chapter we dealt with conjugacy of elements of $AGL(V_2)$ and $AGL(n, F)$ by considering different vectorial or coordinate representations of a given affine transformation. The conjugacy of two elements ϕ_1, ϕ_2 of a group must of course be capable of ate representations of a given affine transformation. The conjugacy of two elements ϕ_1, ϕ_2 of a group must of course be capable of demonstration by finding an element ψ of the group such that $\phi_2 = \psi^{-1}\phi_1\psi$. Obtain such demonstrations of the following results.

(i) If $\phi = \theta\tau \in AGL(V_n(F))$ where $\theta \in GL(V_n(F))$ and $\tau \in T$, and ϕ has a fixed point, then ϕ and θ are conjugate in $AGL(V_n(F))$.

(ii) If $\phi_1 : u \mapsto uA + a$ and $\phi_2 : u \to uA + b$, where $a = [\xi, \eta]$ with $\xi \neq 0$, $b = [1, 0]$, and

$$either\ A = \begin{bmatrix} 1 & 1 \\ 0 & 1 \end{bmatrix}\ or\ A = \begin{bmatrix} 1 & 0 \\ 0 & \lambda \end{bmatrix} (\lambda \neq 0, 1),$$

then ϕ_1, ϕ_2 are conjugate in $AGL(2, \mathbb{R})$.

12.2.   Show that in $\mathscr{A}_2(\mathbb{R})$ every affine transformation of finite order ($\neq 1$) is central or axial, and is representable in a suitable coordinate system by a transformation $u \mapsto uA$ where A is an orthogonal matrix.

12.3.   An ordered set of n points $p_1, \ldots, p_n$, not all collinear, in the affine plane (over an arbitrary field) is said to form an *affine-regular n-gon* if there is an affine transformation ϕ such that $p_i\phi = p_{i+1}$, $i = 1, \ldots, n-1$, $p_n\phi = p_1$. Show that every triangle is affine-regular, and hence show that, for any field F, there is a 2×2 matrix A with elements in F such that A is of multiplicative order 3. Find such a matrix when F is (a) $\mathbb{F}_2$, (b) $\mathbb{F}_3$, (c) $\mathbb{Q}$.

12.4.   Show that in $\mathscr{A}_2(\mathbb{R})$ a 4-gon (a, b, c, d) is affine-regular if and only if it is a proper parallelogram: that is, $(a, b)\|(c, d)$, $(b, c)\|(d, a)$, and a, b, c, d are not collinear.

12.5.   Show that, in $\mathscr{A}_2(F)$ where F is of characteristic 2, in a proper parallelogram (a, b, c, d) we have also $(a, c)\|(b, d)$. Show that a 4-gon is affine-regular if and only if it is a proper parallelogram, and that the vertices of an affine-regular 4-gon taken in *any* order form an affine-regular 4-gon.

Exercises 12.6–12.13 all relate to the geometry $\mathscr{A}_2(\mathbb{R})$.

12.6.   (i) Show that if a, b, c are distinct collinear points, and p is a point not on the line which contains them, then

$$(\Delta pab/\Delta pbc) = (\delta ab/\delta bc).$$

(ii) Show that if a, b, c, d lie on a line, $a \neq b$, $c \neq d$, and p, q lie on a line parallel to but distinct from the first line, then

$$(\Delta pab/\Delta qcd) = (\delta ab/\delta cd).$$

12.7. The points p, a, b are not collinear, and a line which does not contain any of them meets (p, a) in a', (p, b) in b', and (a, b) in c. The lines (a, b'), (b, a') meet in q, and (p, q) meets (a, b) in d. Show that

$$(\delta ac/\delta cb) \div (\delta ad/\delta db) = -1.$$

[Hint: use a suitable coordinate system, e.g. take p, a, b as $(0, 0)$, $(1, 0)$, $(0, 1)$ respectively.]

What happens if c is the mid-point of $\{a, b\}$?

12.8. (i) The lines (p, a), (p, b) are distinct, and $a'\ (\neq p)$, $b'\ (\neq p)$ lie on (p, a), (p, b) respectively. Prove that

$$(\Delta pab/\Delta pa'b') = (\delta pa/\delta pa')(\delta pb/\delta pb').$$

[Hint: use an appropriate affine transformation.]

(ii) Four distinct lines meet in a point p, and two lines, distinct from the first four lines and not containing p, meet the four lines in a, b, c, d and a', b', c', d' respectively. Prove that

$$(\delta ac/\delta cb)(\delta ad/\delta db) = (\delta a'c'/\delta c'b')(\delta a'd'/\delta d'b').$$

(iii) Hence obtain an alternative proof of the result stated in Exercise 12.7.

12.9. The points a, b, c are not collinear, and p does not lie on any of the lines (b, c), (c, a), (a, b). The lines (a, p), (b, c) meet at d; (b, p), (c, a) meet at e; (c, p), (a, b) meet at f. Prove the affine form of *Ceva's Theorem*, namely

$$(\delta bd/\delta dc)(\delta ce/\delta ea)(\delta af/\delta fb) = 1.$$

State and prove a converse of Ceva's Theorem.

12.10. The points a, b, c are not collinear, and $d \in (b, c)$, $d \neq b$, $d \neq c$; $e \in (c, a)$, $e \neq c$, $e \neq a$; $f \in (a, b)$, $f \neq a$, $f \neq b$. Show that, if

$$(\delta bd/\delta dc)(\delta ce/\delta ea)(\delta af/\delta fb) = -1,$$

then d, e, f are collinear.

12.11. We take a fixed triangle (x, y, z), and for any ordered triad of points (a, b, c) we define a (signed) *measure* μ by

$$\mu(a, b, c) := \begin{cases} (\Delta abc/\Delta xyz), & \text{if } a, b, c \text{ are not collinear,} \\ 0, & \text{if } a, b, c \text{ are collinear.} \end{cases}$$

Show that, for any four points p, a, b, c,

$$\mu(p, a, b) + \mu(p, b, c) + \mu(p, c, a) = \mu(a, b, c).$$

Show also that, for any ordered set of points $(a_1, \ldots, a_n)$,

$$\sum_{i=1}^{n} \mu(p, a_i, a_{i+1}) \quad \text{(where } a_{n+1} = a_1)$$

is independent of p, so that we may define the measure μ of $(a_1, \ldots, a_n)$ by

$$\mu(a_1, \ldots, a_n) := \sum_{i=1}^{n} \mu(p, a_i, a_{i+1}).$$

Show that $\mu(a_1, a_2, a_3, a_4) = \mu(a_1, a_2, a_3) + \mu(a_3, a_4, a_1)$.

12.12. Show that, in a vectorial representation, a transformation $\theta \in \mathrm{GL}(V_2)$ is an axial reflexion if and only if $\mathrm{tr}\,\theta = 0$ and $\det \theta = -1$. Hence show that (relative to an arbitrary basis) an axial reflexion θ is represented by one of the following matrices:

$$\begin{bmatrix} \lambda & \mu \\ (1-\lambda^2)/\mu & -\lambda \end{bmatrix} \ (\mu \neq 0), \quad \begin{bmatrix} 1 & 0 \\ c & -1 \end{bmatrix}, \quad \begin{bmatrix} -1 & 0 \\ c & 1 \end{bmatrix}.$$

For each of these matrices, find the equations of the axis and the direction of the reflexion which they represent. (The *direction* of an axial reflexion is the common direction of the lines which join a point and its image.)

12.13. Enumerate the canonical forms of affine transformations of determinant 1, and show that each of these can be expressed as the product of two axial reflexions.

12.14. A (non-empty) locus Γ in $\mathbb{R}^2$ is given by an equation $f(x, y) = 0$, where f is a differentiable function of x and y. Define the *tangent* to Γ at the point (x_0, y_0) on Γ, and show that under an affine transformation which maps Γ to Γ^* and (x_0, y_0) to (x_0^*, y_0^*) the tangent to Γ at (x_0, y_0) is mapped to the tangent to Γ^* at (x_0^*, y_0^*).

An ellipse Γ has centre c, and p, q are points of Γ such that the tangents to Γ at p, q meet at the point t. Prove that the line (c, t) contains the mid-point of $\{p, q\}$. [You may find it helpful to refer back to Chapter 10, where it was shown that any ellipse is affine-equivalent to the unit circle in $\mathbb{R}^2$.]

12.15. Show that, if χ is a collineation of F^2 and m is a line such that $m\chi \| m$, and $n \| m$, then $n\chi \| n$.

Such a collineation is said to preserve the direction of m. Show that a collineation $\neq 1$ which preserves all directions has at most one fixed point. Hence show that a collineation which preserves all directions is an affine transformation.

12.16. A collineation χ of F^2 fixes all points of the line l. Show that it preserves the direction of l.

Show that if χ fixes any point $P \notin l$, then $\chi = 1$.

Show that if χ preserves a direction other than that of l, then all lines with that direction are fixed by χ. Hence show that if χ preserves more than one such direction, then $\chi = 1$.

Thus, if $\chi \neq 1$, either χ preserves just one direction, or χ preserves no direction, other than that of l. Show that χ is uniquely determined when the image P_χ of a point $P \notin l$ is known, and deduce that a collineation which fixes all points of a line l is necessarily an affine transformation.

12.17. Generalise Lemma 12.6 to a statement about collineations of F^n, and show that any collineation of F^n is of the form $u \mapsto (u^\alpha)A + a$, where $\alpha \in \text{Aut } F$ and

$$(x_1, \ldots, x_n)^\alpha := (x_1\alpha, \ldots, x_n\alpha).$$

Verify that the transformations $u \mapsto (u^\alpha)A + a$ form a group, containing as a subgroup the transformations $u \mapsto (u^\alpha)A$. These groups will be denoted by $A\Gamma L(n, F)$, $\Gamma L(n, F)$ respectively. Show that

$$A\Gamma L(n, F)/AGL(n, F) \cong \Gamma L(n, F)/GL(n, F) \cong \text{Aut } F.$$

12.18. A transformation $\theta : V_n(F) \to V_n(F)$ of a vector space is said to be *semilinear* if there is an automorphism $\alpha \in \text{Aut } F$ such that

$$(\lambda x + \mu y)\theta = (\lambda\alpha)(x\theta) + (\mu\alpha)(y\theta)$$

for all $x, y \in V_n(F)$ and all $\lambda, \mu \in F$. Show that, relative to a basis of $V_n(F)$, an invertible semi-linear transformation is represented by an element of $\Gamma L(n, F)$, and that the invertible semi-linear transformations of $V_n(F)$ form a group $\Gamma L(V_n(F)) \cong \Gamma L(n, F)$.

12.19. Show that each of the following fields has trivial automorphism group:
   (i) the finite field $\mathbb{F}_p$ with a prime number p of elements;
   (ii) the field $\mathbb{Q}$ of rational numbers;
   (iii) the field $\mathbb{E}$ of Exercise 11.5.

12.20. Show that the complex field $\mathbb{C}$ has the non-trivial automorphism $z \mapsto \bar{z}$, which maps each element to its complex conjugate, and that this is the only non-trivial automorphism which is a *continuous* map $\mathbb{C} \to \mathbb{C}$, in terms of the standard metric topology of $\mathbb{C}$ given by $d(z_1, z_2) = |z_1 - z_2|$.

Chapter 13

Projective geometry

Although the principles of perspective in pictorial representation were not mastered until $c.1500$, it is now a matter of common knowledge that two parallel horizontal lines are represented in a picture by two lines which meet on the horizon, or 'vanishing line', in the picture (Fig. 13.1). These parallel lines, which do not meet in space, become intersecting lines when projected on the plane of the picture from the point occupied by the eye of the beholder. We can pick out the geometrical essentials of the situation by thinking of the plane Π_1 containing the two parallel lines m_1, n_1, projected on a plane Π_2 (not parallel to Π_1) from a centre O not lying on either plane (see Fig. 13.2). The planes (O, m_1), (O, n_1) containing O and the lines m_1, n_1 respectively meet in a line m through O. This line meets Π_2 in a point P, and the planes (O, m_1), (O, n_1) meet Π_2 in lines m_2, n_2 respectively which pass through P. These are the projections of the lines m_1, n_1 respectively, for if we take any point x_1 on m_1 (say), the line OX_1 will meet Π_2 in a point X_2 of m_2, unless OX_1 does not meet Π_2, which happens if and only if X_1 lies in the plane through O parallel to Π_2. This plane meets Π_1 in a line l_2, and points of l_1 have no image defined by the projection. Further, m does not meet Π_1, and there is no point of Π_1 which corresponds to P. The line m lies in the plane through O parallel to Π_1, which meets Π_2 in a line l_2, and no point of l_2 is the image of a point of Π_1 in the projection. Thus, if $X_1 \in \Pi_1 \setminus l_1$ then OX_1 meets Π_2 in a point $X_2 \notin l_2$, and the mapping $X_1 \mapsto X_2$ is a bijection $\Pi_2 \setminus l_1 \to \Pi_2 \setminus l_2$. Thus the projection establishes only a partial bijective correspondence between the planes Π_1, Π_2.

We get a much clearer picture by considering the set S of lines through O. Given any plane Π not containing O, we have an injective map $\Pi \to S$, which

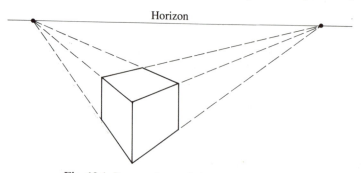

Fig. 13.1. Perspective and the 'vanishing line'.

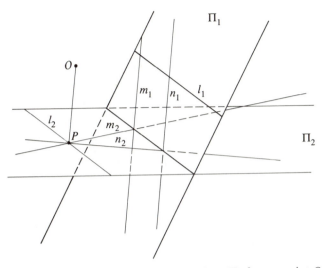

Fig. 13.2. Projection of a plane Π_1 to a plane Π_2 from a point O.

maps any point $P \in \Pi$ to the line $OP \in S$. The lines of S which are not in the image of this map are precisely the lines through O parallel to Π. If we take such a line $l \in S$, and a plane Π^* (not containing O) which meets l, in X (say), then any two lines in Π which are parallel to l will project from O to lines in Π^* which meet in X. We can therefore think of such lines l as representing 'points at infinity' which can be added to Π to make an 'extended plane' in which any two parallel lines of Π will have a point of intersection.

This process of extending the physical plane by adding notional 'points at infinity' was the historical line of development of the idea of a projective plane, but although the terminology of points at infinity is evocative and often useful, we shall obtain a better foundation for the formal definition of a projective plane (which is easily extended to projective space of any dimension) by directing attention to the set S of lines through a point O of 3-space. Now the elements of a real 3-dimensional vector space $V_3(\mathbb{R})$ can be represented by points of physical space, with O representing the zero vector, and the lines through O then represent the 1-dimensional subspaces of $V_3(\mathbb{R})$. This remark provides the motivation for the following formal definitions.

A *projective space* $P_n(F)$ of n dimensions over a field F is the set of 1-dimensional subspaces of a vector space $V_{n+1}(F)$ of dimension $n+1$ over F.

A *projective k-flat* A_k in $P_n(F)$ is the set of points in $P_n(F)$ (:= 1-dimensional subspaces of $V_{n+1}(F)$) which lie in a vector subspace W_{k+1} of V_{n+1}, of dimension $k+1$.

We call $V_{n+1}(F)$ the *underlying vector space* of $P_n(F)$. The 0-flats of $P_n(F)$ are the points of $P_n(F)$, and $P_n(F)$ itself is the unique n-flat of $P_n(F)$. Note also that our definition of k-flats includes the case $k = -1$; there is only one subspace of $V_{n+1}(F)$ of dimension zero, and it contains no 1-dimensional subspaces, hence the unique (-1)-flat is the empty set $\varnothing$. We call k the *projective dimension* of a k-flat A_k, to distinguish it from the (vector space) dimension $k+1$ of the corresponding vector space W_{k+1}.

We shall write P_n, V_{n+1} for $P_{n+1}(F)$, $V_{n+1}(F)$ when no confusion can arise from omission of explicit mention of the field F. A k-flat (for any k, $-1 \leqslant k \leqslant n$) will be called a (projective) subspace of P_n, and the set of all such subspaces will be denoted by $\mathscr{S}(P_n)$. Denoting the set of all vector subspaces of V_{n+1} by $\mathscr{S}(V_{n+1})$, we have a bijection $\pi: \mathscr{S}(V_{n+1}) \to \mathscr{S}(P_n)$, which preserves inclusion. The projective span of a set S of points in P_n is the smallest subspace which contains them. It corresponds in the bijection π to the linear span of the 1-dimensional subspaces of V_{n+1} which are the points of S.

We can coordinatise P_n by choosing a basis $e_0, e_1, \ldots, e_n$ of the underlying vector space V_{n+1}. A given 1-dimensional subspace of V_{n+1} is described by any non-zero vector which it contains. Such a vector will be called a *representative vector* of the corresponding point of P_n. If $u = \sum_0^n \alpha_i e_i$, then the vectors in the 1-dimensional subspace containing u are represented relative to the given basis by the set of ordered $(n+1)$-tuples $\{(\lambda\alpha_0, \lambda\alpha_1, \ldots, \lambda\alpha_n): \lambda \in F\}$. It is convenient to introduce the condensed notation $\{\alpha_0, \alpha_1, \ldots, \alpha_n\}$ for this set, and to call this symbol the *(homogeneous) coordinate* of the point of P_n. If the coordinate system has been fixed, we shall talk (unambiguously) of 'the point $\{\alpha_0, \alpha_1, \ldots, \alpha_n\}$'. It is a consequence of the definitions that

$$\{\alpha_0, \alpha_1, \ldots, \alpha_n\} = \{\lambda\alpha_0, \lambda\alpha_1, \ldots, \lambda\alpha_n\} \quad \text{for any } \lambda \neq 0, \lambda \in F.$$

The $(n+1)$ points of P_n with coordinates $\{1, 0, \ldots, 0\}$, $\{0, 1, \ldots, 0\}$, $\ldots$, $\{0, 0, \ldots, 1\}$ are the 1-dimensional subspaces of V_{n+1} containing $e_0, e_1, \ldots, e_n$ respectively, and are called the *simplex of reference* of the coordinate system of P_n. The projective span of the simplex of reference is the whole of P_n. Conversely, given $(n+1)$ points whose projective span is P_n, then by taking arbitrary representative vectors $e_0, e_1, \ldots, e_n$ we obtain a coordinate system having the given set of points as its simplex of reference. Suppose that for one choice of $e_0, e_1, \ldots, e_n$ we take a point with coordinate $\{\rho_0, \rho_1, \ldots, \rho_n\}$, with no $\rho_i = 0$; this condition is equivalent to saying that any $(n+1)$ vectors out of the set $\{e_0, e_1, \ldots, e_n, f = \sum \rho_i e_i\}$ are linearly independent (and thus span V_{n+1}). We can now adjust the choice of basis of V_{n+1} by taking $e_i^* = \rho_i e_i$, $i = 0, 1, \ldots, n$, as a new basis, and obtain a coordinate system with the same simplex of reference, in which the selected point has coordinate $\{1, 1, \ldots, 1\}$. The point with this coordinate is called the *unit point* of the coordinate system. We have thus shown:

THEOREM 13.1: *Given $(n+2)$ points $p_0, p_1, \ldots, p_n, q$, of which any $(n+1)$ span P_n, there is a coordinate system having $p_0, p_1, \ldots, p_n$ as simplex of reference and q as unit point.*

The group $GL(V_{n+1})$ acts in a natural way on P_n, and maps k-flats to k-flats. However the action is not faithful, since the natural homomorphism $\rho: GL(V_{n+1}) \to Sym(P_n)$ defined by the action has the non-trivial kernel $\Lambda = \{\lambda.1 : \lambda \in F, \lambda \neq 0\}$ consisting of the non-zero scalar multiplications. It is readily checked that the elements of Λ fix every 1-dimensional subspace of V_{n+1}, and hence fix every point of P_n; also, taking any coordinate system, we see easily that the only transformations in $GL(V_{n+1})$ which fix the simplex of reference and the unit point are the non-zero scalar multiplications. The image of ρ is therefore isomorphic to $GL(V_{n+1})/\Lambda$; we shall call this subgroup of $Sym(P_n)$ *the n-dimensional projective group* over F, and denote it by $\Pi_n(F)$. Projective geometry of n dimensions over F is then defined as the pair $\mathscr{P}_n(F) = (P_n(F), \Pi_n(F))$. The reader will verify without difficulty that any coordinate system for $P_n(F)$ sets up an isomorphism $\Pi_n(F) \to PGL(n+1, F)$, where $PGL(n+1, F)$ is the matrix group defined in Example 4.5, and that the geometry $\mathscr{P}_n(F)$ is equivalent to the geometry $(\Sigma_0, PGL(n+1, F))$.

It will be noted that the definition of projective geometry $\mathscr{P}_n(F)$ in terms of a vector space $V_{n+1}(F)$ and its invertible linear transformations is coordinate-free, and we have therefore been able to avoid the difficulties which in the case of affine and Euclidean geometry led us to define these geometries as 'abstract copies' of coordinatised geometries. We could have defined $\mathscr{P}_n(F)$ as a pair (Ω, G) equivalent to $(\Sigma_0, PGL(n+1, F))$, and then explained coordinate systems in terms of self-equivalences of $(\Sigma_0, PGL(n+1, F))$. However, the route which we have adopted is formally less sophisticated and reflects more naturally the ideas which emerged from our preliminary discussion.

In view of that discussion, we expect to be able to embed affine space in a meaningful way in projective space, and we now discuss the relationship between $\mathscr{A}_n(F)$ and $\mathscr{P}_n(F)$. Our assertion is that if we select any $(n-1)$-flat H of P_n, and consider the subgroup G of Π_n which fixes H (as a set, not point-wise), so that G acts on $P_n\backslash H$, then $(P_n\backslash H, G)$ is a geometry equivalent to $\mathscr{A}_n(F)$. This is seen most readily in terms of coordinates. We take the simplex of reference $P_0, P_1, \ldots, P_n$ so that $P_0 \notin H$, $P_i \in H$ for $i = 1, \ldots, n$. Then H consists of the points $\{x_0, x_1, \ldots, x_n\}$ with $x_0 = 0$. The points of $P_n\backslash H$ are those for which $x_0 \neq 0$, and any such point is uniquely representable by a coordinate $\{1, \alpha_1, \ldots, \alpha_n\}$. A transformation in $GL(V_{n+1})$ is represented by a matrix $A = [\lambda_{ij}]$ $(i, j = 0, 1, \ldots, n)$, and maps the point $\{x_0, x_1, \ldots, x_n\}$ to the point $\{x_0^*, x_1^*, \ldots, x_n^*\}$, where $[x_0^*, x_1^*, \ldots, x_n^*] = [x_0, x_1, \ldots, x_n]A$. In particular, $x_0^* = x_0\lambda_{00} + x_1\lambda_{10} + \ldots + x_n\lambda_{n0}$. If the transformation fixes H, we must have $x_0^* = 0$ whenever $x_0 = 0$, hence $\lambda_{10} = \lambda_{20} = \ldots = \lambda_{n0} = 0$, and (since A is invertible) $\lambda_{00} \neq 0$. Now, any non-zero scalar multiple of A is in the same coset of Λ in $GL(V_{n+1})$, and effects on P_n the same transformation as A; we may therefore assume that $\lambda_{00} = 1$. We can therefore write the transformation of $P_n\backslash H$ in the matrix form

$$[1, \alpha_1, \ldots, \alpha_n] \mapsto [1, \alpha_1^*, \ldots, \alpha_n^*] = [1, \alpha_1, \ldots, \alpha_n]A,$$

where $A = [\lambda_{ij}]$ $(i, j = 0, 1, \ldots, n)$ with $\lambda_{00} = 1$, $\lambda_{i0} = 0$ $(i = 1, \ldots, n)$. In equational form this is

$$\alpha_j^* = \sum_{i=1}^{n} \alpha_i \lambda_{ij} + \lambda_{0j} \quad (j = 1, \ldots, n),$$

which can be put back in the matrix form $u \mapsto u^* = uT + a$, where $u^* = [\alpha_1^*, \ldots, \alpha_n^*]$, $u = [\alpha_1, \ldots, \alpha_n]$, $T = [\lambda_{ij}]$ $(i, j = 1, \ldots, n)$ and $a = [\lambda_{01}, \ldots, \lambda_{0n}]$. We now have identified the points of $P_n \backslash H$ with F^n, and the group G with $\mathrm{AGL}(n, F)$, and it is readily checked that these identifications establish the equivalence of $(P_n \backslash H, G)$ with $(F^n, \mathrm{AGL}(n, F))$, and hence with $\mathscr{A}_n(F)$.

The point $\{1, 0, \ldots, 0\}$ of the simplex of reference in P_n is identified with the zero vector $(0, \ldots, 0)$ of F^n. The definition of projective geometry is coordinate-free, and in the argument just given we can take the coordinate system so that *any* point of $P_n \backslash H$ is $\{1, 0, \ldots, 0\}$, and thus becomes the origin of the coordinatisation of $P_n \backslash H$ by F^n. It is clearly possible to *define* affine geometry in terms of projective geometry, and if this route of formal development is adopted, it has the advantage of providing a coordinate-free definition of affine geometry. The same route may be pursued further to yield coordinate-free definitions of similarity and Euclidean geometries (see Exercises 13.13, 13.14, 13.15).

The projective $(n-1)$-flat H is called the *hyperplane at infinity* (in relation to the associated affine space). We can usefully think of n-dimensional affine space as an n-dimensional projective space with a projective $(n-1)$-flat removed. A projective k-flat K which is not contained in H meets H in a projective $(k-1)$-flat; this follows from consideration of the dimensions of intersections of subspaces in the underlying vector spaces (see Exercise 13.1). The intersection of K with the affine space $P_n \backslash H$ is $K^* = K \cap (P_n \backslash H) = K \backslash K \cap H$. This is a k-dimensional projective space with a projective $(k-1)$-flat removed, and is thus in a natural way a k-dimensional affine space. The set of points K^* is indeed an affine k-flat of the affine space $P_n \backslash H$, and every such affine k-flat is of the form $K \backslash K \cap H$, where K is a projective k-flat of P_n not contained in H. In terms of the coordinate system described above, a projective k-flat K not contained in H can be spanned by $(k+1)$ points not in H, with homogeneous coordinates which can be written as $\{1, \alpha_1^{(i)}, \ldots, \alpha_n^{(i)}\}$ $(i = 1, \ldots, k+1)$, and then $K^* = K \backslash K \cap H$ is the affine span of the affine-independent set of points in F^n given by $(\alpha_1^{(i)}, \ldots, \alpha_n^{(i)})$ $(i = 1, \ldots, k+1)$. Conversely, given any affine k-flat K^* in F^n spanned by the affine independent set of points $(\alpha_1^{(i)}, \ldots, \alpha_n^{(i)})$ $(i = 1, \ldots, k+1)$, the projective span K of the points $\{1, \alpha_1^{(i)}, \ldots, \alpha_n^{(i)}\}$ $(i = 1, \ldots, k+1)$ is a projective k-flat such that $K^* = K \backslash K \cap H$.

It is often helpful in discussing affine geometry to think of the affine space as embedded in projective space, so that we can talk about the points at infinity. For example, two affine k-flats are parallel if and only if their points at infinity coincide. An affine transformation is the restriction to $P_n \setminus H$ of a projective transformation which fixes H, and again it is helpful to look at the projective transformation restricted to H. For example, the affine transformation is a translation if and only if the projective transformation fixes all points of H and no other point of P_n.

We can make good pictures of $P_2(\mathbb{R})$ in a physical plane, using the representation of lines through a point O of 3-space by the points in which they meet the plane (Fig. 13.3), as suggested by our initial discussion. The

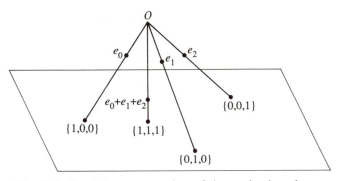

Fig. 13.3. Partial representation of the projective plane.

figure, however much extended, cannot represent all the points of $P_2(\mathbb{R})$, as the lines through O in the plane parallel to the plane of the figure do not meet it. This is not a serious disability, as no figure can be expected to give more than a partial representation of a geometrical situation, and we are content if we are able to show only the features relevant to the particular matters under discussion. We can always choose the plane so that the figure shows any finite number of lines and points which we want to talk about. In particular, if we wish to discuss the real affine plane in terms of the projective plane $P_2(\mathbb{R})$ in which it is embedded, we can draw the picture so that the hyperplane at infinity (in this case a 1-flat, or projective line) is represented by a line l_∞. Parallel lines of the affine plane will then be represented by lines which meet on l_∞.

Such pictures of $P_2(\mathbb{R})$ can of course convey useful ideas about situations in $P_2(F)$, where F is any field, and can indeed be regarded as pictures, in a somewhat loose sense, of $P_2(F)$. Consider for instance the representation of $P_1(\mathbb{R})$ indicated by Fig. 13.4. $P_1(\mathbb{R})$ consists of the lines through O in the plane of the picture, and is represented by the points of the line l, the only unrepresentable point of $P_1(\mathbb{R})$ being the line through O (the 1-dimensional vector space $\langle e_1 \rangle$) which is parallel to l. Any other point of $P_1(\mathbb{R})$ has a representative vector of the form $e_0 + \lambda e_1$, $\lambda \in \mathbb{R}$, and a unique homogeneous coordinate $\{1, \lambda\}$. The points of $P_1(\mathbb{R})$ are thus indexed by the elements $\lambda \in \mathbb{R}$, except for the point $\{0, 1\}$ which we index by the symbol ∞, and the points of $P_1(\mathbb{R})$ are in bijective correspondence with the set $\mathbb{R} \cup \{\infty\}$. The algebraic

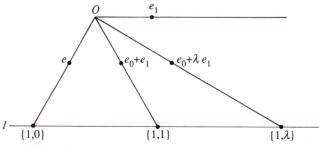

Fig. 13.4. Representation of the projective line.

essentials of the argument are equally valid if $\mathbb{R}$ is replaced by an arbitrary field F, and the picture serves equally well to illustrate the geometrical relationships which support the argument that $P_1(F)$ can be coordinatised by the set $F \cup \{\infty\}$, a result which has already been stated in slightly different terminology in Example 4.5.

Note. The notation $\{\alpha_0, \alpha_1, \ldots, \alpha_n\}$ for the homogeneous coordinate of a point in projective space is not standard. Most writers use $(\alpha_0, \alpha_1, \ldots, \alpha_n)$, relying on an implicit understanding that $(\lambda\alpha_0, \lambda\alpha_1, \ldots, \lambda\alpha_n)$ $(\lambda \neq 0)$ stands for the same point in projective space as $(\alpha_0, \alpha_1, \ldots, \alpha_n)$. In this exposition, we have wished to distinguish clearly between the non-zero vector represented by $(\alpha_0, \alpha_1, \ldots, \alpha_n)$ and the 1-dimensional subspace (point of projective space) spanned by that vector.

Exercises

13.1.   Denoting the span of two subspaces A, B of $P_n(F)$ by (A, B) show that

$$\dim(A, B) = \dim A + \dim B - \dim A \cap B,$$

where dim denotes the projective dimension. Hence show that

$$\max(\dim A, \dim B) \leqslant \dim(A, B) \leqslant \min(n, \dim A + \dim B + 1)$$

$$\min(\dim A, \dim B) \geqslant \dim A \cap B \geqslant \max(-1, \dim A + \dim B - n).$$

Show in particular that in $P_2(F)$ two lines (1-flats) meet in a unique point or coincide and that two points span a unique line or coincide.

13.2.   A set S of points is said to form an *incidence plane* if it has a family of subsets, called lines, with the property that two distinct lines intersect in a single point, and two distinct points are members of one and only one line; and further (to exclude trivial cases) not all points lie on the same line, and every line contains at least three points. Show that, if S is finite, then every line contains the same number $m+1$ of points $(m \geqslant 2)$, that every point belongs to $m+1$ lines and that the set S has

m^2+m+1 points and m^2+m+1 lines. (The number m is called the *order* of the incidence plane.)

Show that incidence planes of orders 2, 3, 4, 5 exist. [Hint: consider $P_2(F)$, where F is a finite field.]

13.3. If S is an incidence plane (as defined in Exercise 2, but not necessarily finite), a bijection $S \to S$ is called a *collineation* if it maps each line onto a line. Show that, if a collineation $\neq 1$ fixes every point of one line, it can fix at most one other point.

Show that, if a collineation of S fixes every point of a line L, but fixes no other point, and maps $p \notin L$ to q, then it is the only collineation with these properties.

Show that, if a collineation of S fixes every point of a line L, and also a point $x \notin L$, and maps p ($\neq x$, $\notin L$) to q, then x, p, q are collinear, and the given collineation is the only one with these properties.

13.4. (i) Find a transformation in $\Pi_2(F)$ which fixes every point $\{0, \beta, \gamma\}$ but fixes no other point, and maps $\{1, 0, 0\}$ to $\{\lambda, 1, 0\}$ ($\lambda \neq 0$).
(ii) Find a transformation in $\Pi_2(F)$ which fixes every point $\{0, \beta, \gamma\}$ and also the point $\{1, 0, 0\}$, and maps $\{\lambda, 1, 0\}$ ($\lambda \neq 0$) to $\{\mu, 1, 0\}$ ($\mu \neq 0$).

13.5. Using the results of Exercises 13.3 and 13.4, show that any collineation of $P_2(F)$ which fixes all points of a line is necessarily a projective transformation (cf. Exercises 12.15, 12.16).

13.6. Show that there is a unique coordinate system for $P_n(F)$ with a given simplex of reference and unit point, and a unique transformation in $\Pi_n(F)$ which takes one ordered set forming a simplex and unit point to another such set.

A coordinate system establishes an equivalence of $\mathscr{P}_n(F)$ with $(PG(n, F), PGL(n+1, F))$. Show that any such equivalence is a coordinate system (that is to say, can be set up in the way described in the text by selecting a suitable basis for $V_{n+1}(F)$). Show also that any two coordinate systems are related by an inner self-equivalence of $(PG(n, F), PGL(n+1, F))$.

13.7. Show that in $P_1(F)$ any ordered set of 3 distinct points satisfies the condition of Theorem 13.1 for a simplex of reference and unit point.

Hence or otherwise show that $\Pi_1(F)$ is *sharply* 3-transitive; that is, given any two triads a, b, c and a^*, b^*, c^*, each of distinct points of $P_1(F)$, there is a unique $\phi \in \Pi_1(F)$ such that $a\phi = a^*$, $b\phi = b^*$, $c\phi = c^*$.

13.8. There is a natural equivalence of $(PG(1, F), PGL(2, F))$ and $(F \cup \{\infty\}, \Phi(F))$, where $\Phi(F)$ is the fractional linear group over F. We may thus talk of coordinatisations of $\mathscr{P}_1(F)$ by $(F \cup \{\infty\}, \Phi(F))$, any two such being related by an inner self-equivalence of $(F \cup \{\infty\}, \Phi(F))$. Use this idea to discuss the conjugacy classes of $\Pi_2(\mathbb{R})$, obtaining canonical representatives of these classes in $\Phi(\mathbb{R})$, in the forms:

(i) $\lambda \mapsto \alpha\lambda$, $(|\alpha| \leqslant 1, \alpha \neq 0)$,

(ii) $\lambda \mapsto \lambda + 1$,

(iii) $\lambda \mapsto (\lambda + \beta)/(1 - \beta\lambda)$, $(\beta \neq 0)$,

(iv) $\lambda \mapsto -1/\lambda$.

To which of these conjugacy classes does the transformation $\lambda \mapsto 1/\lambda$ belong?

13.9. Adapt the argument of Lemma 12.6 to show that a collineation of PG(2, F) is a member of the group of transformations $P\Gamma L(3, F)$ induced on PG(2, F) by transformations of the group $\Gamma L(3, F)$. Show also that $P\Gamma L(3, F) \cong \Gamma L(3, F)/\Lambda$ where $\Lambda \cong F^*$, the multiplicative group of non-zero elements of F.

Extend these results to (PG(n, F), $P\Gamma L(n+1, F)$).

13.10. Use the results of Exercise 13.9 to obtain an alternative proof of the result stated in Exercise 13.5.

13.11. The projective group $\Pi_n(F)$ acts in a natural way as a group of transformations of $\mathscr{S}(P_n(F))$, and we consider the geometry $(\mathscr{S}(P_n), \Pi_n)$. If $V_{n+1}(F)$ is the underlying vector space of $P_n(F)$, and $V'_{n+1}(F)$ is its dual space, we have a dual geometry $(\mathscr{S}(P'_n), \Pi'_n)$ defined in terms of $V'_{n+1}(F)$. Show that any isomorphism $\alpha: V_{n+1} \to V'_{n+1}$ yields an equivalence of geometries

$$\varepsilon_\alpha : (\mathscr{S}(P_n), \Pi_n) \to (\mathscr{S}(P'_n), \Pi'_n).$$

(Remember that ε_α stands for a pair of maps $\mathscr{S}(P_n) \to \mathscr{S}(P'_n)$ and $\Pi_n \to \Pi'_n$.)

Now consider the map $W \mapsto W^0$ of $\mathscr{S}(V_{n+1}) \to \mathscr{S}(V'_{n+1})$, which maps a subspace W of V_{n+1} on its annihilator W^0 in V'_{n+1}. Show that this map yields a canonical equivalence

$$\delta : (\mathscr{S}(P_n), \Pi_n) \to (\mathscr{S}(P'_n), \Pi'_n).$$

Hence show that

$$\delta\varepsilon_\alpha^{-1} : (\mathscr{S}(P_n), \Pi_n) \to (\mathscr{S}(P_n), \Pi_n)$$

is a self-equivalence of $P_n(F)$ which (i) maps k-flats to $(n-k-1)$-flats; (ii) reverses inclusion relations of subspaces.

13.12. Consider a vectorial representation $(V_2(F), AGL(V_2(F)))$ of affine geometry $\mathscr{A}_2(F)$, and let D be the set of directions of lines (D is the set of 1-dimensional subspaces of $V_2(F)$). The group $GL(V_2)$ acts in a natural way as a group of transformations of D, and we define the action of

$\phi = \theta\tau \in \mathrm{AGL}(V_2)$ on $d \in D$ by $d^\phi = d\theta$. Hence $\mathrm{AGL}(V_2)$ can be seen as a group of transformations on $V_2 \cup D$, and we can consider the geometry $(V_2 \cup D, \mathrm{AGL}(V_2))$. Show that this geometry is equivalent to the geometry $(P_2(F), \Pi_2^*(F))$, where $\Pi_2^*(F)$ is the subgroup of $\Pi_2(F)$ which fixes a hyperplane (line l_∞) of $P_2(F)$, and that in this equivalence D corresponds to the projective line l_∞.

[This exercise expresses in formal terms the sort of process by which, in the historical development, the affine plane was extended to the projective plane by adding to it 'points at infinity', one for each set of parallel lines in the affine plane.]

13.13. Let H be a hyperplane of $P_n(\mathbb{R})$, corresponding to a subspace $W_n(\mathbb{R})$ of $V_{n+1}(\mathbb{R})$, and suppose that we have a positive definite symmetric inner product defined on $W_n(\mathbb{R})$. Points A, $B \in H$ will be called *conjugate* if their representing vectors are orthogonal with respect to the inner product. Show that the relation of conjugacy of points in H is well-defined (independent of choice of representing vectors), and that a transformation in $\mathrm{GL}(W_n(\mathbb{R}))$ which preserves conjugacy of points in H is a scalar multiple of a transformation in $\mathrm{O}(W_n)$.

If now G_1 is the group of transformations in $\Pi_n(\mathbb{R})$ which fix H *and* preserve conjugacy of points in H, and $G_1^* = G_1 | (P_n \setminus H)$, show that $(P_n \setminus H, G_1^*)$ is equivalent to the similarity geometry $(\mathbb{R}^n, \Sigma)$ where Σ is the group of transformations $u \mapsto u(\lambda A) + a$, A an orthogonal matrix, $\lambda \in \mathbb{R}$, $\lambda \neq 0$.

13.14. Using the notations of Exercise 13.13, if G is the group of transformations in $\Pi_n(\mathbb{R})$ which fix H, and $G^* = G | (P_n \setminus H)$, so that $(P_n \setminus H, G^*)$ is equivalent to $(\mathbb{R}^n, \mathrm{AGL}(n, \mathbb{R}))$, let G_2^* be the subgroup of G^* which corresponds in the equivalence to the equi-affine subgroup of $\mathrm{AGL}(n, \mathbb{R})$. Show that G_2^* is independent of the coordinate system used to establish the equivalence of $(P_n \setminus H, G^*)$ and $(\mathbb{R}^n, \mathrm{AGL}(n, \mathbb{R}))$.

13.15. Using the notations of Exercises 13 and 14, let $G_3^* = G_1^* \cap G_2^*$. Show that $(P_n \setminus H, G_3^*)$ is equivalent to $(\mathbb{R}^n, \mathrm{I}(n))$, and is thus a Euclidean geometry $\mathscr{E}_n$.

13.16. A non-singular projective conic in $P_2(\mathbb{R})$ is defined as a non-empty point set

$$\Gamma = \{\{x, y, z\} : ax^2 + by^2 + cz^2 + 2fyz + 2gzx + 2hxy = 0\}$$

where

$$\begin{vmatrix} a & h & g \\ h & b & f \\ g & f & c \end{vmatrix} \neq 0.$$

Show that all non-singular projective conics are projectively equivalent.

Show that, if Γ is a non-singular projective conic and H is a projective line in $P_2(\mathbb{R})$, then $\Gamma^* = \Gamma \cap (P_2 \backslash H) = \Gamma \backslash \Gamma \cap H$ is a non-singular affine conic in the affine space $P_2 \backslash H$, and that Γ^* is an ellipse, parabola or hyperbola according as $\Gamma \cap H$ consists of 0, 1, or 2 points respectively.

Chapter 14

Euclidean geometry

We recall that Euclidean geometry $\mathscr{E}_n$ is a G-space (Ω, G) equivalent to $(\mathbb{R}^n, \mathrm{I}(n))$, and has coordinate representations by $(\mathbb{R}^n, \mathrm{I}(n))$, any two of which are related by an inner self-equivalence of $(\mathbb{R}^n, \mathrm{I}(n))$. The geometry $\mathscr{E}_n$ also has vectorial representations by $(V_n(\mathbb{R}), \mathrm{I}(V_n(\mathbb{R})))$ where $V_n(\mathbb{R})$ is a vector space endowed with a positive definite symmetric inner product, and $\mathrm{I}(V_n(\mathbb{R}))$ is the associated isometry group which for any two vectors $x, y \in V_n(\mathbb{R})$ preserves the distance

$$d(x, y) = \|x - y\| = +\{(x - y) \cdot (x - y)\}^{1/2}.$$

By taking any orthonormal basis of $V_n(\mathbb{R})$, a vectorial representation yields a coordinate representation, and if x, y are represented by n-tuples $(\alpha_1, \ldots, \alpha_n)$, $(\beta_1, \ldots, \beta_n) \in \mathbb{R}^n$, then $d(x, y) = +\{\sum(\alpha_i - \beta_i)^2\}^{1/2}$.

Since $\mathrm{I}(V_n) \leqslant \mathrm{AGL}(V_n)$, any geometrical properties of $\mathscr{A}_n(\mathbb{R})$ are geometrical properties of $\mathscr{E}_n$. Thus, for instance, the affine invariant $(\delta ac/\delta ab)$ of three collinear points a, b, c, with $a \neq b$ is necessarily a Euclidean invariant. It is in fact closely connected with the Euclidean invariants $d(a, c)$, $d(a, b)$, for if $(\delta ac/\delta ab) = \lambda$ then $c = (1 - \lambda)a + \lambda b$, giving $(a - c) = \lambda(a - b)$, whence $\|a - c\| = |\lambda| \cdot \|a - b\|$, so that

$$d(a, c)/d(a, b) = |(\delta ac/\delta ab)|.$$

It will be noted that the ratio of the (essentially positive) Euclidean distances only gives the absolute value of the invariant $(\delta ac/\delta ab)$. It is this fact which gives rise to some tiresome technical difficulties in classical Euclidean geometry, which relies principally on the notion of distance. In a non-algebraic development the notion of distance has to be supported by a notion of 'order' or 'between-ness' or 'separation' of points on a line.

As in the case of the affine group, a Euclidean transformation is represented uniquely in a vectorial representation by $\phi = \theta\tau$, where now $\theta \in \mathrm{O}(V_n)$ (the inner-product-preserving transformations in $\mathrm{GL}(V_n)$), and $\tau \in T$ (the translation group). In a coordinate representation, ϕ is represented by a transformation $u \mapsto uA + a$, where A is a real orthogonal $n \times n$ matrix, and $a \in \mathbb{R}^n$. We shall carry out for the Euclidean group the same sort of programme as we conducted for the affine group, investigating the conjugacy classes by seeking canonical representatives, which are obtained by considering the representations of a given transformation in different coordinate systems.

We first consider transformations in $\mathrm{O}(V_n)$, and we start by recalling some basic results about a real vector space V endowed with a positive-definite symmetric inner product. Such an inner product $\pi: V \times V \to \mathbb{R}$ defines by restriction an inner product $\pi|_{U \times U} = \pi^*: U \times U \to \mathbb{R}$ on any subspace U of V, and π^* is clearly positive-definite and symmetric. For any subspace U of V, the vectors in V which are orthogonal to every vector in U form a subspace $U^\perp$, called the *orthogonal complement* of U. Clearly $U \cap U^\perp = \{0\}$, since the zero vector is the only vector orthogonal to itself. Also, $\dim U + \dim U^\perp = \dim V$, as can be seen by taking an orthonormal basis $e_1, \ldots, e_r$ for U, and extending it to an orthonormal basis $e_1, \ldots, e_r, e_{r+1}, \ldots, e_n$ for V, and checking that $U^\perp = \langle e_{r+1}, \ldots, e_n \rangle$. We therefore have $V = U \oplus U^\perp$.

If U is invariant under an orthogonal transformation $\theta: V \to V$, so that $U\theta = U$, then

$$x \in U^\perp \Leftrightarrow x \cdot u = 0 \qquad \text{for all } u \in U$$

$$\Leftrightarrow x\theta \cdot u\theta = 0 \quad \text{for all } u \in U$$

$$\Leftrightarrow x\theta \cdot v = 0 \qquad \text{for all } v \in U\theta = U$$

$$\Leftrightarrow x\theta \in U^\perp.$$

Hence $U^\perp\theta = U^\perp$, and $U^\perp$ is also invariant under θ.

For any subspace W invariant under θ, the restriction $\theta^* = \theta|_W: W \to W$ preserves the inner product on W, and is thus an orthogonal transformation of W.

Before starting our next lemma, we recall the definition and some properties of the *minimum polynomial* $m(t)$ of a linear transformation (not necessarily invertible) $\theta: V_n(F) \to V_n(F)$ of a finite dimensional vector space over an arbitrary field F. For any polynomial $f(t) = a_0 t^r + a_1 t^{r-1} + \cdots + a_r \in F[t]$, we write $f(\theta) = a_0 \theta^r + a_1 \theta^{r-1} + \cdots + a_r 1$. The set of polynomials $\{f(t): f(\theta) = 0\}$ is non-empty and, indeed, contains some non-zero polynomials (this follows from the Cayley–Hamilton theorem that $\chi(\theta) = 0$, where $\chi(t)$ is the characteristic polynomial of θ, but can also be shown by more elementary arguments) and is an ideal of the principal ideal ring $F[t]$. It follows that the set is precisely the set of multiples of a unique monic polynomial $m(t)$. If $m(t)$ has a non-trivial factor $g(t)$, so that $m(t) = g(t)h(t)$, then $g(\theta)$ is not invertible; for if it were, we should have $h(\theta) = (g(\theta))^{-1}m(\theta) = 0$, giving $m(t)|h(t)$. Hence $g(\theta)$ has a non-trivial kernel. We now state

LEMMA 14.1:   *If $\theta \in GL(V_n(\mathbb{R}))$ then $V_n(\mathbb{R})$ has a subspace of dimension 1 or 2 which is invariant under θ.*

Proof.   Let $m(t) \in \mathbb{R}[t]$ be the minimum polynomial of θ. Then the irreducible factors of $m(t)$ in $\mathbb{R}[t]$ are linear or quadratic. We can therefore find $g(\theta) = \theta - c$ or $g(\theta) = \theta^2 + a\theta + b$ with a non-trivial kernel. In the first case, take $x \neq 0$ in $\ker(\theta - c)$, so that $x\theta = cx$. Now $c \neq 0$ since otherwise θ has a non-trivial kernel, whereas $\theta \in GL(V_n(\mathbb{R}))$ and is invertible. Hence the 1-dimensional subspace $U = \langle x \rangle$ is invariant under θ, and $U\theta = U$. In the second case, take $x \neq 0$ in $\ker(\theta^2 + a\theta + b)$ and let $y = x\theta$. The vectors x, y are linearly independent, for otherwise $y = kx$, and we have $0 = x(\theta^2 + a\theta + b) = (k^2 + ak + b)x$, whence $k^2 + ak + b = 0$ and $t - k$ is a factor of the (irreducible) polynomial $t^2 + at + b$. Also,

$$y\theta = x\theta^2 = -a(x\theta) - bx = -ay - bx,$$

and this, together with $x\theta = y$, tells us that the 2-dimensional subspace $U = \langle x, y \rangle$ is mapped into itself by θ. Since θ has trivial kernel, we have $U\theta = U$.

We can now state and prove

THEOREM 14.2:   *If $\theta \in O(V_n(\mathbb{R}))$, then $V_n(\mathbb{R})$ $(= V)$ is a direct sum $V = W_1 \oplus \ldots \oplus W_s$ of mutually orthogonal subspaces, each of which is of dimension 1 or 2 and invariant under θ.*

Proof.   We find (by Lemma 14.1) a subspace W_1 of V invariant under θ, and write $V = W_1 \oplus V_1$ where $V_1 = W_1^{\perp}$. We then consider $\theta_1 = \theta | V_1$, and find a subspace W_2 of V_1 invariant under θ_1 (and hence invariant under θ), and write $V_1 = W_2 \oplus V_2$ where V_2 is the orthogonal complement of W_2 in V_1. Every element of V_2 is not only orthogonal to every element of W_2; it is also (in virtue of belonging to V_1) orthogonal to every element of W_1. The subspaces W_2 and V_2 are invariant under θ_1, and hence under θ. Continuing in this way, we obtain the required decomposition of V.

The decomposition process described in the proof of Theorem 14.2 can be made more precise if we always select by preference a 1-dimensional invariant subspace. In that case, the 2-dimensional invariant subspaces obtained in the decomposition can contain no 1-dimensional invariant subspace.

We now construct an orthonormal basis for $V_n(\mathbb{R})$ by taking an orthonormal basis for each of the component subspaces. The orthogonal transformation induced by θ on a component subspace is then represented by an orthogonal matrix. The only orthogonal 1×1 matrices are $[1]$ and $[-1]$. The 2×2 orthogonal matrices are (see Exercise 2.14) of the forms

$$\begin{bmatrix} \cos\alpha & \sin\alpha \\ -\sin\alpha & \cos\alpha \end{bmatrix}, \begin{bmatrix} \cos\alpha & \sin\alpha \\ \sin\alpha & -\cos\alpha \end{bmatrix}.$$

These represent transformations with an invariant 1-dimensional subspace if and only if they have a real eigenvalue. All matrices of the second form have eigenvalues 1, -1. Those of the first form have a real eigenvalue if and only if α is a multiple of π. The transformations on the 2-dimensional component subspaces are therefore represented by matrices R_α, $\alpha \in \mathbb{R}/2\pi\mathbb{Z}$, $\alpha \neq 0, \pi$. If we now arrange the summands in an appropriate order, θ is represented by a matrix

$$\text{diag}\{1, \ldots, 1, -1, \ldots, -1, R_{\alpha_1}, \ldots, R_{\alpha_s}\} \quad (\alpha_i \neq 0, \pi, i = 1, \ldots, s).$$

Now, two 1-dimensional subspaces on which the restriction of θ is the identity can be combined in the decomposition to form a 2-dimensional invariant subspace on which θ is the identity, represented by the matrix $I = R_0$. Likewise, two 1-dimensional subspaces on which the restriction of θ is -1 can be combined to form a 2-dimensional invariant subspace on which θ is represented by the matrix $-I = R_\pi$. If we write

$$K = \begin{bmatrix} 1 & 0 \\ 0 & -1 \end{bmatrix}$$

(which we adopt as a standard notation), we obtain the following result.

THEOREM 14.3:   *If $\theta \in O(V_n(\mathbb{R}))$ there is an orthonormal basis of $V_n(\mathbb{R})$ relative to which θ is represented by one of the following matrices:*
(i) $\mathrm{diag}(1, R_{\alpha_1}, \ldots, R_{\alpha_m})$ *or* $\mathrm{diag}(-1, R_{\alpha_1}, \ldots, R_{\alpha_m})$ *if* $n = 2m+1$;
(ii) $\mathrm{diag}(R_{\alpha_1}, \ldots, R_{\alpha_m})$ *or* $\mathrm{diag}(K, R_{\alpha_2}, \ldots, R_{\alpha_m})$ *if* $n = 2m$.

    In each case of Theorem 14.3, matrices of the first form have determinant $+1$, and represent a transformation $\theta \in SO(V_n(\mathbb{R}))$; those of the second form have determinant -1. The representing matrices which we have obtained are not unique, because the ordering of the matrices R_{α_i} is clearly arbitrary. Also, for a 2-dimensional space, the transformation represented relative to the orthonormal basis e_1, e_2 by the matrix R_α is represented relative to the orthonormal basis e_1, $-e_2$ by $R_{-\alpha}$. The proof that modulo change of sign and order of the α_i these representations are unique, and provide canonical representatives of the conjugacy classes of $O(n)$, is left to the reader (see Exercise 14.9).
    Theorem 14.3 tells us that the eigenvalues of $\theta \in O(V_n(\mathbb{R}))$ are the eigenvalues of the representing matrices, and are therefore 1, -1 or eigenvalues of a matrix R_α ($\alpha \neq 0, \pi$), which are $e^{i\alpha}$, $e^{-i\alpha}$. Thus, all the eigenvalues of θ are of unit modulus. A possible approach to the study of the orthogonal group is by way of the eigenvalues and eigenvectors of an orthogonal transformation. However, the eigenvalues are (in general) complex numbers. If we work in coordinate terms, so that we think of a real orthogonal matrix A acting on $\mathbb{R}^n$, we cannot find an eigenvector in $\mathbb{R}^n$ for a complex eigenvalue of A. We can, however, find a solution in the complex field $\mathbb{C}$ for the coordinates of an eigenvector satisfying the equation $xA = \lambda x$. If we do this, we are making use of the natural embeddings of $\mathbb{R}^n$ in $\mathbb{C}^n$, and of $GL(n, \mathbb{R})$ in $GL(n, \mathbb{C})$. Using such arguments, we can work back to results about the action of A on $\mathbb{R}^n$ (see Exercises 14.6–9). This process is less easy to describe in basis-free terms if we are thinking of a transformation θ of a vector space $V_n(\mathbb{R})$; a description in basis-free terms of a natural embedding of $V_n(\mathbb{R})$ in a vector space $V_n(\mathbb{C})$ requires a knowledge of the tensor product. We have therefore provided in the text a treatment which avoids consideration of vector spaces over the complex field.

It is convenient at this stage to introduce the exponential function $\exp A$ of a matrix A. For a real or complex $n \times n$ matrix A, consider the series

$$I + A + \frac{1}{2!} A^2 + \cdots + \frac{1}{r!} A^r + \cdots .$$

Let

$$A = [a_{ij}], \quad \frac{1}{r!} A^r = [a_{ij}^{(r)}],$$

and suppose that $M \geqslant |a_{ij}|$ for all i, j $(1 \leqslant i, j \leqslant n)$. We prove easily by induction that $|a_{ij}^{(r)}| \leqslant n^{r-1} M^r / r!$, so writing $a_{ij}^{(0)} = \delta_{ij}, a_{ij}^{(1)} = a_{ij}$ and using the fact that the infinite series $\sum_{r=1}^{\infty} n^{r-1} M^r / r!$ converges, it follows that $\sum_{r=0}^{\infty} a_{ij}^{(r)}$ converges absolutely, to c_{ij} say, and we can define $\exp A$ as the matrix $[c_{ij}]$. We can further show (copying the proof for the exponential function of a real or complex variable) that *provided A and B commute*,

$$\exp A . \exp B = \exp (A + B),$$

and in particular that $(\exp A)^m = \exp(mA)$ for any integer $m \geqslant 0$. Also $(\exp A)(\exp(-A)) = I$, whence $(\exp A)^{-1} = \exp(-A)$, and $(\exp A)^m = \exp(mA)$ holds also for any integer $m < 0$. Finally, we note that, since $(T^{-1}AT)^r = T^{-1}A^rT$, we have $\exp(T^{-1}AT) = T^{-1}(\exp A)T$ for any non-singular $n \times n$ matrix T.

In particular, if

$$J := \begin{bmatrix} 0 & 1 \\ -1 & 0 \end{bmatrix}$$

then, since $J^2 = -I$, we have

$$\exp(\alpha J) = I + \alpha J - \frac{1}{2!} \alpha^2 I - \frac{1}{3!} \alpha^3 J + \cdots = (\cos \alpha)I + (\sin \alpha)J = R_\alpha,$$

and consequently any special orthogonal transformation is represented relative to a suitable orthonormal basis by a matrix

$$\operatorname{diag}(\exp(\alpha_1 J), \ldots, \exp(\alpha_m J)) = \exp(\operatorname{diag}(\alpha_1 J, \ldots, \alpha_m J)) \quad \text{if } n = 2m,$$

or

$$\operatorname{diag}(1, \exp(\alpha_1 J), \ldots, \exp(\alpha_m J)) = \exp(\operatorname{diag}(0, \alpha_1 J, \ldots, \alpha_m J))$$
$$\text{if } n = 2m + 1.$$

In both cases, the transformation is represented by a matrix $\exp S$, where S is a skew-symmetric matrix $(S' = -S)$. Now, relative to any other orthonormal basis, the transformation will be represented by $T^{-1}(\exp S)T$, where T is an orthogonal matrix representing the change of basis, and we have $T^{-1}(\exp S)T = \exp(T^{-1}ST) = \exp(T'ST)$, and since $(T'ST)' = T'S'T = -T'ST$, the matrix $T'ST$ is skew-symmetric. Thus any special orthogonal matrix A can be expressed in the form $A = \exp S$, where S is skew-symmetric. Conversely, given any real skew-symmetric matrix S, we have $(\exp S)' = \exp S' = \exp(-S) = (\exp S)^{-1}$, showing that $\exp S$ is orthogonal. To show that $\exp S$ is special orthogonal, consider $\exp(tS)$, where t is a real parameter. The entries in this matrix are given by convergent power series in t (with infinite radius of convergence) and are thus continuous functions of t. Hence $\det(\exp(tS))$ is also a continuous function of t which, since $\exp(tS)$ is orthogonal, can only take the value $+1$ or -1. But for $t = 0, \exp(tS) = I_n$ and $\det I_n = 1$, and hence $\det(\exp(tS)) = 1$ for all values of t, in particular for $t = 1$.

Let $S \neq 0$ be any skew-symmetric 3×3 real matrix. We can then calculate the matrix $A \in SO(3)$ given by $A = \exp S$. We write S in the form

$$S = \begin{bmatrix} 0 & -v & \mu \\ v & 0 & -\lambda \\ -\mu & \lambda & 0 \end{bmatrix}$$

Let $\omega = [\lambda, \mu, v]$, and $\alpha = +(\lambda^2 + \mu^2 + v^2)^{1/2}$. We then have $\omega S = 0$, and

$$S^2 = \begin{bmatrix} \lambda^2 & \lambda\mu & \lambda v \\ \mu\lambda & \mu^2 & \mu v \\ v\lambda & v\mu & v^2 \end{bmatrix} - (\lambda^2 + \mu^2 + v^2)I = \omega'\omega - \alpha^2 I$$

whence

$$S^3 = -\alpha^2 S,$$
$$S^4 = -\alpha^2 S^2 = -\alpha^2 \cdot \omega'\omega + \alpha^4 I,$$
$$S^5 = \alpha^4 S,$$
$$S^6 = \alpha^4 \omega'\omega - \alpha^6 I,$$

$$\cdots$$

and

$$\exp S = \left(1 - \frac{\alpha^2}{2!} + \frac{\alpha^4}{4!} - \cdots\right)I + \left(1 - \frac{\alpha^2}{3!} + \frac{\alpha^4}{5!} - \cdots\right)S$$

$$+ \left(\frac{1}{2!} - \frac{\alpha^2}{4!} + \frac{\alpha^4}{6!} - \cdots\right)(\omega'\omega),$$

giving $A = (\cos\alpha)I + (\alpha^{-1}\sin\alpha)S + \alpha^{-2}(1 - \cos\alpha) \cdot (\omega'\omega)$. Since $\omega S = 0$ and $\omega(\omega'\omega) = (\omega\omega')\omega = \alpha^2\omega$, we have

$$\omega A = (\cos\alpha)\omega + (1 - \cos\alpha)\omega = \omega,$$

so the vector ω is fixed by the rotation represented by A. If x is orthogonal to ω, we have $x\omega' = 0$, and

$$xA = \cos\alpha \cdot x + \sin\alpha(\alpha^{-1} \cdot xS).$$

Now, from $\omega S = 0$ we have $S\omega' = 0$ and $(xS)\omega' = 0$, whence xS is orthogonal to ω. Also, the scalar $(xS)x' = 0$, since $xSx' = (xSx')' = -xSx'$, whence xS is orthogonal to x. If x is a unit vector, $(xS)(xS)' = -xS^2x' = \alpha^2(xx)' = \alpha^2$, and the vectors $\alpha^{-1}\omega$, x, $\alpha^{-1}xS$ are an orthonormal triad. Denoting these by e_1, e_2, e_3 respectively, we have

$$e_1 A = e_1,$$

$$e_2 A = (\cos \alpha)e_2 + (\sin \alpha)e_3,$$

$$e_3 A = (-\sin \alpha)e_2 + (\cos \alpha)e_3,$$

so that relative to this basis, the rotation represented by A takes the canonical form $\mathrm{diag}(1, R_\alpha)$ for its representing matrix. The vector ω is in the direction of the axis, and of magnitude equal to the angle of rotation.

If we displace a rigid body in physical space, keeping one point of the body fixed, then the effect of the displacement is to apply to the particles of the body an isometry of the space which is represented, relative to an orthonormal basis of $V_3(\mathbb{R})$ with origin at the fixed point, by an orthogonal matrix A, so that the particle initially in a position represented by the vector x_0 is moved to a point represented by the vector x_1, where $x_1 = x_0 A$. The body does not, however, reach the final position instantaneously, but over a period of time through intermediate positions. Suppose for simplicity that the displacement takes place over unit time; we then have a time-dependent orthogonal matrix $A(t) = [a_{ij}(t)]$, the position of the particle at time t being given by $x(t) = x_0 A(t)$, with $x(0) = x_0$, $x(1) = x_1$ and $A(0) = I$, $A(1) = A$. Now, the particle can only move with finite velocity $\dot{x}(t)$, given by $\dot{x}(t) = x_0 \dot{A}(t)$, where $\dot{A}(t) = [\dot{a}_{ij}(t)]$, showing that the entries in $A(t)$ must be differentiable functions of t. This implies continuity of all $a_{ij}(t)$, and hence of $\det A(t)$, as functions of t, and since $\det A(0) = 1$ we have $\det A(t) = 1$ for all $t \in [0, 1]$, and in particular the matrix A is a *special* orthogonal matrix.

These ideas transfer easily to Euclidean space of any dimension n. A *motion with fixed point* is defined by a matrix $A(t) = [a_{ij}(t)]$, where t is a real parameter in $[0, 1]$, and $A(0) = I$, $A(t) \in SO(n)$, and the entries $a_{ij}(t)$ are all differentiable functions of t for $t \in [0, 1]$. If $x(t) = x_0 A(t)$, then $x(0) = x_0$, and $\dot{x}(t) = x_0 \dot{A}(t) = x(t)(A(t))^{-1}\dot{A}(t)$, which we may write as $\dot{x} = x(A'\dot{A})$. Now $(A'\dot{A})' = \dot{A}'A$, and by differentiating the relation $A'A = I$, we have $\dot{A}'A + A'\dot{A} = 0$, so $A'\dot{A}$ is a skew-symmetric matrix $S(t)$, and we have the relation $\dot{x} = xS$. This matrix $S(t)$ is called the *angular velocity matrix* (or simply, the *angular velocity*) of the motion at time t. If the matrix $A(1) = \exp S_1$, then the matrix $A(t) = \exp(tS_1)$ defines a motion in which

$$A'(t) = \exp(tS_1') = \exp(-tS_1),$$

and

$$\dot{A}(t) = (\exp(tS_1))S_1,$$

giving $A'(t)\dot{A}(t) = S_1$, which is independent of t. We therefore have a motion with constant angular velocity S_1. It should be noted that the angular velocity S_1 is not uniquely determined by the matrix $A(1)$, because the numbers α_i, which occur in the canonical form of $A(1)$, are only determined modulo 2π.

We now examine the conjugacy classes of the Euclidean group in dimensions 2 and 3. We recall from the discussion of affine transformations in Chapter 12 that for a transformation with a fixed point, we may take the origin at that point and obtain a representation by a transformation in $O(n)$. If a transformation is fixed-point-free, then it must have 1 as an eigenvalue. The only matrix of the form R_α which has eigenvalue 1 is $R_0 = I$. Hence the fixed-point-free transformations may be represented in one of the forms $u \mapsto u + a$, $u \mapsto uK + a$. The first of these is a translation, and we may choose the orthonormal basis so that the first basis vector is a positive multiple of a; the transformation then takes the form

$$[x, y] \mapsto [x^*, y^*] = [x, y] + [\xi, 0], \quad \xi > 0.$$

A transformation of the second kind may be written as $[x, y] \mapsto [x, -y] + [\xi, \eta]$, and is fixed-point-free unless $\xi = 0$, when it has the line of fixed points given by $y = \frac{1}{2}\eta$. By moving the origin to a point on this line, the transformation is represented in the form

$$[x, y] \mapsto [x^*, y^*] = [x, -y] + [\xi, 0].$$

For a fixed-point-free transformation of this type, with $\xi \neq 0$, we can, by reversing the x-axis (a process equivalent to conjugation by the transformation $[x, y] \mapsto [-x, y]$), ensure that $\xi > 0$.

We can now list in equational form the standard representations of the various types of transformations in I(2).

Fixed-point-free

    F.1 Translation $x^* = x + \xi,\ y^* = y, \quad \xi > 0.$

    F.2 Glide reflexion $x^* = x + \xi,\ y^* = -y, \quad \xi > 0.$

Central

    C.1 Rotation $x^* = \cos \alpha \cdot x - \sin \alpha \cdot y,$

                                  $y^* = \sin \alpha \cdot x + \cos \alpha \cdot y, \quad \alpha \in (0, \pi).$

    C.2 Central reflexion $x^* = -x,\ y^* = -y.$

Axial

    A.1 Reflexion $x^* = x,\ y^* = -y.$

Identity

    I $x^* = x,\ y^* = y.$

These transformations, with the stated restrictions on the parameters ξ, α, are a canonical set of representatives of the conjugacy classes of I(2). The transformations C.1, C.2, A.1, I with a fixed point have distinct sets of eigenvalues. For a translation, ξ is the distance between any point and its image, and is thus the same in any coordinate system. Likewise, for a glide reflexion ξ is the distance between any point on the invariant line and its image. Thus no two of the transformations listed belong to the same conjugacy class of I(2).

The central reflexions and the identity may be regarded as special cases of rotations (with $\alpha = \pi, 0$ respectively). Every transformation in I(2) of determinant $+1$ is the product of two (axial) reflexions, and the reflexions generate the whole of I(2). (See Exercise 14.11.)

The transformations in I(2) of determinant $+1$ form a subgroup, which will be denoted by SI(2), and which consists of the rotations (including the identity and central reflexions) and the translations. The transformations in SI(2) are of the form $u \mapsto uA + a$, where A is a special orthogonal matrix. The group I(2) contains SI(2) as a normal subgroup of index 2, and the transformations F.1, C.1, C.2, I which are listed above are a canonical set of representatives of the conjugacy classes in I(2) of the elements of SI(2). They are not a complete set of representatives for the conjugacy classes of SI(2), because the orthogonal matrices R_α, $R_{-\alpha}$ are *not* conjugate in SO(2), and in C.1 we must take $\alpha \in (0, \pi) \cup (\pi, 2\pi)$ to get all the conjugacy classes.

This last point is one which deserves further discussion. If we had set out to investigate the conjugacy classes of SI(2), it would have been appropriate to consider not $\mathscr{E}_2$ (which is equivalent to $(\mathbb{R}^2, I(2))$), but a more restricted geometry equivalent to $(\mathbb{R}^2, SI(2))$. Now, the appropriate coordinate systems for such a geometry are related by self-equivalences of $(\mathbb{R}^2, SI(2))$, and any such self-equivalence is associated with a transformation in SI(2). These coordinate systems are therefore more restricted than those which we used for Euclidean geometry $\mathscr{E}_2$. In terms of vector spaces, we can not derive a coordinate system by arbitrary choice of an orthonormal basis, but must use only orthonormal bases related to each other by a special orthogonal transformation (of determinant $+1$). Thus, in particular, if e_1, e_2 is a legitimate basis for $V_2(\mathbb{R})$, the basis e_1, $-e_2$ is not legitimate. The orthonormal bases (in any dimension) divide into two sets, two bases in the same set being related by an orthogonal transformation of determinant $+1$, and two in different sets being related by an orthogonal transformation of determinant -1. When picturing such bases in a plane (or, for 3-dimensional Euclidean geometry, in 3-space), we can distinguish the two sets as 'right-handed' and 'left-handed', but neither set is superior to, or more 'natural' than, the other. The universal convention in cartesian geometry is to use right-handed sets of axes, so that in plane geometry the y-axis is obtained by rotating the x-axis anti-clockwise through a right angle. If, as is generally the case in cartesian geometry, we select one coordinate system and stay with it, an agreement as to how the selected system should be depicted is convenient, but we should not be misled by habitual usage into thinking that the practice is anything more than a convention.

We conclude this chapter with a discussion of Euclidean transformations in 3 dimensions. By an argument which must now be familiar, any such transformation with a fixed point is representable in a suitable coordinate system by a transformation $u \mapsto uA$, where A (by Theorem 14.3) has the form $\mathrm{diag}(\pm 1, R_\alpha)$. Transformations of determinant $+1$ are represented by matrices $\mathrm{diag}(1, R_\alpha)$, and such transformations are called *rotations*. The rotations include the identity with representing matrix $I = \mathrm{diag}(1, R_0)$. Every rotation $\neq 1$ has a line of fixed points, which is called the *axis* of the rotation; for the identity, every point is fixed. The rotations include the *axial reflexions* with representing matrix $\mathrm{diag}(1, R_\pi) = \mathrm{diag}(1, -1, -1)$. A set of canonical representatives of rotations in the Euclidean group is given by limiting α to the range $[0, \pi]$. The rotations all belong to the restricted Euclidean group $(\cong \mathrm{SI}(3))$ and in that group the range $\alpha \in [0, 2\pi)$ provides canonical representatives. In this latter situation, there is a unique $\alpha \in \mathbb{R}/2\pi\mathbb{Z}$ which specifies the conjugacy class of a rotation, and this is called the *angle* of the rotation.

The transformations with fixed points and determinant -1 are representable by matrices $\mathrm{diag}(-1, R_\alpha)$. These include the matrix $\mathrm{diag}(-1, R_0) = \mathrm{diag}(-1, 1, 1)$. A transformation so representable has a plane of fixed points, and is called a *plane reflexion*. The matrix $\mathrm{diag}(-1, R_\alpha)$ is the product (in either order) of $\mathrm{diag}(-1, R_0)$ and $\mathrm{diag}(1, R_\alpha)$, and thus represents the product of a plane reflexion and a rotation with axis orthogonal to the plane of the reflexion. Such a transformation is called a *rotatory reflexion*, a term which includes the (pure) plane reflexion, corresponding to $\alpha = 0$. It should be noted that in 3 dimensions the *central reflexions*, with representing matrix $\mathrm{diag}(-1, -1, -1)$, are rotatory reflexions, and not rotations. The fact that $-1 \in \mathrm{SO}(n)$ for n even but $-1 \notin \mathrm{SO}(n)$ for n odd is one of a number of significant ways in which $\mathrm{O}(n)$ for n even behaves differently from $\mathrm{O}(n)$ for n odd.

Transformations $\phi = \theta\tau$ without a fixed point must have 1 as an eigenvalue of θ. Hence θ is either a rotation (if $\det \theta = 1$) or a plane reflexion (if $\det \theta = -1$). If $\theta \neq 1$ is a rotation, we find that θ has a fixed line parallel to the axis of θ, and by moving the origin to a point on this line, and choosing axes suitably, such a transformation, which is called a *screw*, can be represented uniquely by equations

$$x^* = x + \xi, \qquad\qquad \xi > 0,$$
$$y^* = (\cos \alpha)y - (\sin \alpha)z,$$
$$z^* = (\sin \alpha)y - (\cos \alpha)z, \quad \alpha \in (0, \pi].$$

If $\theta = 1$, ϕ is simply a translation, and can be represented uniquely by equations

$$x^* = x + \xi, \; y^* = y, \; z^* = z, \quad \xi > 0.$$

This can be regarded as a special case of a screw, with $\alpha = 0$. Indeed, the rotations can also be regarded as special cases of a screw, with $\xi = 0$. The rotations, translations and screws, are all the transformations of determinant $+1$ and thus the whole of the restricted Euclidean group. In a number of physical applications, it is the restricted group which is of interest. For instance, in displacing a rigid body we are using transformations of the restricted group, and many of the classical treatises on mechanics include a thorough discussion of the geometrical theory of screws.

Finally we have the transformation $\phi = \theta\tau$ without fixed points, where θ is a plane reflexion. Here we find that ϕ has a fixed plane parallel to the plane of fixed points of θ, and by moving the origin to a point on the fixed plane of ϕ and choosing the axes suitably, we show that ϕ can be represented uniquely by equations

$$x^* = -x, \; y^* = y+\eta, \; z^* = z, \quad \eta > 0.$$

The transformation is the product of a plane reflexion and a translation in a direction lying in the plane of the reflexion. Such a transformation is called a *glide reflexion*.

Exercises

14.1. Two lines in $\mathscr{E}_n$, given in a vectorial representation by (p, q) and (r, s), are said to be perpendicular, and we write $(p, q) \perp (r, s)$, if $(p-q)\cdot(r-s) = 0$. A *unit triangle* is a triangle (x, y, z) with $(x, y) \perp (x, z)$ and $d(x, y) = d(x, z) = 1$. Show that all unit triangles are Euclidean-equivalent, and that for $n > 2$ all unit triangles are equivalent under the restricted group of special isometries $\mathrm{SI}(V_n)$.

14.2. For any triangle (a, b, c) in $\mathscr{E}_2$ we define the area $\triangle abc$ by

$$\triangle abc := \tfrac{1}{2}|(\triangle abc/\triangle xyz)|.$$

where (x, y, z) is any unit triangle. Show that this is a good definition.

We say that p is the foot of the perpendicular from a to (b, c) if $p \in (b, c)$ and $(a, p) \perp (b, c)$. Show that this is a good definition.

Show that $\triangle abc = \tfrac{1}{2}d(a, p).d(b\,c)$.

14.3. Suppose that the lines (p, q), (r, s) in Euclidean space of $n > 2$ dimensions are *skew*, that is, they neither intersect nor are parallel. Prove that, if $x \in (p, q)$ and $y \in (r, s)$, then $d(x, y)$ attains its minimum value when $(x, y) \perp (p, q)$ and $(x, y) \perp (r, s)$, and that these conditions determine x, y uniquely.

14.4. In a vectorial representation of $\mathscr{E}_n$, *a half-line with end-point a* is defined as a set of points

$$(a; p) = \{x \mid x = (1-\lambda)a + \lambda p, \; \lambda \in \mathbb{R}, \; \lambda \geqslant 0\}$$

where $p \neq a$. Show that if $n = 2$, there is a rotation with representing matrix R_α which maps the half-line $(a; p)$ to the half-line $(a; q)$, and that the angle between $(a; p)$ and $(a; q)$ can be uniquely defined as

$$\angle((a; p), (a; q)) := \alpha \in \mathbb{R}/2\pi\mathbb{Z}.$$

Show that

$$\angle((a; p), (a; q)) = - \angle((a; q), (a; p)).$$

Show that, if (p, q, r) is any triangle in $\mathcal{E}_2$, then

$$\angle((p; q), (p; r)) + \angle((q; r), (q; p)) + \angle((r; p), (r; q)) = \pi.$$

14.5.   Discuss the possible definitions of
>   (i) the angle between two lines in the Euclidean plane,
>   (ii) the angle between two half-lines with common end-point in Euclidean space of dimension >2.

In discussing an orthogonal transformation of $\mathbb{R}^n$, it is often convenient to use the natural embedding of $\mathbb{R}^n$ as a subset of $\mathbb{C}^n$. A transformation $\theta: \mathbb{R}^n \to \mathbb{R}^n$ given in matrix form by $u \mapsto uA$ where A is a real orthogonal matrix extends naturally to $\theta*: \mathbb{C}^n \to \mathbb{C}^n$ given by the same matrix transformation (but with u now in $\mathbb{C}^n$), and $\theta*|\mathbb{R}^n = \theta$. The passage to the complex field is useful, because the eigenvalues of the real matrix A are (in general) complex, and for a complex eigenvalue we can always find an eigenvector in $\mathbb{C}^n$ (but not in $\mathbb{R}^n$). For a vectorial representation of Euclidean space, we can obtain a similar embedding of $V_n(\mathbb{R})$ in $V_n(\mathbb{C})$, but to describe this embedding in basis-free terms we have to appeal to the tensor product, defining $V_n(\mathbb{C})$ as $\mathbb{C} \otimes_\mathbb{R} V_n(\mathbb{R})$ with its natural structure of a vector space over $\mathbb{C}$. Because we elected to work with a vectorial representation, and could not rely on familiarity with the tensor product, this embedding was not used in this chapter. The following exercises indicate how the embedding can be exploited if we work with a coordinate representation.

14.6.   The $n \times n$ matrix A is real orthogonal, and $\lambda \in \mathbb{C}$ is an eigenvalue, so that there is $x \in \mathbb{C}^n$, $x \neq 0$ such that $xA = \lambda x$. By considering $xAA'\bar{x}'$, where $\bar{x}$ denotes the n-tuple obtained by taking complex conjugates of all members of the n-tuple x, show that $|\lambda| = 1$. Hence show that the eigenvalues of A which are not equal to $+1$ or -1 occur in complex conjugate pairs $e^{i\alpha}$, $e^{-i\alpha}$.

14.7.   Supposing that A (as in Exercise 14.6) has complex (non-real) eigenvalues $e^{i\alpha}$, $e^{-i\alpha}$, there is $z \in \mathbb{C}^n$, $z \neq 0$ such that $zA = e^{-i\alpha}z$. Write $z = u + iv$, $u, v \in \mathbb{R}^n$, and show that

$$uA = (\cos \alpha)u + (\sin \alpha)v,$$

$$vA = (-\sin \alpha)u + (\cos \alpha)v.$$

Hence show that, if $\theta: \mathbb{R}^n \to \mathbb{R}^n$ is given by $x \mapsto xA$, then
>   (i) $\langle u, v \rangle$ is a 2-dimensional subspace of $\mathbb{R}^n$ invariant under θ;
>   (ii) $u \cdot u = v \cdot v$;
>   (iii) $u \cdot v = 0$.

14.8.   From Exercises 14.6 and 14.7 derive results for $\mathbb{R}^n$ similar to those stated in Theorems 14.2 and 14.3 for $V_n(\mathbb{R})$.

14.9.   Show, either in terms of the treatment in the text or in terms of the treatment developed in Exercises 14.6–8, that the elements $\alpha_i \in \mathbb{R}/2\pi\mathbb{Z}$ which occur in the representation of a given orthogonal transformation θ by a matrix (see Theorem 14.3) are uniquely determined, up to changes of sign and order of the α_i.

14.10. Show that an axial reflexion in the Euclidean plane is completely determined by its axis, and find (in matrix form) the reflexion of $\mathbb{R}^2$ having as axis the line with equation $x \cos \alpha + y \sin \alpha = p$.

14.11. Show that every transformation in SI(2) is a product of two reflexions, and that every transformation in I(2) is a product of at most three reflexions.

14.12. The mapping $\mathbb{R}^2 \to \mathbb{C}$ given by $(x, y) \mapsto x + iy = z$ is a bijection, and yields a coordinatisation of the Euclidean plane by the complex numbers. Show that SI(2) is represented by the group of transformations $z \mapsto az + b$, with $a, b, \in \mathbb{C}$ and $|a| = 1$, and that I(2) is represented by these transformations together with the transformations $z \mapsto a\bar{z} + b$, $a, b \in \mathbb{C}$, $|a| = 1$.

Chapter 15

Finite groups of isometries

We start our discussion by proving two important lemmas about finite groups of affine transformations. These lemmas, which depend essentially on the finiteness of the group, clear a great amount of ground in preparation for the subsequent investigation.

LEMMA 15.1: *If G is a finite subgroup of $\mathrm{AGL}(V_n(\mathbb{R}))$, then there is a point $c \in V_n(\mathbb{R})$ which is fixed by every element of G.*

Proof. Let $a \in V_n(\mathbb{R})$, and let $\{\phi_1, \ldots, \phi_m\}$ be the set of elements of G. Write

$$c := \frac{1}{m}(a\phi_1 + \cdots + a\phi_m).$$

If $\phi \in G$, we have $\phi = \theta\tau_b$, $\theta \in \mathrm{GL}(V_n)$, $\tau_b \in T$, and

$$c\phi = \left(m^{-1}\left(\sum_{i=1}^{m} a\phi_i\right)\right)\theta\tau_b$$

$$= \left(m^{-1}\left(\sum_{i=1}^{m} a\phi_i\theta\right)\right)\tau_b$$

$$= \left(m^{-1}\sum_{i=1}^{m} a\phi_i\theta\right) + b$$

$$= m^{-1}\left(\sum_{i=1}^{m} (a\phi_i\theta + b)\right)$$

$$= m^{-1}\sum_{i=1}^{m} a\phi_i\phi.$$

But if $\phi \in G$ is fixed, and ϕ_i runs through all the elements of G, then $(\phi_i\phi)$ also runs through all the elements of G. Hence

$$c\phi = m^{-1}\sum_{j=1}^{m} a\phi_j = c,$$

and this holds for all $\phi \in G$.

Lemma 15.1 clearly holds for any field F (not merely for $\mathbb{R}$) provided that m has a multiplicative inverse in F. This is always the case if char $F = 0$, and the condition is satisfied by fields of non-zero characteristic provided that m is not a multiple of char F.

Lemma 15.1 tells us that any finite subgroup of $AGL(V_n(\mathbb{R}))$ is conjugate to a (finite) subgroup of $GL(V_n(\mathbb{R}))$, and also that any finite subgroup of $I(V_n)$ is conjugate (in $I(V_n)$) to a finite subgroup of $O(V_n)$. This means that our search for finite subgroups of $AGL(V_n)$ or of $I(V_n)$ can be restricted to $GL(V_n)$ or to $O(V_n)$, respectively.

LEMMA 15.2: *If G is a finite subgroup of $GL(V_n(\mathbb{R}))$, then there is a positive-definite symmetric inner product on $V_n(\mathbb{R})$ which is preserved by every element of G.*

Proof. Take any positive-definite symmetric inner product on $V_n(\mathbb{R})$, and denote the product of $x, y \in V_n(\mathbb{R})$ by $x * y$. Let $\{\theta_1, \ldots, \theta_m\}$ be the set of elements of G, and write

$$x \cdot y := (x\theta_1 * y\theta_1) + \cdots + (x\theta_m * y\theta_m) = \sum_{i=1}^{m} x\theta_i * y\theta_i,$$

for all $x, y \in V_n(\mathbb{R})$. It is easily verified that the product $x \cdot y$ is symmetric and bilinear, from the corresponding properties of $x * y$. Also $x \cdot y \geq 0$ since $x\theta_i * x\theta_i \geq 0$, and $x \cdot x = 0$ only if $x\theta_i * x\theta_i = 0$ for all $\theta_i \in G$, in particular for $\theta_i = 1 \in G$, which implies that $x * x = 0$, and hence $x = 0$. The product $x \cdot y$ is therefore positive-definite. Also, for any $\theta \in G$ we have

$$x\theta \cdot y\theta = \sum_{i=1}^{m} x\theta\theta_i * y\theta\theta_i = \sum_{j=1}^{m} x\theta_j * y\theta_j = x \cdot y,$$

since when θ_i runs through all the elements of G, so does $\theta_j = \theta\theta_i$.

Lemma 15.2 tells us that, given any finite subgroup G of $GL(V_n(\mathbb{R}))$, we can find an inner product preserved by every element of G, and taking a basis for $V_n(\mathbb{R})$ which is orthonormal with respect to this inner product, each element of G is represented by an orthogonal matrix. Hence G is conjugate in $GL(V_n(\mathbb{R}))$ to a subgroup of $GL(V_n(\mathbb{R}))$ which is isomorphic to a finite subgroup of $O(n)$. Relative to an arbitrary basis of V_n the elements θ_i of G are represented by matrices of the form $T^{-1}A_i T$, where T is non-singular and the matrices A_i ($i = 1, \ldots, m$) are orthogonal.

The problem of finding finite subgroups of the real affine group or of the Euclidean group has therefore been reduced to that of finding finite subgroups of $O(n)$. The rest of the chapter is devoted to this problem, in the cases $n = 2$ and $n = 3$. We shall present the discussion in terms of $O(V_n)$, rather than the matrix group $O(n)$, in order to retain the free choice of orthonormal basis.

The case $n = 2$ gives little difficulty. If G is a finite subgroup of $O(V_2)$ then either $G \leqslant SO(V_2)$ and all its elements are rotations, or $H := G \cap SO(V_2)$

is a normal subgroup of G of index 2, and $G = H \cup \sigma H$, where $\sigma \in O(V_2) \backslash SO(V_2)$ is an axial reflexion.

If $G \leqslant SO(V_2)$, then relative to any orthonormal basis the elements of G are represented by matrices R_θ $(\theta \in \mathbb{R}/2\pi\mathbb{Z})$. The group G is isomorphic to a multiplicative group of such matrices, and as previously noted (see p. 137) any finite group of matrices R_θ is cyclic, of order k (say), and generated by R_α where $\alpha = 2\pi/k$. Hence $G \cong Z_k$, for some k.

If $G \nleqslant SO(V_2)$, then $G \cap SO(V_2) = H \cong Z_k$ for some k. The group H is generated by a rotation ρ of order k. If σ is any axial reflexion, then we can choose an orthonormal basis for V_2 which represents σ, ρ by J, R_α respectively, where, recall

$$J = \begin{bmatrix} 0 & 1 \\ -1 & 0 \end{bmatrix}, \qquad R_\alpha = \begin{bmatrix} \cos\alpha & \sin\alpha \\ -\sin\alpha & \cos\alpha \end{bmatrix},$$

and from the easily verified relation

$$J^{-1}R_\alpha J = R_{-\alpha} = (R_\alpha)^{-1}$$

we conclude that $\sigma^{-1}\rho\sigma = \rho^{-1}$. The group G is thus generated by σ, ρ which satisfy the relations

$$\sigma^2 = 1, \quad \rho^k = 1, \quad \sigma^{-1}\rho\sigma = \rho^{-1},$$

and is the dihedral group D_k. We may therefore state

THEOREM 15.3: (i)   *Any finite subgroup of* $SO(V_2)$ *is cyclic.*
(ii)   *Any finite subgroup of* $O(V_2)$ *is either cyclic or dihedral.*

We turn now to the more difficult but more interesting case of $n = 3$. The discussion for $n = 2$ suggests that it will be profitable to concentrate in the first instance on finite subgroups of $SO(V_3)$. The elements of $SO(V_3)$ are rotations, and every rotation of V_3 has an axis. If G is a finite subgroup of $SO(V_3)$, all of whose elements have the same axis, then G is easily identified with a subgroup of $SO(V_2)$. However, there are certainly finite subgroups of $SO(V_3)$ which are more genuinely 3-dimensional. If we consider any regular polyhedron such as the cube, the rotations which fix it will form a group. If the cube is represented in $V_3(\mathbb{R})$, with the zero vector at the centre of the cube, this group of rotations is a finite subgroup of $SO(V_3)$. We might search for finite subgroups of $SO(V_3)$ by looking for regular polyhedra and investigating their rotational symmetry groups. The best sort of regularity requires the polyhedron to be built up of faces, all of which are regular polygons with m sides, and to have n such polygons meeting at a vertex. These are called the *Platonic polyhedra*, and they attracted much attention from the later classical Greek mathematicians. It has been plausibly suggested that Euclid's *Elements* was designed as an introduction to the study of these figures. It has long been known that there are ony five solid figures of this kind, namely those shown in Fig. 15.1. To these we

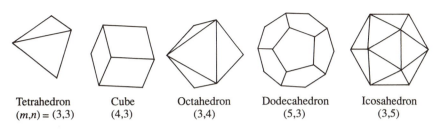

Tetrahedron   Cube   Octahedron   Dodecahedron   Icosahedron
$(m,n) = (3,3)$ (4,3)   (3,4)   (5,3)   (3,5)

Fig. 15.1. The Platonic polyhedra.

must add the *dihedron* and its dual. The dihedron with $n = 2$ and any $m \geqslant 3$ looks like a regular m-gon cut out of thin cardboard, but is thought of as a closed polyhedron with two distinct faces. It is often convenient to think of the regular polyhedra as figures on a sphere. For the solid polyhedra we obtain such a representation by inscribing the polyhedron in a sphere, and then projecting the polyhedral surface on the surface of the sphere by radial projection from the centre of the sphere. In the analogous representation of the dihedron, the m vertices are on a great circle, and the two faces are represented by the two hemispheres into which this great circle divides the sphere. The *dual dihedron* has two vertices, which are opposite to each other on the sphere, and n edges which are halves of great circles through those two points and which divide the sphere into equal segments (like segments of an orange).

Some geometrical information about these figures is given in Table 15.1.

Table 15.1. Combinatorial facts about the regular polyhedron.

	m	n	Number of:			
			Vertices	Edges	Faces	
Dihedron	k	2	k	k	2	$(k \geqslant 3)$
Dual dihedron	2	k	2	k	k	$(k \geqslant 3)$
Tetrahedron	3	3	4	6	4	
Cube	4	3	8	12	6	
Octahedron	3	4	6	12	8	
Dodecahedron	5	3	20	30	12	
Icosahedron	3	5	12	30	20	

Some information about the rotational symmetry groups of these figures is easily obtained. The possible axes are lines through the centre which also pass through a vertex, or the mid-point of an edge, or the centre of a face. The rotations with a given axis which belong to the rotational symmetry group form a cyclic subgroup of the symmetry group, and the order of this cyclic group is called the order of the axis. Each axis of order r accounts for $(r-1)$ elements $\neq 1$ of the symmetry group, so we can easily calculate the order of the symmetry group. Note that for the dihedron with k odd an axis through a vertex also passes through the mid-point of an edge, for the dual dihedron with k odd an axis through the mid-point of a face also passes through the mid-point of an edge and that for the tetrahedron an axis through a vertex also passes through the centre of a face. In all other cases, the polyhedron has central symmetry (that is to say, is fixed by the central reflexion), and the possible axes contain two vertices, or two mid-points of edges, or two centres of faces (Table 15.2).

Table 15.2. Rotational symmetries of the regular polyhedron.

	Number of axes of orders:					Order of the symmetry group	
	2	3	4	5	k		
Dihedron	k	—	—	—	1	$2k$	$(k \geqslant 3)$
Dual dihedron	k	—	—	—	1	$2k$	$(k \geqslant 3)$
Tetrahedron	3	4	—	—	—	12	
Cube	6	4	3	—	—	24	
Octahedron	6	4	3	—	—	24	
Dodecahedron	15	10	—	6	—	60	
Icosahedron	15	10	—	6	—	60	

The similarity which appears between the groups of the cube and octahedron, and between the groups of the dodecahedron and icosahedron, is not accidental. These pairs of groups are in fact isomorphic. The centres of the faces of a cube are the vertices of an octahedron, and the centres of the faces of an octahedron are the vertices of a cube. The rotational symmetry group of a cube therefore fixes an octahedron, and the rotational symmetry group of an octahedron fixes a cube. This shows that the groups of the cube and octahedron are isomorphic. Likewise the centres of the faces of a dodecahedron (icosahedron) are the vertices of an icosahedron (dodecahedron), and the groups of the dodecahedron and icosahedron are isomorphic. These are instances of *duality*: the cube and octahedron are dual polyhedra, the dodecahedron and icosahedron are dual to each other, the dihedron and dual dihedron with k edges are dual, and the tetrahedron is self-dual.

One well known finite rotation group has escaped our listing. This is the rotation group of a rectangle (with adjacent sides not equal), which is of order 4, the three elements $\neq 1$ being of order 2. It comes under the heading of the dihedron, as the case $k = 2$, but to obtain a sensible meaning of 'dihedron' for

$k = 2$, we must take a spherical representation. In such a representation, the dihedron has 2 diametrically opposite vertices, 2 edges which are the halves into which a great circle through the two vertices is divided by the vertices, and 2 faces which are the halves into which the spherical surface is divided by the great circle. This figure has the same rotation group as the rectangle.

All the distinct non-trivial finite rotation groups which we have found are listed in Table 15.3. It is easy to see that the rotational symmetry group of a dihedron is generated by a rotation ρ of order k, and a rotation σ of order 2, and that we have $\rho^k = 1, \sigma^2 = 1, \sigma^{-1}\rho\sigma = \rho^{-1}$. The symmetry group is thus in fact isomorphic to the dihedral group D_k with which we are already familiar. (This of course explains why the dihedral group is so named.)

Table 15.3. Finite rotation groups of three-dimensional space.

	Order	
Cyclic	k	(all $k \geqslant 2$)
Dihedral	$2k$	(all $k \geqslant 2$)
Tetrahedral	12	
Octahedral	24	
Icosahedral	60	

The question which now arises is whether our list contains all the finite subgroups of $SO(V_3)$. This is in fact so, but before we can be satisfied on this point, we must undertake a systematic search for such finite subgroups to assure ourselves that none has been missed. In conducting this investigation, it is convenient to think of $SO(V_3)$ (and $O(V_3)$) as operating on the *unit sphere S*, where

$$S := \{u \mid u \in V_3, \|u\| = 1\}.$$

Any vector $v \in V_3$ may be written as $v = \lambda u$, where $\lambda \in \mathbb{R}$ and $\|u\| = 1$, and for any $\theta \in O(V_3)$ we have $v\theta = \lambda(u\theta)$, so that the action of θ on V_3 is completely described by its action on the unit sphere. Any element of $SO(V_3)$ is a rotation, and has an axis which meets the sphere in two diametrically opposite points p, $-p$ of S; these points are called the *poles* of the rotation, and are the only fixed points on S of a rotation $\neq 1$. An element of $SO(3)$ which fixes more than two points of S (or fixes two points which are not diametrically opposite) is necessarily the identity.

Now consider a finite subgroup G of $SO(V_3)$; each of its elements $\neq 1$ has a pair of poles p, $-p$, and the set of elements of G with a given axis is a (finite) subgroup of G. Now this subgroup is cyclic, of order $m_p \,(= m_{-p})$, say; for the elements of $SO(V_3)$ with a given axis form a group isomorphic to $SO(V_2)$, and any finite subgroup of $SO(V_2)$ is cyclic. Hence with every pole is associated an order m_p.

If p is a pole of $\psi \in G$, then $p\phi$ is a pole of $\phi^{-1}\psi\phi \in G$, for all $\phi \in G$. Hence G acts on the set Ω of poles of its elements, and we may consider the G-space (Ω, G). The points of Ω are the only points of S with non-trivial stabilisers under the action of G. For any $p \in \Omega$, $\text{Stab}(p)$ is a cyclic group of order m_p, and hence the number of elements in the orbit of p is n/m_p, where $n = |G|$. Every pole p is a pole of $m_p - 1$ elements $\neq 1$ of G. If p^* is in the orbit of p, we have $p^* = p\phi$ for some $\phi \in G$, and

$$\text{Stab}(p^*) = \phi^{-1}(\text{Stab}(p))\phi \cong \text{Stab}(p),$$

so that $m_{p^*} = m_p$. Thus counting up over all the poles in the orbit of p, we have $n(m_p - 1)/m_p$ elements $\neq 1$ of G. If we count up over all the orbits, we shall have counted every element $\neq 1$ of G exactly twice, as it will have been counted in association with each of its two poles.

We therefore have

$$2(n-1) = \sum n(m_p - 1)/m_p,$$

where each orbit contributes one summand to the right-hand side of the equation, and m_p denotes the common order of the poles in that orbit. We rewrite this relation in the form

$$2\left(1 - \frac{1}{n}\right) = \sum\left(1 - \frac{1}{m_p}\right).$$

Remembering that n, $m_p \geq 2$, the left-hand side lies in the range $[1, 2)$ and each term of the right-hand side lies in the range $[\frac{1}{2}, 1)$, and hence there are either *two* or *three* orbits.

If there are two orbits, the relation may be written as

$$\frac{2}{n} = \frac{1}{m_p} + \frac{1}{m_q}.$$

Now, m_p is the order of a subgroup of G, so $m_p | n$; but if $m_p \leq n/2$, we have $1/m_q \leq 0$, so we can only have $m_p = m_q = n$. There is just one pole in each of the orbits, and Ω consists of a pair of diametrically opposite points. All elements $\neq 1$ of G have the same axis, and hence G is cyclic, and $G \cong Z_n$.

If there are three orbits, the relation may be written as

$$1 + \frac{2}{n} = \frac{1}{m_p} + \frac{1}{m_q} + \frac{1}{m_r}.$$

If the smallest of m_p, m_q, m_r is ≥ 3, we have $1 + (2/n) \leq 1$; hence one, m_p say, is 2, and we have

$$\frac{1}{2} + \frac{2}{n} = \frac{1}{m_q} + \frac{1}{m_r}.$$

If m_q (say) is 2, then the relation is satisfied by $m_r = k$, $n = 2k$ for all $k \geqslant 2$. If neither m_q nor m_r is 2, then one (m_q say) must be 3; for if m_q, m_r are both $\geqslant 4$, we have $(2/n) \leqslant 0$. We then have

$$\frac{1}{m_r} = \frac{1}{2} - \frac{1}{3} + \frac{2}{n} > \frac{1}{6},$$

giving $m_r < 6$, which restricts the possible values of m_r to $m_r = 3, 4, 5$. We can now calculate n for all these possible values of m_p, m_q, m_r, and find that they do in fact give integer values for n, divisible by m_p, m_q and m_r. The possible numerical solutions for the three-orbit case are set out in Table 15.4.

Table 15.4. Solutions of the equation for the three-orbit case.

| Case | m_p | m_q | m_r | n | Number of elements in the orbit | | |
					p	q	r
(i)	2	2	k	$2k$	k	k	2	($k \geqslant 2$)
(ii)	2	3	3	12	6	4	4
(iii)	2	3	4	24	12	8	6
(iv)	2	3	5	60	30	20	12

Now, we have already shown that rotation groups with these numerical properties exist, for cases (i)–(iv) are realised by the dihedral, tetrahedral, octahedral, and icosahedral groups respectively. We still need to satisfy ourselves that they can be realised in no other way: thus for example we must show that a finite group of rotations which falls under case (i) is the rotational symmetry group of some dihedron with k vertices. The geometrical figures which we seek are provided by orbits of poles.

In case (i), there is a 2-element orbit of poles of order k. If $k > 2$, these are the only poles of order k, and such poles come in diametrically opposite pairs. Hence the two poles must lie on the same axis, and have the same stabiliser, which must be a cyclic group of order k generated by a rotation ϕ about that axis. Any other rotation $\neq 1$ in the group G is of order 2 and must interchange the two poles of order k; it must therefore be a rotation of angle π about an axis in the equatorial plane. Hence all poles other than the two of order k must lie on the equator of the sphere. If p is such a pole, the k distinct points $p\phi^r$ ($r = 0, 1, \ldots, k-1$) are in the orbit of p and hence are the whole of the orbit of p. These k poles are the vertices of a regular k-gon. If now we consider the dihedron having these vertices, it is clearly fixed by all elements of G, hence G is contained in the rotational symmetry group of the dihedron.

But these two groups have the same order $2k$, hence G coincides with the rotational symmetry group of the dihedron.

If $k = 2$, we have 6 poles forming 3 orbits, each with 2 points. The group G is of order 4, and necessarily abelian; in fact, as all its elements $\neq 1$ are of order 2, we know that $G \cong Z_2 \times Z_2$. If a pole p is fixed by $\theta \in G$, and $\phi \neq 1$ is any other element of G, then $(p\phi)\theta = (p\theta)\phi = p\phi$. Hence $p\phi$ is fixed by θ, and $p\phi \neq p$, so $p\phi = -p$. Hence the axes of θ and ϕ are orthogonal, and G consists of the rotations through π about three mutually orthogonal axes, together with the identity. A suitable dihedron is obtained by taking as its vertices any pair of diametrically opposite poles, and as edges the halves of the great circle containing these poles and one other pair of poles.

In cases (ii), (iii), and (iv) we look at the smallest orbit of poles, containing 4, 6, and 12 points respectively. In case (ii) it is relatively easy to show that the set of four points must be the vertices of a regular tetrahedron. Each point is a pole of order 3, and a rotation $\neq 1$ which fixes one of them must permute the other three cyclically (since it cannot fix all of them). The three points are therefore the vertices of an equilateral triangle, and this holds for every subset of three points in the set of four. The four points are therefore the vertices of a regular tetrahedron. The group G fixes the tetrahedron and is therefore contained in its rotational symmetry group; and as the two groups are both of order 12, G coincides with the rotational symmetry group of the tetrahedron.

We need a rather more elaborate argument to deal with cases (iii) and (iv). For any finite set S of points on the unit sphere there is a minimum distance d between two distinct points of S, and two points of S which are at this minimum distance from each other will be called *adjacent*. If now S is an orbit of poles, and $p, q \in S$ are adjacent, then so are $p\phi, q\phi$ for all $\phi \in G$. Since G acts transitively on S, every point of S is adjacent to some other point of S. In fact, if p is a pole of order m, then G contains a rotation ψ of order m which fixes p, and if q is adjacent to p, so are all the m points $q\psi^r$ ($r = 0, 1, \ldots, m-1$). We shall establish the pattern of the set S of points in the orbit of poles by investigating the adjacency relationships of points of S. Our discussion of case (ii) showed that in the set of 4 points of S every point is adjacent to every other, which is just another way of saying that the four points are the vertices of a regular tetrahedron.

In case (iii), we have S consisting of 6 poles of order 4. A pole p has 4 adjacent poles $r, s, t, u = r, r\psi, r\psi^2, r\psi^3$, where $\psi^4 = 1$. These lie at the vertices of a square, and $d(r, r\psi^2) = \sqrt{2} \cdot d(r, r\psi)$, so $r\psi^2$ is not adjacent to r. There are 4 poles adjacent to r, and these must be $p, s \, (= r\psi)$, $u \, (= r\psi^3)$ and the sixth pole v. Likewise, s is adjacent to p, r, t, v; t is adjacent to p, s, u, v; and u is adjacent to p, r, t, v. The adjacency relations are exhibited in Fig. 15.2 and show that the points p, r, s, t, u, v are the vertices of a regular octahedron.

In case (iv), S consists of 12 poles of order 5. Each pole p has 5 adjacent poles $q, r, s, t, u \, (= q, q\psi, q\psi^2, q\psi^3, q\psi^4$, where $\psi^5 = 1)$. These lie at the vertices of a regular pentagon, and $d(q, s) = d(q, t) > d(q, r) = d(q, u)$. The pole q is not

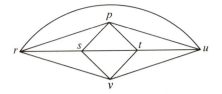

Fig. 15.2. Adjacency relations of 6 poles of order 4.

adjacent to s or t. It is either adjacent to both r and u or to neither of them. In the latter case, we know that for any pole no two of the five adjacent poles are adjacent to each other. This situation however soon leads us to a contradiction. The five poles adjacent to q must be p, q_1, q_2, q_3, q_4, and the five poles adjacent to r must be p, r_1, r_2, r_3, r_4. Since there are only 12 poles in all, the sets $\{q_1, \ldots, q_4\}$ and $\{r_1, \ldots, r_4\}$ must have at least two elements in common, and the regular pentagons $\{p, q_1, \ldots, q_4\}$ and $\{p, r_1, \ldots, r_4\}$ must coincide, which is impossible since q, r are distinct and not diametrically opposite. Hence q is adjacent to r and u, and it is now easy to work out the adjacency relationships for all the 12 points of S, which are indicated in Fig. 15.3 and which ensure that the 12 points are the vertices of a regular icosahedron.

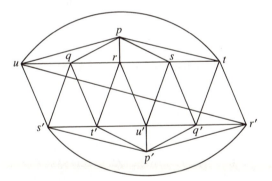

Fig. 15.3. Adjacency relations of 12 poles of order 5.

Thus, in case (iii) the group G fixes an octahedron and is therefore contained in, and from consideration of its order coincides with, the rotational symmetry group of this octahedron. Similarly, in case (iv) the group G is the rotational symmetry group of an icosahedron. We can now state

THEOREM 15.4: *Any finite subgroup of* $\mathrm{SO}(V_3)$ *is either* (i) *cyclic,* (ii) *dihedral,* (iii) *tetrahedral,* (iv) *octahedral, or* (v) *icosahedral.*

We now examine further the structure of the tetrahedral, octahedral, and icosahedral symmetry groups. We identify these groups with permutation groups by finding sets of geometrical objects on which the group acts faithfully. If we can find such a set of n objects, then we have a monomorphism $G \rightarrow S_n$ which identifies G with a subgroup of S_n. From consideration of the order of G, we see that we must have $n \geqslant 4$ for the tetrahedral and octahedral groups, and $n \geqslant 5$ for the icosahedral group. We shall therefore look for a set of 4 objects in the cases of the tetrahedral and octahedral groups, and a set of 5 objects in the case of the icosahedral group.

For the tetrahedral group, we may take the set of 4 vertices of the tetrahedron. Since no element $\neq 1$ of G fixes each of the vertices, G acts faithfully on this set, and G is isomorphic to a subgroup of S_4. The only such subgroup of order 12 is in fact the alternating group A_4; but without appealing to this result, it can be seen that the elements $\neq 1$ of G acting on the 4 vertices produce permutations which are either 3-cycles or the product of two disjoint 2-cycles, and are therefore even permutations. This means that the monomorphism $G \rightarrow S_4$ is in fact a monomorphism into A_4, and from consideration of the order of G, $G \cong A_4$.

For the octahedral group, we take the set of 4 axes through the centres of the faces. An element $\neq 1$ of G which fixes two of these axes must be a rotation through π about the axis orthogonal to both of them, and hence there is no element $\neq 1$ of G which fixes each of the 4 axes. Hence G acts faithfully on the set of 4 axes, and we have a monomorphism $G \rightarrow S_4$, which (since G and S_4 are both of order 24) must be an isomorphism. We thus have $G \cong S_4$.

For the icosahedral group, the choice of a set of 5 objects may not be immediately obvious. There are however several such sets, all closely related geometrically, on which G acts faithfully. We shall content ourselves with the description of one of them. The 30 mid-points of edges (which lie on 15 axes) fall into 5 sets of 6, the points in each set being the vertices of a regular octahedron. To see this, we need only verify that, given one of the 15 axes, there are just two others which together with the given axis form a mutually orthogonal set.

The first picture in Fig. 15.4 shows a view of the icosahedron, looking along the axis (p, p'). The axis (q, q') is orthogonal to this axis. At each of x and y is a

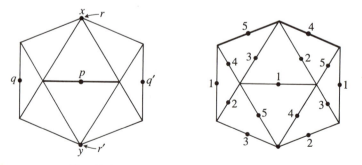

Fig. 15.4. Octahedral sets in an icosahedron.

fifth (non-visible) edge parallel to the line of sight, and the mid-points r, r' of these edges are diametrically opposite. The axes (p, p'), (q, q'), (r, r') are a mutually orthogonal set and the points p, p', q, q', r, r' are an octahedral set. The second picture in Fig. 15.4 is the same view of the icosahedron with the visible mid-points of edges numbered so that those in the same octahedral set carry the same number. With the help of this picture, we can examine the effect on the 5 octahedral sets of typical rotations of orders 2, 3, and 5 in the icosahedral group G, and see that they produce permutations of the 5 sets which are respectively a product of two 2-cycles, a 3-cycle, and a 5-cycle. These are all even permutations, and no element $\neq 1$ of G fixes each of the octahedral sets; hence we have a monomorphism $G \rightarrow A_5$, and (since G and A_5 are both of order 60) this shows that $G \cong A_5$.

We have now completed our investigation of the finite subgroups of $SO(V_3)$, and turn to the consideration of finite subgroups of $O(V_3)$ which are not subgroups of $SO(V_3)$. Such a group G contains elements of determinant -1, and the homomorphism $G \rightarrow \{1, -1\} \cong Z_2$ given by $\theta \mapsto \det \theta$ is onto, with kernel $H \leqslant SO(V_3)$. The subgroup H is normal of index 2 in G, and $G = H \cup \sigma H$, where $\sigma \in G \backslash H$.

The central reflexion, which as a linear transformation of V_3 is the transformation -1, will be denoted by σ_0. This transformation is in the centre of $O(V_3)$. Two cases now arise, depending on whether G does or does not contain σ_0.

First, suppose that $\sigma_0 \in G$. In this case, we may write $G = H \cup \sigma_0 H$, and since σ_0 commutes with all elements of $O(V_3)$ we have

$$G = H \times \{1, \sigma_0\} \cong H \times Z_2.$$

Secondly, suppose that $\sigma_0 \notin G$. We write $G = H \cup \sigma H$, and consider $G^* = H \cup \sigma_0 \sigma H$. Now G^* is a group (since σ_0 commutes with everything), and is moreover a group of rotations, and is therefore one of the known finite subgroups of $SO(V_3)$. Also, G is isomorphic to G^* under the mapping $G \rightarrow G^*$ given by

$$\psi \mapsto \psi, \quad \sigma\psi \mapsto \sigma_0\sigma\psi, \quad \text{for all } \psi \in H.$$

Now this shows that H can not be octahedral or icosahedral, as we should then have a rotation group of order 48 or 120, and the only such rotation groups are cyclic or dihedral. It is a trivial fact that a subgroup of a cyclic group is cyclic, and it is almost trivial that a subgroup of a dihedral group is either dihedral or cyclic. The same argument tells us that if H is tetrahedral, then G^* can only be octahedral. Further, if G^* is tetrahedral, octahedral, or icosahedral, it can not contain a cyclic subgroup H of index 2, as this would imply the existence in G^* of elements of orders 6, 12, 30 respectively. Similarly, if G^* contains a dihedral subgroup of index 2, this implies the existence in G^* of elements of orders 3, 6, 15 respectively, which rules out this possibility immediately when G^* is octahedral or icosahedral. We have to look a little more closely to see that a tetrahedral group does not contain a dihedral subgroup $\cong D_3$. Such a subgroup of A_4 would have to contain all the three elements (12) (34), (13) (24), (14) (23) of order 2 in A_4; but these elements with the identity form a group, and D_3 (of order 6) can have no subgroup of order 4. We are therefore left with the following possibilities (Table 15.5).

Table 15.5. Possibilities for finite subgroups of $O(V_3)$ not contained in $SO(V_3)$ and not containing σ_0.

H		$G^* (\cong G)$	
Cyclic,	$\cong Z_k$	Cyclic,	$\cong Z_{2k}$
		Dihedral,	$\cong D_k$
Dihedral,	$\cong D_k$	Dihedral,	$\cong D_{2k}$
Tetrahedral,	$\cong A_4$	Octahedral,	$\cong S_4$

It remains to show that all these possibilities can be realised by subgroups G of $O(V_3)$ which are not rotation groups. To deal with the first three cases, select any axis, and define

$\sigma_1 :=$ rotatory reflexion of angle π/k, with the given axis;
$\sigma_2 :=$ reflexion in a plane containing the given axis;
$\rho_1 :=$ rotation of angle $2\pi/k$, with the given axis;
$\rho_2 :=$ rotation of angle π with axis orthogonal to the given axis.

We now have

$$\langle \sigma_1 \rangle \cong Z_{2k}, \quad \text{with rotation subgroup } \langle \sigma_1^2 \rangle = \langle \rho_1 \rangle \cong Z_k$$
$$\langle \rho_1, \sigma_2 \rangle \cong D_k, \quad \text{with rotation subgroup } \langle \rho_1 \rangle \cong Z_k$$
$$\langle \sigma_1, \rho_2 \rangle \cong D_{2k}, \quad \text{with rotation subgroup } \langle \sigma_1^2, \rho_2 \rangle = \langle \rho_1, \rho_2 \rangle \cong D_k.$$

Finally, we observe that the full symmetry group (that is, the stabiliser in $O(V_3)$) of a regular tetrahedron is $\cong S_4$, as it acts faithfully on the set of 4 vertices, and induces all permutations of the 4 vertices. (A 2-cycle is induced by reflexion in a plane containing an edge, and all 2-cycles can be obtained in this way.) The rotation subgroup of this group is the rotational symmetry group of the tetrahedron, which is isomorphic to A_4. We may now state

THEOREM 15.5: *Any finite subgroup of $O(V_3)$ is either*
(i)   *a subgroup of $SO(V_3)$,*
(ii) *generated by a subgroup $SO(V_3)$ and the central reflexion σ_0,*
(iii) *cyclic,*
(iv) *isomorphic to a dihedral group, or*
(v)   *isomorphic to S_4.*

Our information on the structures of finite subgroups of $SO(V_3)$ and of $O(V_3)$ is set out in Table 15.6.

Table 15.6. The finite subgroups of $SO(V_3)$ and of $O(V_3)$.

	$O(V_3)$	
$SO(V_3)$	(a)	(b)
Z_k	$Z_k \times Z_2$	Z_{2k}
		D_k
D_k	$D_k \times Z_2$	D_{2k}
A_4	$A_4 \times Z_2$	S_4
S_4	$S_4 \times Z_2$	
A_5	$A_5 \times Z_2$	

We conclude the chapter by looking at the full symmetry groups of the Platonic polyhedra. If the polyhedron has central symmetry, the full symmetry group contains the central reflexion σ_0, and is one of the groups listed in column (a) of the table; otherwise we obtain a group listed in column (b). Thus without difficulty we obtain Table 15.7.

Table 15.7. Symmetry groups of the Platonic polyhedra.

Figure	Rotational symmetry group	Full symmetry group
Dihedron (k odd)	D_k	D_{2k}
Dihedron (k even)	D_k	$D_k \times Z_2$
Tetrahedron	A_4	S_4
Octahedron	S_4	$S_4 \times Z_2$
Icosahedron	A_5	$A_5 \times Z_2$

Exercises

15.1. Describe, as a figure on the unit sphere, the dihedron with $k = 1$ (that is, a polyhedron with one vertex, one edge, and two faces). Find the rotational and full symmetry groups of this figure.

15.2. Find the structure of the group $\langle \sigma_1, \sigma_2 \rangle$, where σ_1, σ_2 are the transformations defined on p. 184.

15.3.   The eight faces of a double pyramid on a square base are congruent isosceles (not equilateral) triangles. Find the rotational and full symmetry groups of this figure.

15.4.   Describe geometrically the isometry which permutes cyclically the four vertices of a regular tetrahedron.

15.5.   The rhombic dodecahedron has 12 faces, each of which is a rhombus with diagonals of lengths a and $a\sqrt{2}$. Satisfy yourself that such a closed polyhedron exists, and draw a picture of it.
   Find the rotational and full symmetry groups of the rhombic dodecahedron.

15.6.   Show that it is possible to pick out of the 20 vertices of a regular dodecahedron a set of 8 which are the vertices of a cube, and that this can be done in 5 different ways, each vertex of the dodecahedron being used twice. (This is a set of 5 objects which can be used to show that the rotational symmetry group of the dodecahedron is isomorphic to A_5.)

15.7.   Show that the 30 edges of an icosahedron fall into 5 sets, each set consisting of 3 pairs of parallel edges, the three directions being mutually orthogonal. (This is another set of 5 figures on which the icosahedral group acts faithfully.)

15.8.   Show that the distance between two non-adjacent vertices of a regular pentagon of side a is $a\tau$, where $\tau = \frac{1}{2}(1 + \sqrt{5})$.
   Hence show that two opposite edges of an icosahedron are the short sides of a rectangle whose sides are in the ratio $\tau:1$.
   Show that the points with coordinates (in a rectangular cartesian system)

$$(0, \pm 1, \pm\tau), \quad (\pm\tau, 0, \pm 1), \quad (\pm 1, \pm\tau, 0)$$

are the vertices of a regular icosahedron.

   [The 'golden number' τ satisfies the equation $\tau^2 - \tau - 1 = 0$. A rectangle with sides a, $a\tau$ is called a 'golden rectangle'. The removal of a square of side a from such a rectangle leaves another golden rectangle, of sides $a\tau^{-1}$, a (since $\tau - 1 = \tau^{-1}$). The result of Exercise 15.8 is the 'twelfth almost incomprehensible property' of the number τ listed by Luca Pacioli in his treatise *De Divina Proportione* published in 1509.]

15.9.   Draw a picture of three golden rectangles formed by the vertices of an icosahedron.
   Find the length of an edge of an icosahedron inscribed in the unit sphere.

15.10. Draw a picture of the orthogonal projection of an icosahedron on a plane containing two opposite edges. Indicate on your picture any distances equal to τ, the edge length of the icosahedron being taken as 1.

15.11. The rotational symmetry group of the icosahedron acts in a natural way on the set of 6 axes through the vertices. Show that the action is faithful and transitive, and establishes a monomorphism $A_5 \to A_6$ (cf. p. 90).

15.12. The application to a regular tetrahedron inscribed in a sphere of a central reflexion (with centre at the centre of the sphere) produces a second tetrahedron. Show that the 8 vertices of the two tetrahedra are the vertices of a cube.

Show that both tetrahedra have the same rotational symmetry group, and that this is a normal subgroup of the rotational symmetry group of the cube.

15.13. Show that the subgroup of the icosahedral group which fixes one of the octahedra described on pp. 182–3 is tetrahedral. Find two tetrahedra of which this subgroup is the rotational symmetry group. (Describe the two tetrahedra with the help of a picture such as those on p. 182.)

15.14. Find the structures of all subgroups of the icosahedral group, showing in each case that such a subgroup exists.

15.15. Find the subgroup of the icosahedral group which is generated by an element of order 2 and an element of order 3, distinguishing the possible cases.

Chapter 16

Complex numbers and quaternions

Complex numbers first appeared in the early 17th century as a way of expressing the 'impossible' roots of an algebraic equation with real coefficients with the help of the 'imaginary' square root i of -1. The representation of a complex number $a + ib$ by the point of the Euclidean plane with coordinates (a, b) in a rectangular cartesian coordinate system, known to us as the Argand diagram, appeared at the end of the 18th century, and was of vital importance to the development of the theory of functions of a complex variable which took place in the 19th century. Although the word 'imaginary' persists in the terminology of 'real part' and 'imaginary part', complex numbers ceased to be mysterious, and as more confidence was gained in handling them it was possible not only to use the Argand representation in reverse, as a way of assigning a complex coordinate to each point of the Euclidean plane, but also to think of extending the notion of complex numbers to 'hyper-complex' numbers which would serve a similar purpose in Euclidean space of higher dimension. It is these developments which concern us here.

First we give some examples of the use of complex coordinates in plane geometry. In terms of such coordinates, a direct similarity transformation is given by $z \mapsto az + b$ ($a, b, \in \mathbb{C}$, $a \neq 0$); the full similarity group includes the axial reflexions, and we obtain this group by adding the reflexion $z \mapsto \bar{z}$ to the generators, obtaining the group of all transformations $z \mapsto az + b$ and $z \mapsto a\bar{z} + b$. We shall be interested in the more restricted group of *similitudes*, which are transformations of the form $z \mapsto \lambda z + a$ ($\lambda \in \mathbb{R}$, $\lambda \neq 0$, $a \in \mathbb{C}$). A similitude will be called *positive* or *negative* according as $\lambda > 0$ or $\lambda < 0$. A similitude with $\lambda = 1$ is a translation; every other similitude has a fixed point with coordinate $(1 - \lambda)^{-1}a$. A similitude takes a circle Γ_1 to a circle Γ_2 and takes the centre O_1 of Γ_1 to the centre O_2 of Γ_2. If Γ_1, Γ_2 have unequal radii r_1, r_2 there are two similitudes which map Γ_1 to Γ_2. If Γ_1, Γ_2 have the same centre, this must be a fixed point, and taking it as the origin, the two similitudes are $z \mapsto \pm \lambda z$ where $\lambda = r_2/r_1$. If the centres have coordinates 0 and c respectively, the two similitudes are $z \mapsto \pm \lambda z + c$. The fixed points of these have coordinates $(1 - \lambda)^{-1}c$, $(1 + \lambda)^{-1}c$; these are called the centres of similitude of Γ_1, Γ_2. The inverse transformations mapping Γ_2 to Γ_1 have the same fixed point. Both points lie on the line joining the centres O_1, O_2, the centre of the positive similitude lies outside, and that of the negative similitude lies inside, the segment $O_1 O_2$. If a centre of similitude lies on one of the circles, it necessarily lies on the other, and as it also lies on the line of centres, the two circles touch at that point (internal or external contact according as the similitude is positive or negative).

Now we tackle the geometry of the triangle. We denote points by capital letters and their complex coordinates by the corresponding small letters. The origin is taken at the centre S of the circumcircle of the triangle ABC. The midpoints of the sides are D, E, F, so that $d = \frac{1}{2}(b+c)$, $e = \frac{1}{2}(c+a)$, $f = \frac{1}{2}(a+b)$ (see Fig. 16.1). Now, $|a| = |b| = |c| = R$; we shall make extensive use of the

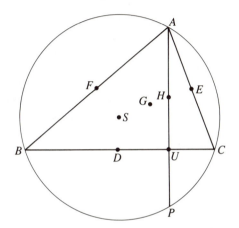

Fig. 16.1. A triangle and its circumcircle.

observation that, if $|a| = |b|$ then $(a+b) = \lambda(ab)^{1/2}$, $a-b = \lambda i(ab)^{1/2}$, where $\lambda \in \mathbb{R}$. (There is an ambiguity of sign in $(ab)^{1/2}$, but the statement is valid whichever square root is taken.) Let the perpendicular from A to BC meet BC in U and the circumcircle in P (V, W and Q, R are similarly defined on the altitudes through B, C). We claim that $p = -bc/a$; for this gives $|p| = R$, whence P is on the circumcircle and $a-p = a+(bc/a) = \lambda(bc)^{1/2}$, and $b-c = \kappa i(bc)^{1/2}$ (λ, $\kappa \in \mathbb{R}$), whence AP is perpendicular to BC. Now let $h := a+b+c$; then $h-a = b+c = \rho(bc)^{1/2}$ ($\rho \in \mathbb{R}$), so H lies on AP. Similarly H lies on the other altitudes BQ, CR; hence H is the orthocentre of the triangle. The centroid G of the triangle has coordinate $g = \frac{1}{3}(a+b+c)$, and consequently S, G, H are collinear; this line is the *Euler line* of the triangle.

Now consider the similitude $z \mapsto \frac{1}{2}(z+a+b+c) = \frac{1}{2}(z+h)$. This has fixed point H. The point on the circumcircle diametrically opposite to A has coordinate $-a$, and is mapped to D. Hence the circumcircle is mapped to a circle through D (and similarly, through E, F). The origin S is mapped to the point N where $n = \frac{1}{2}(a+b+c)$, so that G (the midpoint of SH) is the centre of the circle through D, E, F. The radius of the circle is $\frac{1}{2}R$. The midpoints A', B', C' of HA, HV, HC also lie on the circle. Finally, we claim that U is the midpoint of HP, by showing that the midpoint given by

$$u = \frac{1}{2}(h+p) = \frac{1}{2}\left(a+b+c-\frac{bc}{a}\right)$$

lies on BC; for we have

$$u-d = \frac{1}{2}\left(a-\frac{bc}{a}\right)$$

and $b-d = \frac{1}{2}(b-c)$, and both are real multiples of $i(bc)^{1/2}$. We have thus shown that the nine points D, E, F, A', B', C', U, V, W all lie on the circle, which is therefore known as the *nine-point circle*. The other centre of similitude of the circumcircle and nine-point circle is the fixed point of $z \mapsto \frac{1}{2}(-z+a+b+c)$, with fixed point given by $\frac{1}{3}(a+b+c) = g$.

We also observe that H may lie outside the triangle ABC. This occurs only if the triangle is obtuse-angled, in which case H lies outside the circumcircle. if the triangle is right-angled, at A say, then H coincides with A. Thus the triangle is acute-angled, right-angled or obtuse-angled according as $|h| < R, |h| = R$ or $|h| > R$. We always have $|h| < |a| + |b| + |c| = 3R$ and $|g| = \frac{1}{3}|h| < R$, a result which also follows from the fact that G always lies inside the triangle ABC.

If the tangents to the circumcircle at B and C meet at T, then $t-b = \lambda ib$, $t-c = \mu ic$, where $\lambda, \mu \in \mathbb{R}$, and since $|t-b| = |t-c|$, we have $|\mu| = |\lambda|$, and since $\mu = \lambda$ implies $c-b = \lambda i(b-c)$, we must have $\mu = -\lambda$, and $2t-(b+c) = \lambda i(b-c) = \kappa(b+c)$ for some $\kappa \in \mathbb{R}$, and $2t = (1+\kappa)(b+c)$. Hence, T lies on SD, and from the similarity of the triangles SBD, STB we have $SB/SD = ST/SB$, and $|t||d| = R^2$. If $d = \rho e^{i\theta}$ then

$$t = (R^2/\rho)e^{i\theta} = R^2/(\rho e^{-i\theta}) = R^2/\bar{d} = 2R^2/(\bar{b}+\bar{c}).$$

Now $b\bar{b} = c\bar{c} = R^2$, so we have $t = 2/(b^{-1}+c^{-1}) = 2bc/(b+c)$.

The results obtained so far are not at all difficult to achieve by arguments of elementary Euclidean geometry. The theorem which follows is more difficult, and provides more convincing evidence of the usefulness of complex coordinates.

THEOREM 16.1 [FEUERBACH'S THEOREM]:   *The nine-point circle of a triangle touches the incircle and the three excircles.*

Proof. Let Γ be a circle which touches the sides of a triangle with vertices, V_1, V_2, V_3 at T_1, T_2, T_3 on V_2V_3, V_3V_1, V_1V_2 respectively (see Fig. 16.2). The triangle $T_1T_2T_3$ cannot be right-angled, since if it is right-angled at T_1 (say), then T_2, T_3 are diametrically opposite points of Γ, and the tangents to Γ at T_2, T_3 are parallel. If $T_1T_2T_3$ is acute-angled, then Γ is the incircle of the triangle $V_1V_2V_3$; if $T_1T_2T_3$ is obtuse-angled, Γ is an excircle.

Let U_1, U_2, U_3 be the midpoints of T_2T_3, T_3T_1, T_1T_2, and let W_1, W_2, W_3 be the midpoints of V_2V_3, V_3V_1, V_1V_2. Take the origin at the centre of Γ, and

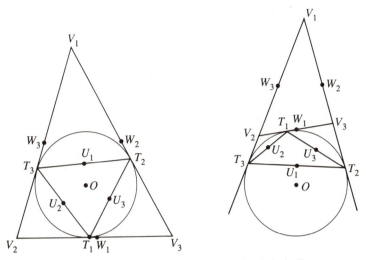

Fig. 16.2. Triangle $V_1V_2V_3$ with inscribed circle Γ.

take the radius of Γ as the unit of distance. We have (from a previous result)

$$v_1\bar{u}_1 = 1, \quad u_1 = \tfrac{1}{2}(t_2+t_3), \quad \bar{u}_1 = \tfrac{1}{2}(\bar{t}_2+\bar{t}_3) = \tfrac{1}{2}(t_2^{-1}+t_3^{-1}),$$

whence $v_1 = (\bar{u}_1)^{-1} = 2t_2t_3/(t_2+t_3)$. This gives

$$w_1 = t_3t_1/(t_3+t_1)+t_1t_2/(t_1+t_2)$$
$$= (t_1^2t_2+t_1^2t_3+2t_1t_2t_3)/(t_3+t_1)(t_1+t_2).$$

We now aim to express this conveniently in terms of t_1 and the elementary symmetric polynomials

$$p = t_1+t_2+t_3, \quad q = t_2t_3+t_3t_1+t_1t_2, \quad r = t_1t_2t_3.$$

(We must remember that p, q, r, t_1 are not independent, but satisfy the relation $t_1^3 - pt_1^2 + qt_1 - r = 0$.) We first make the denominator symmetric, writing

$$w_1 = (t_1q+r)(p-t_1)/\{(p-t_1)(p-t_2)(p-t_3)\},$$

and have $(p-t_1)(p-t_2)(p-t_3) = p^3-p^2p+pq-r = pq-r$. Also, from the relation noted above, $t_1^2(p-t_1) = qt_1-r$, so the numerator can be written as $(qt_1+r)(qt_1-r)t_1^{-2} = q^2-r^2t_1^{-2}$. Now we have $\bar{p} = \bar{t}_1+\bar{t}_2+\bar{t}_3 = t_1^{-1}+t_2^{-1}+t_3^{-1} = q/r$, so $pq-r = r(|p|^2-1)$, giving

$$w_i = \{(r/t_i^2)-(q^2/r)\}/(1-|p|^2) \quad \text{for } i = 1, 2, 3.$$

If $s_i := r/t_i^2$, then $|s_i| = 1$, and s_i represents a point S_i on Γ. Writing $\lambda = (1-|p|^2)^{-1}$, the transformation $z \mapsto \lambda(z-(q^2/r))$ is a similitude (since $\lambda \in \mathbb{R}$) which maps S_1, S_2, S_3 on Γ to W_1, W_2, W_3 on the nine-point circle of the triangle $V_1V_2V_3$. The fixed point of the similitude has coordinate

$q^2/(r|p|^2)$, and since $|r| = 1$ and $|q| = |\bar{p}r| = |p|$, this coordinate is of unit modulus and the point is on Γ. It follows that Γ touches the nine-point circle at this point.

The quaternion algebra, devised by W. R. Hamilton and published in 1844, was the outcome of an attempt to construct for three-dimensional Euclidean space a system of 'numbers' which would copy the relationship of complex numbers with 2-dimensional space. Thus he sought in the first instance to represent a point with coordinate (a, b, c) in a rectangular cartesian system by a 'number' $u = a + ib + jc$, and to define a multiplication of such numbers which would extend that of the complex numbers. This suggested $i^2 = -1$ and $j^2 = -1$ (to secure reduction to the complex numbers for points $(a, b, 0)$ and $(a, 0, c)$). If we define the conjugate $\bar{u}$ of u as $\bar{u} = a - ib - jc$, we obtain

$$u\bar{u} = \bar{u}u = a^2 + b^2 + c^2 + (ij + ji)bc,$$

which suggests $ij + ji = 0$, which assuming commutativity, gives $ij = ji = 0$. Hamilton aimed to define a multiplication with the property that

$$(a + ib + jc)(a' + ib' + jc') = A + iB + jC,$$

where $A^2 + B^2 + C^2 = (a^2 + b^2 + c^2)(a'^2 + b'^2 + c'^2)$, but this is not the case if $ij = ji = 0$. It was necessary to abandon commutativity, and write $ij = -ji = k$, giving a product of the form $(a + ib + jc)(a' + ib' + jc') = A + iB + jC + kD$. At first, Hamilton hoped to identify k with an element $\alpha + i\beta + j\gamma$ corresponding to a point (α, β, γ) in the three-dimensional space, but he was eventually forced to the conclusion that this was impossible, and that a fourth dimension was necessary to accommodate k. In fact, Hamilton had achieved for four dimensions what he had aimed to do for three. We now give a sightly more formal account of the quaternion algebra.

Let $V_4(\mathbb{R})$ be a real vector space of dimension 4, on which a positive definite symmetric bilinear form is defined, and let e, i, j, k be an orthonormal basis. We now define a multiplication of elements of V_4, stipulating bilinearity, that is to say

$$(\lambda x + \mu y)z = \lambda xz + \mu yz, \quad z(\lambda x + \mu y) = \lambda zx + \mu zy.$$

The multiplication is then defined for all elements by the multiplication of the basis elements, which is given by the following table.

	e	i	j	k
e	e	i	j	k
i	i	$-e$	k	$-j$
j	j	$-k$	$-e$	i
k	k	j	$-i$	$-e$

The multiplication of all elements is associative if (and only if) the multiplication of basis elements is associative. To check that this is so may appear to be a formidable task, involving the proof of 64 relations of the form $(ab)c = a(bc)$, with $a, b, c \in \{e, i, j, k\}$, but it is much curtailed by the observation that e is a unit element, and the relation is immediate if a or b or c is e, so reducing the number to the 27 relations involving only i, j, k. We now observe that the multiplication table is unchanged by cyclic permutation of i, j, k, so it is sufficient to prove it for all cases in which $a = i$, reducing the number to 9, which we leave the reader to check.

With this definition of multiplication (and the pre-existing definition of addition in V_4), the elements of V_4 become a ring of *quaternions*, which (in tribute to Hamilton, and/or because $\mathbb{Q}$ has already been assigned to the field of rationals) is denoted by $\mathbb{H}$. If associativity of multiplication is assumed, the basic relations can be condensed to $i^2 = j^2 = k^2 = -ijk = -e$ where e is a unit element, for we can then deduce the whole multiplication table.

The quaternion algebra $\mathbb{H}$ has several points of resemblance to the complex numbers $\mathbb{C}$. Thus for a quaternion $q = \lambda e + \alpha i + \beta j + \gamma k$, we define the conjugate $\bar{q}$ of q by $\bar{q} = \lambda e - \alpha i - \beta j - \gamma k$, and obtain

$$q\bar{q} = \bar{q}q = (\lambda^2 + \alpha^2 + \beta^2 + \gamma^2)e = |q|^2 e,$$

where $|q| = +(\lambda^2 + \alpha^2 + \beta^2 + \gamma^2)^{1/2}$ is the modulus of q. Thus every non-zero quaternion has an inverse $q^{-1} = \bar{q} \cdot |q|^{-2}$. If $|q| = 1$, we call q a *unit quaternion*; for a unit quaternion, $q^{-1} = \bar{q}$. We also see quite easily that altering the signs of i, j, k in q_1 and q_2, and reversing the order of terms in the product $q_1 q_2$, simply alters the signs of i, j, k in the result, so that $\bar{q}_2 \bar{q}_1 = \overline{q_1 q_2}$. From this it follows that $|q_1 q_2|^2 \cdot (q_1 q_2)^{-1} = |q_1|^2 |q_2|^2 q_2^{-1} q_1^{-1}$, and since $(q_1 q_2)^{-1} = q_2^{-1} q_1^{-1}$ we have $|q_1|^2 \cdot |q_2|^2 = |q_1 q_2|^2$, from which we derive an algebraic identity of the form

$$(\lambda_1^2 + \alpha_1^2 + \beta_1^2 + \gamma_1^2)(\lambda_2^2 + \alpha_2^2 + \beta_2^2 + \gamma_1^2) = (\Lambda^2 + A^2 + B^2 + \Gamma^2)$$

expressing the product of two sums of four squares as a sum of four squares.

For any $q \in V_4$, we define a *right multiplication map* $\rho_q := V_4 \to V_4$ by $x \mapsto xq$. This is a linear map, and if $q = \lambda e + \alpha i + \beta j + \gamma k$, then ρ_q is represented, relative to the basis e, i, j, k, by the matrix

$$A = \begin{bmatrix} \lambda & \alpha & \beta & \gamma \\ -\alpha & \lambda & -\gamma & \beta \\ -\beta & \gamma & \lambda & -\alpha \\ -\gamma & -\beta & \alpha & \lambda \end{bmatrix}.$$

We have $A'A = (\lambda^2 + \alpha^2 + \beta^2 + \gamma^2)I$, whence $(\det A)^2 = (\lambda^2 + \alpha^2 + \beta^2 + \gamma^2)^4$, and since $\det A$ has the term λ^4, it follows that $\det A = (\lambda^2 + \alpha^2 + \beta^2 + \gamma^2)^2$. If $|q| = 1$, we have $A'A = I$ and $\det A = 1$, hence $A \in \mathrm{SO}(4)$, and $\rho_q \in \mathrm{SO}(V_4)$. We can also define a left multiplication map λ_q by $x \mapsto q^{-1}x$ and if $|q| = 1$, we find similarly that $\lambda_q \in SO(V_4)$ with representing matrix

$$
\begin{bmatrix}
\lambda & -\alpha & -\beta & -\gamma \\
\alpha & \lambda & -\gamma & \beta \\
\beta & \gamma & \lambda & -\alpha \\
\gamma & -\beta & \alpha & \lambda
\end{bmatrix}
$$

Right (or left) multiplication by a unit vector will be called right (or *left*) *rotations*. The right rotations ρ_q (and similarly the left rotations λ_q) form a subgroup of $SO(V_4)$; the two groups are both isomorphic to the multiplicative group U of unit quaternions, but are different subgroups of $SO(V_4)$. We see from examination of the representing matrices that the only common members of the two groups are the identity $x \mapsto x$ (right or left multiplication by e) and $x \mapsto -x$ (right or left multiplication by $-e$).

The quaternion algebra which we have constructed on the elements of V_4 clearly depends on the choice of the orthonormal basis e, i, j, k, for a different choice of the first element e gives a different unit element in the algebra. We therefore need to investigate how the algebra depends on the choice of basis. We recall that there are two distinguishable systems of similarly oriented orthonormal bases. For economy of language, we shall describe two bases of the same orientation as *compatible*. Two bases are compatible if and only if there is a transformation $\sigma \in SO(V_4)$ which maps one on the other.

LEMMA 16.2:   *Two compatible bases with the same first element e define identical quaternion algebras on V_4.*

Proof.   Let (e, i, j, k) and (e, u, v, w) be compatible bases. It is sufficient to show that, in the algebra constructed on the first basis we have $u^2 = v^2 = w^2 = uvw = -e$. Now, (i, j, k) and (u, v, w) are compatible bases of the 3-dimensional space $\langle e \rangle^\perp$, so $u = \alpha i + \beta j + \gamma k$, $v = \alpha' i + \beta' j + \gamma' k$, and $w = \alpha'' i + \beta'' j + \gamma'' k$, where

$$
T = \begin{bmatrix}
\alpha & \beta & \gamma \\
\alpha' & \beta' & \gamma' \\
\alpha'' & \beta'' & \gamma''
\end{bmatrix} \in SO(3).
$$

We have immediately $u^2 = -(\alpha^2 + \beta^2 + \gamma^2)e = -e$, and similarly $v^2 = w^2 = -e$. A somewhat more arduous calculation, using the orthogonality of the rows of T, yields $u(vw) = -(\det T)e = -e$.

Thus for compatible bases, the quaternion algebra is uniquely specified by the first basis element, and we can without ambiguity call the algebra 'the e-algebra'.

LEMMA 16.3:   *If (e, i, j, k) and (f, u, v, w) are compatible bases for V_4, then a right (or left) rotation in the f-algebra is a right (or left) rotation in the e-algebra.*

Proof.   We shall use our normal notation for products in the e-algebra. The product of vectors a, b in the f-algebra will be denoted by $a * b$, and the inverse of a in the f-algebra will be denoted by $a^\dagger$. Let f^{-1} be the inverse of f in the e-algebra. Then $x \mapsto xf^{-1}$ is a right rotation, and $(e, uf^{-1}, vf^{-1}, wf^{-1})$ is a basis compatible with (f, u, v, w) and therefore with (e, i, j, k). This map is therefore an isomorphism of the f-algebra to the e-algebra and $a * b = c$ is equivalent to $(af^{-1})(bf^{-1}) = cf^{-1}$ whence $a * b = af^{-1}b$. Now let ρ_a^*, λ_a^* denote right and left rotations (multiplications by the unit vector a) in the f-algebra. We have $x * a = x(f^{-1}a)$, so $\rho_a^* = \rho_{f^{-1}a}$, and since f^{-1} is a unit vector this is a right rotation in the e-algebra. To deal with a left rotation in the f-algebra, we need to calculate $a^\dagger * x$. Now $a^\dagger * a = f$, so $(a^\dagger f^{-1})(af^{-1}) = e$, $a^\dagger = fa^{-1}f$. Hence $(a^\dagger * x)f^{-1} = (a^\dagger f^{-1})(xf^{-1}) = (fa^{-1})(xf^{-1})$, and $a^\dagger * x = (fa^{-1})x = (af^{-1})^{-1}x$, giving $\lambda_a^* = \lambda_{af^{-1}}$.

Finally, we compare the algebras on incompatible bases. There are two sets of compatible bases, and we can take any two bases, one from each set. It will be convenient to consider the pair of orthonormal bases (e, i, j, k) and $(e, i, j, -k)$. Using normal notations for the first basis and denoting multiplication in the second basis by $*$, we have

$$i * j = -j * i = -k, \quad j * (-k) = k * j = i, \quad (-k) * i = i * k = j,$$

whence

$$i * j = ji, \quad j * i = ij, \quad j * k = kj, \quad k * j = jk, \quad k * i = ik, \quad i * k = ki.$$

Thus multiplication of all non-commuting pairs of basis elements e, i, j, k is reversed on passing from one algebra to the other, and for any vectors a, b we have $a * b = ba$. If a^{-1} denotes the inverse of a in the first algebra, we have $a * a^{-1} = a^{-1}a = e$ and $a^{-1} * a = aa^{-1} = e$, so a^{-1} is also the inverse of a in the second algebra. Thus the map $\rho_a^* : x \mapsto x * a$ is the same as $\lambda_{a^{-1}} : x \mapsto ax$, and the map $\lambda_a^* : x \mapsto a^{-1} * x$ is the same as $\rho_{a^{-1}} : x \mapsto xa^{-1}$. We have thus proved

LEMMA 16.4:   *Two incompatible bases define the same pair of left and right rotation groups, but with the names 'left' and 'right' interchanged.*

The fact that the left and right rotation groups do not depend on the choice of basis indicates their geometrical importance. It must be possible to identify this pair of subgroups of $SO(V_4)$ in a way which is independent of any specific basis, or even of any quaternion algebra.

If we examine the expression obtained for the representing matrix (relative to the basis (e, i, j, k)) of ρ_q, where $q = \lambda e + \alpha i + \beta j + \gamma k$, with $\lambda^2 + \alpha^2 + \beta^2 + \gamma^2 = 1$, we see that this is in canonical form if and only if $\beta = \gamma = 0$. We can then write q uniquely in the form $q = \cos\theta\,.\,e - \sin\theta\,.\,i$, $\theta \in \mathbb{R}/2\pi\mathbb{Z}$, and the representing matrix is diag $(\bar{R}_{-\theta},\ R_\theta)$. The representing matrix of λ_q is then also in canonical form, and is diag$(R_\theta,\ R_\theta)$. Clearly, we may select any $\theta \in \mathbb{R}/2\pi\mathbb{Z}$ and define q as above so that ρ_q, λ_q have these representing matrices. Although a given transformation $\sigma \in \mathrm{SO}(V_4)$ may be represented (depending on choice of canonical basis) by any of the matrices diag$(R_{\pm\psi},\ R_{\pm\phi})$, $(R_{\pm\phi}, R_{\pm\psi})$, we can now give the following basis-free description of the aggregate of left and right rotations.

LEMMA 16.5:   *The left and right rotations are the elements $\sigma \in \mathrm{SO}(V_4)$ with canonical forms* diag$(R_\phi,\ R_\psi)$, *where $\phi = \pm\psi$.*

We can now prove the important

LEMMA 16.6:   *Every transformation in $\mathrm{SO}(V_4)$ can be expressed as the product of a left rotation and a right rotation.*

Proof.   Let $\sigma \in \mathrm{SO}(V_4)$ have canonical form diag(R_α, R_β). Then (relative to the canonical basis) we have a left rotation λ_{q_1}, with representing matrix diag$(R_\theta,\ R_\theta)$, where $\theta = \frac{1}{2}(\alpha+\beta)$, and a right rotation ρ_{q_2}, with representing matrix diag$(R_{-\phi},\ R_\phi)$, where $\phi = \frac{1}{2}(\beta-\alpha)$. The representing matrix of $\lambda_{q_1}\rho_{q_2}$ is then diag$(R_\alpha,\ R_\beta)$, so that $\sigma = \lambda_{q_1}\rho_{q_2}$.

We shall denote by U, and describe as *the group of unit quaternions*, the group of unit vectors of V_4 under the multiplication defined by the quaternion algebra on an orthonormal basis e, i, j, k of V_4. The group U is isomorphic to the group of right rotations under the mapping $q \mapsto \rho_q$, and the right rotations have a faithful representation by 4×4 real matrices, given by

$$\rho_q \mapsto T_q = \begin{bmatrix} \lambda & \alpha & \beta & \gamma \\ -\alpha & \lambda & -\gamma & \beta \\ -\beta & \gamma & \lambda & -\alpha \\ -\gamma & -\beta & \alpha & \lambda \end{bmatrix},$$

where $q = \lambda e + \alpha i + \beta j + \gamma k$, $\lambda^2 + \alpha^2 + \beta^2 + \gamma^2 = 1$. The matrix T_q can be partitioned into four 2×2 submatrices, all of the form

$$\begin{bmatrix} a & b \\ -b & a \end{bmatrix},$$

and such matrices are an algebra isomorphic to the complex numbers $\mathbb{C}$, under the mapping

$$\begin{bmatrix} a & b \\ -b & a \end{bmatrix} \mapsto a + \varepsilon b$$

(where ε denotes the complex square root of -1). Putting all this together, we have an isomorphism $U \to GL(2, \mathbb{C})$ given by

$$q \mapsto A_q := \begin{bmatrix} u & v \\ -\bar{v} & \bar{u} \end{bmatrix}$$

where $u = \lambda + \varepsilon\alpha$, $v = \beta + \varepsilon\gamma$, and $\bar{u}, \bar{v}$ denote the complex conjugates of u, v. We have

$$\det A_q = u\bar{u} + v\bar{v} = \lambda^2 + \alpha^2 + \beta^2 + \gamma^2 = 1,$$

and

$$A_q^{-1} = \begin{bmatrix} \bar{u} & -v \\ \bar{v} & u \end{bmatrix} = \bar{A}_q',$$

whence $\bar{A}_q' A_q = 1$. Matrices with this second property are called *unitary*, those with unit determinant are *special unitary*; the 2×2 special unitary matrices form a multiplicative subgroup of $GL(2,\mathbb{C})$, which is denoted by $SU(2)$. We have thus proved

LEMMA 16.7: $U \cong SU(2)$.

The product group $U \times U$ is the group of ordered pairs (q_1, q_2) of unit quaternions, with component-wise multiplication $(q_1, q_2)(q_1', q_2') = (q_1 q_1', q_2 q_2')$. An action of $U \times U$ on V_4 is given by $x^{(q_1, q_2)} = q_1^{-1} x q_2$. The transformation $x \mapsto q_1^{-1} x q_2$ of V_4 is in $SO(V_4)$, so the action defines a homomorphism $U \times U \to SO(V_4)$, which (by Lemma 16.6) is surjective. The kernel of the homomorphism consists of the pairs (q_1, q_2) such that $q_1^{-1} x q_2 = x$ for all $x \in V_4$. It is sufficient to examine the basis elements $x = e, i, j, k$ to see that the members of the kernel are given by $q_1 = q_2 = \pm e$. The kernel is thus the 2-element group $\{(e, e), (-e, -e)\} \cong Z_2$. We have thus proved

LEMMA 16.8: $(U \times U)/\{(e, e), (-e, -e)\} \cong SO(V_4) \cong SO(4)$.

Now consider the space $V_3 = \langle i, j, k \rangle$; this is the orthogonal complement in V_4 of the 1-dimensional space $\langle e \rangle$. Any transformation in $SO(V_4)$ which fixes e will fix V_3 and induce on V_3 (by restriction) a transformation in $SO(V_3)$. Conversely, any transformation in $SO(V_3)$ can be extended to a transformation in $SO(V_4)$ by combining it with the identity on the orthogonal complement $\langle e \rangle$ of V_3. Thus $SO(V_3)$ is precisely the restriction to V_3 of the transformations in $SO(V_4)$ which fix e. The condition that the transformation $x \mapsto q_1^{-1} x q_2$ fixes e is simply $q_1 = q_2$. We therefore have an action of U on V_3 given by $x^q = q^{-1} x q$, and an associated surjective homomorphism $U \to SO(V_3)$. The kernel consists of the elements of $q \in U$ such that $q^{-1} x q = x$ for all $x \in V_3$, and the cases $x = i, j, k$ are sufficient to show that the only solutions are $q = e, -e$. These show that the kernel is the 2-element group $\{e, -e\}$, which is isomorphic to Z_2. We thus have

LEMMA 16.9: $U/\{e, -e\} \cong SO(V_3) \cong SO(3)$.

We have shown (Lemma 16.8) that $U \cong SU(2)$, and in this isomorphism $e, -e \in U$ correspond to $I, -I \in SU(2)$. We therefore have $U/\{e, -e\} \cong SU(2)/\{+I\}$ which is the projective group $PSU(2)$, giving the following result.

LEMMA 16.10: $SO(3) \cong PSU(2)$.

Also, $PSO(4)$ is the projective group $SO(4)/\{I, -I\}$, and we have a surjective homomorphism $U \times U \to PSO(4)$, the kernel of which consists of pairs (q_1, q_2) such that $q_1^{-1} x q_2 = x$ or $-x$ for all $x \in V_4$. This kernel is the Klein 4-group

$$\{(e, e), (e - e), (-e, e), (-e, -e)\} = \{e, -e\} \times \{e, -e\},$$

and

$$(U \times U)/(\{e, -e\} \times \{e, -e\}) \cong U/\{e, -e\} \times U/\{e, -e\},$$

giving the following result.

LEMMA 16.11: $PSO(4) \cong SO(3) \times SO(3)$.

This last result provides interesting information on the structure of $SO(4)$, which is peculiar to dimension 4. (In fact, $PSO(n)$ is simple for $n = 3$ and all $n \geqslant 5$.) Knowing the finite subgroups of $SO(3)$, Lemma 16.11 enables us to make a similarly complete classification of the finite subgroups of $PSO(4)$ and of $SO(4)$. A full account of these finite groups will be found in du Val *Homographies, quaternions and rotations* (Clarendon Press, Oxford 1964). Hamilton's original objective was to develop a tool for handling rotations in 3-dimensional Euclidean space. Such applications are most conveniently conducted by fixing the vector e in V_4, and considering the orthogonal space $\langle e \rangle^\perp = V_3$. Further, we shall agree to complete the basis for V_4 by taking a basis i, j, k for V_3 which is a right-handed orthonormal set. Then any two bases e, i, j, k and e, i', j', k' set up in this way will yield (by Lemma 16.2) identical quaternion algebras. We are now only interested in vectors in V_3, and the role of e is merely to serve as the unit element of the quaternion algebra; and since λe operates in the algebra just like the scalar λ, it is normal practice to suppress the symbol e, writing a quaternion in the form $q = \lambda + \alpha i + \beta j + \gamma k$, and the quaternion relations as $i^2 = j^2 = k^2 = ijk = -1$. We then talk of a quaternion q as the sum of a scalar part λ and a vector part $\alpha i + \beta j + \gamma k$. We also know that the product of two quaternions is independent of the choice of right-handed bases. We can therefore write a quaternion in the form $\lambda + u$ $(u \in V_3)$ and carry out manipulation in the quaternion algebra without commitment to a specific basis, or we can choose the basis to fit the particular situation under discussion.

If $u, v \in V_3$, $u = \alpha i + \beta j + \gamma k$, $v = \alpha' i + \beta' j + \gamma' k$, then the quaternion product is

$$uv = -(\alpha\alpha' + \beta\beta' + \gamma\gamma') + (\beta\gamma' - \beta'\gamma)i + (\gamma\alpha' - \gamma'\alpha)j + (\alpha\beta' - \alpha'\beta)k$$

which we can write in the condensed form

$$uv = -(u \cdot v) + u \times v,$$

where $u \cdot v$ is the fundamental bilinear product in V_3. The *vector* or *cross product* $u \times v$ is doubtless also familiar to the reader. Despite the enthusiasm of some advocates, the quaternion calculus never achieved wide acceptance as a tool for handling situations in Euclidean 3-space, but the vector calculus based on the *scalar* (or *dot*) *product* and the *vector* (or *cross*) *product* which was developed by N. H. Gibbs in the 1880s, eventually had better success, and is widely used in classical applied mathematics and certain areas of geometry. We may of course define the cross product in terms of coordinates without any appeal to quaternion theory, but then it is necessary to show that the definition is independent of the coordinate system employed. The advantage in defining it as the vector part of the quaternion product uv is that it then comes with a guarantee of independence of the choice of right-handed basis. Also, the relation $uv = -u \cdot v + u \times v$ can be used to advantage in proving the basic identities involving the dot and cross products because we can appeal to the properties (in particular, associativity) of quaternion multiplication.

The vector calculus finds useful application in spherical geometry. Let us consider the unit sphere, with centre at the origin 0 of the vector space V_3. Points on the sphere will be denoted by capital letters X, Y, ... and the associated unit vectors by the corresponding small letters $x, y, \ldots$. A plane through 0 meets the sphere in a *great circle*. Two great circles meet in two antipodal points X and $\bar{X}$, with representing vectors $x, -x$. (This notation reflects the fact that the quaternion conjugate $\bar{x}$ of a vector x is $-x$.) Two distinct non-antipodal points X, Y lie on just one great circle, which they divide into two segments. The length of the shorter of these is $< \pi$, and this is taken as the (spherical) distance $d(X, Y)$ between the two points. This shorter segment, directed from X to Y, is denoted by XY. A great circle has two (antipodal) poles, the points where the line through 0 orthogonal to the plane of the great circle meets the sphere. We associate one of these poles with each direction of the great circle. If X, Y are two distinct non-antipodal points on the great circle, and the direction is that of the segment XY then the associated pole of the directed great circle is Z', where $x \times y = \sin \alpha . z'$, and $\alpha = d(X, Y)$ is by definition the unique $\alpha \in (0, \pi)$ given by $\cos \alpha = x \cdot y$. The same great circle with the opposite direction has the antipodal pole $\bar{Z}'$.

A spherical triangle XYZ is defined by three distinct points X, Y, Z which do not lie on a great circle (which implies that no two are antipodal). These are the *vertices* of the triangle, and the segments YZ, ZX, XY are the *sides* of the triangle. The lengths of these sides will be denoted by a, b, c respectively. The *angle* at the vertex X is defined as the Euclidean angle A between the (directed) arcs XY, XZ. The poles of YZ, ZX, XY are denoted by X', Y', Z' respectively. The poles of XY, XZ are Z', $\bar{Y}'$, so we have $A = d(Z', \bar{Y}') = \pi - d(Y', Z')$. The triangle $X'Y'Z'$ is the *polar triangle* of the triangle XYZ. The side-lengths a', b', c' of $X'Y'Z'$ thus satisfy the relations

$$a' + A = b' + B = c' + C = \pi.$$

For any triangle, x, y, z are linearly independent, and the triple scalar product $[x, y, z] \neq 0$. The triangle XYZ is said to be positively oriented if $[x, y, z] > 0$. Having agreed to construct our quaternions on right-handed orthonormal bases of V_3, this condition means that if the spherical triangle is viewed from outside the sphere, the order XYZ of the vertices is anti-clockwise. If XYZ is positively oriented, then it is the polar triangle of $X'Y'Z'$, and we have the relations

$$a + A' = b + B' = c + C' = \pi$$

between the angles A', B', C' of $X'Y'Z'$, and the side-lengths a, b, c of XYZ.

The three great circles of which the sides of the triangle XYZ are segments divide the surface of the sphere into eight triangles, namely XYZ, $\bar{X}ZY$, $Z\bar{Y}X$, $YX\bar{Z}$ and the triangles antipodal to these. The triangles XYZ, $\bar{X}ZY$ are said to be colunar, since together they make up the *lune* bounded by the two great semi-circles $XY\bar{X}$, $\bar{X}ZX$. The angles of $\bar{X}ZY$ are A, $\pi - C$, $\pi - B$, and its side lengths are a, $\pi - c$, $\pi - b$.

The results we have obtained for $SO(V_n)$ in the cases $n = 2, 3, 4$ do not easily suggest a common pattern which will serve as a guide to similar results for higher values of n. In fact, the algebraic pattern provided by the cases $n = 2, 4$ is in this respect misleading, and it is the case $n = 3$ which provides the clue which leads to the Clifford algebra, which is applicable in any dimension. This algebra was outlined by W. K. Clifford in a brief note published in 1876, and followed by a longer paper on the structure of the algebra in 1878, but he was unable to develop the theory at all fully before his death at the early age of 33 in 1879, and the algebra was curiously neglected for almost fifty years until taken up again by Brauer and Weyl in the 1920s. Thereafter, its importance was soon recognised, and it was generalised to deal with the group of linear transformations which stabilise a non-singular quadratic form on a vector space $V_n(F)$ over an arbitrary field F. We cannot here go into all these developments, for which the reader must refer to more advanced texts (Artin, Bourbaki, Dieudonné, Chevalley). We can however describe the algebra in the case of positive definite forms on a real vector space $V_n(\mathbb{R})$, and relate it to the results already obtained for $n = 2, 3, 4$.

Let $e_1, \ldots, e_n$ be an orthonormal basis of $V_n(\mathbb{R})$. We consider the algebra C_n generated by a unit element 1 and $e_1, \ldots, e_n$, subject to the relations $e_i^2 = -1$, $e_i e_j = -e_j e_i$ for $i \neq j$ $(i, j = 1, \ldots, n)$. By means of these relations, any product of the e_i may be reduced to a product (with $\pm$ sign) of distinct basis elements in any order. There are thus 2^n such products (corresponding to the 2^n subsets of $\{1, \ldots, n\}$) which provide a linear basis for the algebra. For example, if $n = 3$, the elements of the algebra are the linear combinations (over $\mathbb{R}$) of the elements

$$1, \quad e_1, e_2, e_3, \quad e_2e_3, e_3e_1, e_1e_2, \quad e_1e_2e_3.$$

These elements will be called *even* or *odd* according as the number of terms in the product is even or odd. An element of the algebra which is a linear combination of even basis elements is also called *even*, and since the product of two even basis elements is also even, the even elements form a sub-algebra of C_n, which we denote by C_n^+. Thus C_3^+ is the algebra with basis $1, e_2e_3, e_3e_1, e_1e_2$. Now we have

$$(e_2e_3)^2 = e_2e_3e_2e_3 = -e_2e_2e_3e_3 = -1$$

and similarly

$$(e_3e_1)^2 = (e_1e_2)^2 = -1$$

and also

$$(e_2e_3)(e_3e_1)(e_1e_2) = e_2(e_3e_3)(e_1e_1)e_2 = e_2e_2 = -1;$$

hence C_3^+ is the familiar quaternion algebra, and to relate it to previous results we write $e_2e_3 = i$, $e_3e_1 = j$, $e_1e_2 = k$. We also write $e_1e_2e_3 = \omega$, and observe that $e_1\omega = \omega e_1 = -i$, $e_2\omega = \omega e_2 = -j$, $e_3\omega = \omega e_3 = -k$. Since ω commutes with e_1, e_2, e_3, it commutes with any element of C_3. Also

$$\omega^2 = e_1e_2e_3e_1e_2e_3 = 1,$$

and

$$\omega i = i\omega = -e_1, \quad \omega j = j\omega = -e_2, \quad \omega k = k\omega = -e_3.$$

It follows that the following two statements are equivalent:

$$q^{-1}(xi + yj + zk)q = x*i + y*j + z*k,$$

$$q^{-1}(xe_1 + ye_2 + ze_3)q = x*e_1 + y*e_2 + z*e_3;$$

for each is converted to the other on multiplication by $-\omega$. The second form of the relation, which has been produced by the Clifford algebra, has the pleasing feature that it has separated the quaternion operator q from the vector space $V_3(\mathbb{R})$ on which it operates. It has moreover raised the question of whether a transformation in $SO(V_n(\mathbb{R}))$ can always be expressed in the form $x \mapsto x* = s^{-1}xs$, where $s \in C_n^+$. We examine this question for $n = 2$ and $n = 4$.

The algebra C_2 has the linear basis $1, e_1, e_2, e_1e_2$. We write $e_1e_2 = \varepsilon$, and obtain the elements of C_2^+ in the form $\alpha + \beta\varepsilon$, where $\alpha, \beta \in \mathbb{R}$, and $\varepsilon^2 = -1$, showing that C_2^+ is simply the complex field $\mathbb{C}$. The transformation $\sigma_s : u \mapsto u* = s^{-1}us$ is the same as the transformation σ_t, where $t = s/|s|$, so we may assume that s is a complex number of unit modulus, and thus of the form $\cos\theta + \sin\theta \,.\, \varepsilon$. The transformation then takes the form

$$(\cos\theta - \sin\theta \,.\, \varepsilon)(xe_1 + ye_2)(\cos\theta + \sin\theta \,.\, \varepsilon) = x^*e_1 + y^*e_2.$$

Now $e_1\varepsilon = -\varepsilon e_1$, so multiplying on the right by $-e_1$ we have

$$(\cos\theta - \sin\theta \,.\, \varepsilon)(x + y\varepsilon)(\cos\theta - \sin\theta \,.\, \varepsilon) = x^* + y^*\varepsilon,$$

and (since the terms in the product now commute), this reduces to

$$x^* + y^*\varepsilon = (\cos 2\theta - \sin 2\theta \,.\, \varepsilon)(x + y\varepsilon),$$

which gives a rotation through an angle -2θ. It is of course clear that s and $-s$ produce the same transformation, so that a given rotation in $SO(V_2)$ of angle ϕ is produced by two complex numbers $e^{-\frac{1}{2}\phi}$, $e^{\pi - \frac{1}{2}\phi}$, and we have a surjective homomorphism $S \to SO(V_2)$ from the circle group S of complex numbers of unit modulus, with kernel $\{1, -1\}$.

The algebra C_4^+ has the linear basis

$$1, e_3e_4, e_4e_2, e_2e_3, e_2e_1, e_3e_1, e_4e_1, e_1e_2e_3e_4.$$

We write $e_3e_4 = i$, $e_4e_2 = j$, $e_2e_3 = k$, and $e_1e_2e_3e_4 = \omega$. The elements $1, i, j, k$ are a basis for a quaternion algebra. Since $\omega e_i = -e_i\omega$, $i = 1, \ldots, 4$, it follows that ω commutes with every element of the even algebra, and for the products with i, j, k we have $i\omega = \omega i = e_2e_1$, $j\omega = \omega j = e_3e_1$, $k\omega = \omega k = e_4e_1$. The elements of C_4^+ may therefore be written in the form $q + \omega r$, where q, r are quaternions and $\omega^2 = 1$; following Clifford, we shall call these elements *biquaternions*. If now we write $\xi = \frac{1}{2}(1 + \omega)$, $\eta = \frac{1}{2}(1 - \omega)$, so that $\xi + \eta = 1$, $\xi - \eta = \omega$, we have $q + \omega r = q_1\xi + q_2\eta$ where $q_1 = q + r$, $q_2 = q - r$, and $\xi^2 = \xi$, $\eta^2 = \eta$, $\xi\eta = \eta\xi = 0$, and ξ, η commute with all biquaternions. Since

$$(q_1\xi + q_2\eta)(q_1'\xi + q_2'\eta) = (q_1q_1')\xi + (q_2q_2')\eta = 1$$

if and only if $q_1q_1' = q_2q_2' = 1$, a biquaternion $s = q_1\xi + q_2\eta$ has an inverse, namely $s^{-1} = q_1^{-1}\xi + q_2^{-1}\eta$, if and only if q_1 and q_2 are non-zero quaternions. We now wish to know for which invertible biquaternions s does the transformation $x \mapsto s^{-1}xs$ give a map $V_4 \to V_4$? A necessary and sufficient condition is that $s^{-1}e_is \in V_4$ for $i = 1, \ldots, 4$. We note that $e_i\xi = \eta e_i$, $e_i\eta = \xi e_i$, and e_1 commutes with quaternions, so we have

$$(q_1^{-1}\xi + q_2^{-1}\eta)e_1(q_1\xi + q_2\eta) = (q_1^{-1}\xi + q_2^{-1}\eta)(q_2\xi + q_1\eta)e_1$$

$$= (q_1^{-1}q_2\xi + q_2^{-1}q_1\eta)e_1 = (q_3\xi + q_3^{-1}\eta)e_1$$

$$= \{\tfrac{1}{2}(q_3 + q_3^{-1}) + \tfrac{1}{2}(q_3 - q_3^{-1})\omega\}e_1,$$

where $q_3 = q_1^{-1}q_2$. Now, the image of e_1 under the transformation is a linear combination of $e_1, \ldots, e_4$ if and only if (i) $q_3 + q_3^{-1}$ is scalar, and (ii) $q_3 - q_3^{-1}$ has zero scalar part. The first of these means that *either* $|q_3| = 1$ *or* q_3 *is scalar*, the second that *either* $|q_3| = 1$ *or* q_3 *has no scalar part*. Together, they mean that $|q_3| = 1$, or equivalently that $|q_1| = |q_2|$. We leave it to the reader to verify that this condition also ensures that the images of e_2, e_3, e_4 are in V_4. We also note that the transformation is not affected if we replace q_1, q_2 by $\lambda^{-1}q_1, \lambda^{-1}q_2$, where λ is the common value of $|q_1|$ and $|q_2|$, so that we may suppose that $s = q_1\xi + q_2\eta$, where q_1, q_2 are unit quaternions; we shall call such an element s a unit biquaternion. Our transformation then takes the form

$$(\bar{q}_1\xi + \bar{q}_2\eta)(te_1 + xe_2 + ye_3 + ze_4)(q_1\xi + q_2\eta) = t^*e_1 + x^*e_2 + y^*e_3 + z^*e_4,$$

which on multiplication on the right by $-e_1$ becomes

$$(\bar{q}_1\xi + \bar{q}_2\eta)(t + xe_1e_2 + ye_1e_3 + ze_1e_2)(q_2\xi + q_1\eta) = \\ t^* + x^*e_1e_2 + y^*e_1e_3 + z^*e_1e_4.$$

Now, $t + xe_1e_2 + ye_1e_3 + ze_1e_4 = t - \omega v$, where $v = xi + yj + zk$, and $t - \omega v = (t-v)\xi + (t+v)\eta = \bar{u}\xi + u\eta$, where $u = t + xi + yj + zk$. The equation thus becomes

$$(\bar{q}_1\xi + \bar{q}_2\eta)(\bar{u}\xi + u\eta)(q_2\xi + q_1\eta) = \bar{u}^*\xi + u^*\eta,$$

which is equivalent to the two reltions $\bar{q}_1\bar{u}q_2 = \bar{u}^*$, $\bar{q}_2uq_1 = u^*$, which are mutually equivalent, and of the form previously obtained for a transformation in $SO(V_4)$.

The multiplication of basis vectors by which we have defined the Clifford algebra C_n implies that, for any vector $x \in V_n$, $x^2 = -(x \cdot x)$. Conversely, from this property we can deduce that for orthogonal vectors x, y we have $xy + yx = 0$; for

$$xy + yx = (x+y)^2 - x^2 - y^2 = -\{(x+y) \cdot (x+y) - x \cdot x - y \cdot y\} = -2x \cdot y,$$

and we have $e_i^2 = -1$, $e_ie_j + e_je_i = 0$ $(i \neq j)$ for any orthonormal set $e_1, \ldots, e_n$. The specification $x^2 = -x \cdot x$ therefore provides a basis-free definition of C_n. Also, any product of an even or odd number of vectors is a linear combination of products of an even or odd number of basis vectors, respectively, so the definition of the even subalgebra C_n^+ is also independent of the basis. Also, the element $\omega = e_1 \ldots e_n$ is the same for any two similarly oriented orthonormal bases $e_1, \ldots, e_n$ and $f_1, \ldots, f_n$; for $f_i = \sum \alpha_{ij}e_j$, where $A = [\alpha_{ij}] \in SO(n)$, and we see that

$$f_1 \ldots f_n = \prod_{i=1}^{n} \left(\sum_{j=1}^{n} \alpha_{ij}e_j \right) = \det A . e_1 \ldots e_n = e_1 \ldots e_n.$$

Now, the map σ_s given by $x\sigma_s = s^{-1}xs$ is the identity for all $x \in V_n$ if and only if s commutes with all $x \in V_n$ (and hence with all elements of C_n). Since e_i commutes with e_i but anti-commutes with all other e_j, it follows that ω

commutes with all $x \in V_n$ if n is odd, and anti-commutes with all $x \in V_n$ if n is even. On the other hand, any product of distinct e_i of degree $\geqslant 0$ and $< n$ will commute with some basis element and anti-commute with some other (depending on whether the basis element in question does or does not occur in the product). It follows that σ_s is the identity if and only if n is odd and $s = \lambda + \mu\omega$ ($\lambda, \mu \in \mathbb{R}$) or n is even and $s \in \mathbb{R}$. If $s \in C_n^+$, we must in either case have $s \in \mathbb{R}$.

We denote by G the multiplicative group of invertible elements $s \in C_n$ such that $s^{-1}xs \in V_n$ for all $x \in V_n$, and by ϕ the map $\phi: G \to O(V_n)$ which maps such an element s on the orthogonal transformation $\sigma_s: x \mapsto s^{-1}xs$. Writing $G^+ = G \cap C_n^+$, and $\phi|_{C_n^+} = \phi^+: G^+ \to O(n)$, we have shown that $\ker \phi^+ = \mathbb{R}^*$, the multiplicative group of non-zero elements of $\mathbb{R}$. We shall now show that $\operatorname{im} \phi^+ = SO(V_n)$. Consider the map σ_y, where $y \in V_n$. We have $y\sigma_y = y^{-1}(y)y = y$, so σ_y fixes all vectors in the 1-dimensional subspace $\langle y \rangle$. If $x \in \langle y \rangle^\perp$, we have $x \cdot y = 0$ and $xy + yx = 0$, whence $x\sigma_y = y^{-1}xy = -y^{-1}yx = -x$, so all vectors in $\langle y \rangle^\perp$ are reversed. Now let $\alpha: V_n \to V_n$ be the central reflexion, defined by $x\alpha = -x$ for all $x \in V_n$. Then $\alpha\sigma_y$ reverses vectors in $\langle y \rangle$ and fixes vectors in $\langle y \rangle^\perp$, and is thus a hyperplane reflexion. Now, every transformation in $SO(V_n)$ is a product of an even number of hyperplane reflexions, and every other transformation in $O(V_n)$ is a product of an odd number of such reflexions. Thus, if $\theta \in SO(V_n)$, we have $\theta = \alpha\sigma_{y_1} \ldots \alpha\sigma_{y_{2k}}$ and since α commutes with all linear transformations and α^2 is the identity, $\theta = \sigma_s$ where

$$s = y_1 \ldots y_{2k} \in G^+.$$

If, on the other hand, $\theta \in O(V_n) \setminus SO(V_n)$ we have $\theta = \alpha\sigma_{y_1} \ldots \alpha\sigma_{y_{2k+1}} = \alpha\sigma_t$, where $t = y_1 \ldots y_{2k+1}$. Now if n is even, $\alpha = \sigma_\omega$ and $\theta = \sigma_{\omega t}$; but ωt is an odd element and therefore not in G^+. If x is odd, there is no element $u \in G$ such that $\alpha = \sigma_u$; for u would have to anti-commute with all the elements e_i, and every basis element of C_n commutes with some e_i. Thus in all cases, $\theta \in O(V_n) \setminus SO(V_n)$ implies $\theta \notin \operatorname{im} \phi^+$, and we have $\operatorname{im} \phi^+ = SO(V_n)$.

Finally, we can define conjugates in C_n with the help of the transformation β which reverses the order of any product of vectors, so that in the standard basis for C_n, elements of degree k are unchanged by β if $k \equiv 0$ or 1 (mod 4), and are reversed in sign if $k \equiv 2$ or 3 (mod 4). For any element $s \in C_n$, we may define a *norm* $N(s)$ by $N(s) := s \cdot s^\beta$. The norm so defined is not necessarily a scalar, but for $s \in G^+$ we can write $s = y_1 \ldots y_{2k}$, $y_i \in V_n$, and then $N(s) = y_1 \ldots y_{2k}y_{2k} \ldots y_1$. Since $y_i^2 = -(y_i \cdot y_i)$, a negative scalar, the norm $N(s)$ is in this case a positive scalar, and since $N(\lambda s) = \lambda^2 N(s)$ for any scalar λ, we have two elements $\pm \lambda s$, where $\lambda^2 = 1/N(s)$, such that $N(\lambda s) = N(-\lambda s) = 1$, and $\sigma_{\lambda s} = \sigma_{-\lambda s} = \sigma_s$. The elements of unit norm in G^+ form a subgroup G_0^+, for we have $N(st) = (st)(st)^\beta = (st)(t^\beta s^\beta) = (ss^\beta)(tt^\beta) = N(s)N(t)$ (since tt^β is a scalar). We therefore obtain a surjective homomorphism $G_0^+ \to SO(V_n)$, with kernel $\{1, -1\} \cong Z_2$. The group G_0^+ is the 'classical' *spin group* Spin(n).

Exercises

16.1. Let ABC be a plane triangle represented by complex numbers a, b, c with $|a| = |b| = |c| = R$, and let P be a point on its circumcircle, so that $|p| = R$. For $|q| = R$ define

$$z := a + b + c + p + q + \frac{abc}{pq}.$$

(i) Show that as Q varies on the circumcircle (but A, B, C, P remain fixed) z describes a segment of a straight line of length $2R$.

(ii) Show that when $q = -a$, Z lies on BC and $PZ \perp BC$.

The line containing the locus of Z is called the *Simson line* of p with respect to the triangle ABC. From the symmetry of the expression for z it is clear that Z is the intersection of the Simson lines (with respect to ABC) of P and Q.

(iii) Show that the Simson line of P passes through the midpoint of PH (where H is the orthocentre of ABC).

(iv) Show that the Simson lines of diametrically opposite points are perpendicular and meet on the nine-point circle of ABC.

(v) Finally, by putting $q = p$, investigate the envelope of the Simson line with respect to ABC of a variable point on its circumcircle. (The envelope is a 3-cusped hypocycloid with centre at H.)

16.2. If the vector product $u \times v$ of u, $v \in \mathbb{R}^3$, where $u = (a_1, a_2, a_3)$, $v = (b_1, b_2, b_3)$, is defined by

$$u \times v := (a_2 b_3 - a_3 b_2, a_3 b_1 - a_1 b_3, a_1 b_2 - a_2 b_1),$$

show by direct calculation that, if $A = [\lambda_{ij}]$ is an orthogonal matrix of determinant 1, then

$$(uA) \times (vA) = (u \times v)A.$$

Obtain an alternative proof of this result by showing first that if u, v are linearly dependent, the result holds. Then suppose that u, v are linearly independent, and let $w = (c_1, c_2, c_3) \in \mathbb{R}^3$ be such that $\{u, v, w\}$ is a linearly independent set. Let

$$X = \begin{bmatrix} a_1 & a_2 & a_3 \\ b_1 & b_2 & b_3 \\ c_1 & c_2 & c_3 \end{bmatrix}$$

and let $XA = Y$. Now consider $(Y')^{-1}$.

16.3. Using quaternion multiplication and the relation $uv = -(u \cdot v) + u \times v$, obtain the following results, for u, v, $w \in V_3(\mathbb{R})$:

(i) $u \times v = -(v \times u)$;

(ii) $\|u \times v\| = \|u\| . \|v\| . |\sin \gamma|$, where $\cos \gamma = (u . v)/\|u\| . \|v\|$;

(iii) $u \times v = 0$ if and only if u, v are linearly dependent;

(iv) $u \times v = \frac{1}{2}(uv - vu)$, $u . v = -\frac{1}{2}(uv + vu)$;

(v) $(u \times v) \times w = (u . w)v - (v . w)u$;

(vi) $(u \times v) \times w + (v \times w) \times u + (w \times u) \times v = 0$;

(vii) $u . (v \times w) = v . (w \times u) = w . (u \times v)$.

16.4. Using the notation established in the text, prove the following formulae for a spherical triangle:

(i) $\cos a = \cos b \cos c + \sin b \sin c \cos A$;

(ii) $\dfrac{\sin A}{\sin a} = \dfrac{\sin B}{\sin b} = \dfrac{\sin C}{\sin c} = \dfrac{2\delta}{\sin a \sin b \sin c}$ where $\delta = \frac{1}{2}|[x, y, z]|$.

Deduce the following formulae, where $s = \frac{1}{2}(a + b + c)$:

(iii) $\delta = \{\sin s \sin(s - a)\sin(a - b)\sin(s - c)\}^{1/2}$

(iv) $\sin \frac{1}{2}A = \{\sin(s - b)\sin(s - c)/\sin b \sin c\}^{1/2}$;

(v) $\cos \frac{1}{2}A = \{\sin s \sin(s - a)/\sin b \sin c\}^{1/2}$

(vi) $\tan \frac{1}{2}A = \{\sin(s - b)\sin(s - c)/\sin s \sin(s - a)\}^{1/2}$.

16.5. By considering the area of the lunes which contain the spherical triangle XYZ, show that the area Δ of the triangle is $A + B + C - \pi$. Show also that $\pi < A + B + C < 3\pi$ and $0 < a + b + c < 2\pi$.

16.6. Show that, if Δ_0 is the area of the plane triangle XYZ and p is the perpendicular distance from O to the plane XYZ, then $p\Delta_0 = \delta$.

Let S_0 be the circumcentre of the plane triangle and R_0 its circumradius. Let OS_0 meet the sphere at S. Show that the distances of S from X, Y and Z are all equal. We call S the circumcentre, and the spherical distance $R = d(S, x)$ the circumradius of the spherical triangle XYZ. Show that

$$\tan R = R_0/p = 2 \sin \tfrac{1}{2}a \sin \tfrac{1}{2}b \sin \tfrac{1}{2}c/\delta.$$

16.7. Assuming that XYZ is positively oriented, show that the great circle through S and X' is orthogonal to YZ and passes through the midpoint of YZ, and is also the internal bisector of the angle at X' of the polar triangle $X'Y'Z'$. Show also that a circle with centre S and (spherical) radius $\pi/2 - R$ touches the sides of $X'Y'Z'$.

Hence show that the triangle XYZ has an in-circle of radius r, given by $\tan r = \delta/\sin s$.

16.8. Assuming that none of the sides a, b, c or angles A, B, C of the triangle XYZ is equal to $\pi/2$, show that the great circle through X and X' is orthogonal to YZ and to $Y'Z'$, and that the three great circles through X and X', Y and Y', Z and Z' are concurrent.

Show that if all sides of the triangle XYZ are $< \pi/2$, or if all are $> \pi/2$, then one of the points of concurrence lies inside the polar triangle $X'Y'Z'$.

16.9. By Lemma 16.9 (and the discussion preceding it) every rotation $\rho \in SO(V_3)$ is of the form $x \mapsto q^{-1}xq$ where q is a unit quaternion (determined up to a factor ± 1 by ρ). A unit quaternion is of the form $q = \cos\frac{1}{2}\alpha - \sin\frac{1}{2}\alpha \, . \, u$ where u is a unit vector in V_3.

   (i) Show that if $\rho : x \mapsto q^{-1}xq$, where q is as above, then ρ is a rotation through angle α, with axis in the direction of u.

   (ii) Show that if $\rho = \rho_1\rho_2$, where ρ_1 corresponds to the unit quaternion $\cos\frac{1}{2}\alpha - \sin\frac{1}{2}\alpha \, . \, u$ and ρ_2 to the unit quaternion $\cos\frac{1}{2}\beta - \sin\frac{1}{2}\beta \, . \, v$, then ρ has angle $\pm\theta$ given by

$$\cos\tfrac{1}{2}\theta = \cos\tfrac{1}{2}\alpha \cos\tfrac{1}{2}\beta - (u \cdot v)\sin\tfrac{1}{2}\alpha \sin\tfrac{1}{2}\beta.$$

Show further that if $\beta = \alpha$ and the unit vectors u, v may be chosen arbitrarily then

$$\cos\theta \text{ takes all values in } \begin{cases} [-1, 1] \text{ if } \cos\alpha < 0, \\ [\cos 2\alpha, 1] \text{ if } \cos\alpha \geqslant 0. \end{cases}$$

16.10. Using the notation of Exercise 16.9, show that if u, v are orthogonal and $\alpha = \beta = \pi/2$ then $\cos\theta = -\frac{1}{2}$. Find, in terms of an orthonormal basis u, v, w, a unit vector x such that ρ has axis in the direction of x.

   Give an alternative treatment of this problem by considering the actions of rotational symmetries of a regular octahedron on its set of vertices.

16.11. Show that rotations $\rho_1, \rho_2 \in SO(V_3)$, through angles α, β respectively, are conjugate in $SO(V_3)$ if and only if $\cos\alpha = \cos\beta$. Deduce that a non-trivial normal subgroup of $SO(V_3)$ contains all rotations through angle α for some non-zero $\alpha \in \mathbb{R}/2\pi\mathbb{Z}$, and hence that it contains a rotation through angle θ for every θ such that $\cos\alpha \leqslant \cos\theta \leqslant 1$ (see Exercise 16.9 (ii)).

   Hence prove that $SO(V_3)$ is a simple group (i.e. has no non-trivial proper normal subgroup).

16.12. Show that the eight elements $\pm 1, \pm i, \pm j, \pm k$ form a subgroup H of the multiplicative group of unit quaternions. (This group is called the quaternion group, and is denoted by H in honour of Hamilton.)

   Show that H has only one subgroup of order 2, which is the centre $Z(H)$ of H, and that every subgroup of H is normal.

   Show that $H/Z(H) \cong Z_2 \times Z_2$.

16.13. Show that the 24 elements $\pm 1, \pm i, \pm j, \pm k, \frac{1}{2}(\pm 1 \pm i \pm j \pm k)$, form a subgroup G of the multiplicative group of unit quaternions. Show also that the centre $Z(G)$ of G is of order 2, and that $G/Z(G) \cong A_4$.

   Show that G is not isomorphic to $A_4 \times Z_2$.

16.14. Find a group G of quaternions of order 48, such that $G/Z(G) \cong S_4$.

16.15. Classify the conjugacy classes of finite subgroups of the multiplicative group of unit quaternions.

16.16. Show that the transformations $\sigma \in$ SO(4) which are left or right rotations are precisely those for which the minimum polynomial is irreducible in $\mathbb{R}[t]$.

16.17. Let J_{2n} denote the $2n \times 2n$ matrix diag$(J, J, \ldots, J)$ where

$$J = \begin{bmatrix} 0 & 1 \\ -1 & 0 \end{bmatrix},$$

and denote by Sp$(2n, F)$ the *symplectic group* of matrices $A \in$ GL$(2n, F)$ such that $AJ_{2n}A' = J_{2n}$. Verify that Sp$(2, F) =$ SL$(2, F)$.

   Let $\phi: M_n(\mathbb{C}) \to M_{2n}(\mathbb{R})$ be the mapping which replaces any entry $a + ib$ in a complex matrix by the 2×2 matrix

$$\begin{bmatrix} a & b \\ -b & a \end{bmatrix}.$$

Show that ϕ applied to GL$(n, \mathbb{C})$ yields an injective homomorphism GL$(n, \mathbb{C}) \to$ GL$(2n, \mathbb{R})$. By considering the image under ϕ of the complex unitary group U(n), show that U$(n) \cong$ O$(2n) \cap$ Sp$(2n, \mathbb{R})$.

16.18. We consider the set $M_n(\mathbb{H})$ of $n \times n$ matrices with quaternion entries, and in particular the matrices $A \in M_n(\mathbb{H})$ such that $A\bar{A}' = I$ (where $\bar{A}$ now denotes the *quaternion* conjugate). Using the mapping

$$q \mapsto \begin{bmatrix} u & v \\ -\bar{v} & \bar{u} \end{bmatrix}$$

of a quaternion element $q = \lambda + \alpha i + \beta j + \gamma k$ to a 2×2 complex matrix with $u = \lambda + \alpha\varepsilon$, $\sigma = \beta + \gamma\varepsilon$ (where ε is the complex unit with $\varepsilon^2 = -1$ and $\bar{u}$, $\bar{v}$ denote *complex* conjugation), show that the matrices A form a group U$(n, \mathbb{H})$, and that U$(n, \mathbb{H}) \cong$ U$(2n) \cap$ Sp$(2n, \mathbb{C})$.

16.19. Let $G = G_1 \times G_2$, the (external) product of G_1, G_2 consisting of ordered pairs (g_1, g_2), $g_1 \in G_1$, $g_2 \in G_2$ with the multiplication $(g_1, g_2)(g_1', g_2') = (g_1 g_1', g_2 g_2')$. Let H be a subgroup of G,

$$H_1 = \{g_1 \in G_1 : (g_1, g_2) \in H \text{ for some } g_2 \in G_2\},$$

$$K_1 = \{g_1 \in G_1 : (g_1, 1) \in H\}.$$

Show that $G_1 \geqslant H_1 \trianglerighteq K_1$. Let

$$H_2 = \{g_2 \in G_2 : (g_1, g_2) \in H \text{ for some } g_1 \in G_1\}$$

and

$$K_2 = \{g_2 \in G_2 : (1, g_2)) \in H\},$$

so that $G_2 \geqslant H_2 \trianglerighteq K_2$. Show that $K_1 \times K_2 \leqslant H_1 \times H_2$, and that

$$\frac{H_1}{K_1} \cong \frac{H_2}{K_2}.$$

Show also that if H is finite, so are H_1, H_2, and

$$|H| = |H_1| \cdot |K_2| = |H_2| \cdot |K_1|.$$

16.20. If H is a subgroup of PSO(4), show that a subgroup of SO(4) which maps to H under the standard homomorphism SO(4) → PSO(4) either
   (i) contains the central reflexion $-I$, and is isomorphic to $H \times Z_2$, or
   (ii) does not contain $-I$, and is isomorphic to H.

16.21. In the notation of Exercise 16.19, let

$$G_1 = G_2 = SO(3), \quad H_1 = H_2 \cong A_5, \quad K_1 = K_2 = \{1\},$$

and suppose that we are given an automorphism of A_5. Show that this information defines a subgroup of PSO(4) which is isomorphic to A_5.
   Show that each of the following subgroups of $SO(V_4)$ is isomorphic to A_5:
   (i) the stabiliser of a unit vector e and an icosahedron in the orthogonal space $\langle e \rangle^\perp$;
   (ii) the stabiliser of a simplex in V_4.
(A simplex is the analogue of the tetrahedron in V_3; it has 5 vertices equidistant from each other. We are supposing that the vertices of the icosahedron and of the simplex are all at unit distance from the origin.)
   Show that these two types of subgroup are geometrically distinct, that is to say, they are not conjugate in $SO(V_4)$.
   If we follow through the construction indicated, the type of subgroup of SO(4) which is obtained depends on the automorphism of A_5 used in the construction; in fact, an inner automorphism of A_5 yields one type and an outer automorphism yields the other. Can you say which?

Chapter 17

Inversive geometry

Our main objective in this chapter is the further study of the space $\Omega = \mathbb{C} \cup \{\infty\}$ under the action of the group M of Möbius transformations, which we shall write as

$$z \mapsto z^* = \frac{\alpha z + \beta}{\gamma z + \delta}, \quad \alpha, \beta, \gamma, \delta \in \mathbb{C}, \alpha\delta \neq \beta\gamma,$$

without mentioning explicitly the interpretation of the transformation for $z = \infty$ or $z = -\delta/\gamma$. This group of transformations was introduced in Chapter 2 (Example 2.4), and the geometry (Ω, M) appeared in Chapter 13 as the geometry of the complex projective line.

In the course of the discussion we shall be led to consider the larger group P generated by the Möbius transformations together with the transformation κ of complex conjugation, given by $z \mapsto \bar{z}$. This larger group will (for reasons to be explained) be called the *inversive group*, and *inversive geometry* will be the geometry (Ω, P).

First, we obtain a good representation of the space Ω in Euclidean space. The representation of $\mathbb{C}$ in $\mathbb{R}^2$, using the bijection $\xi + i\eta \mapsto (\xi, \eta)$, will be familiar. In order to extend this representation to $\mathbb{C} \cup \{\infty\}$, we need to add to $\mathbb{R}^2$ a further point ω. A sensible way of doing this is suggested by the stereographic projection of $\mathbb{R}^2$ on the unit sphere.

$$S := \{(\lambda, \mu, \nu) \in \mathbb{R}^3 \mid \lambda^2 + \mu^2 + \nu^2 = 1\}.$$

Under this projection, $(\xi, \eta) \in \mathbb{R}^2$ is mapped on $(\lambda, \mu, \nu) \in S$, where (λ, μ, ν) is the point of intersection with S (other than $(0, 0, 1)$) of the line joining $(0, 0, 1)$ and $(\xi, \eta, 0)$. In Fig. 17.1, $(0, 0, 1)$, $(\xi, \eta, 0)$ and (λ, μ, ν) are represented by n, p, and p^* respectively. The stereographic projection maps $\mathbb{R}^2$ bijectively on the punctured sphere $S \setminus \{n\}$. We thus have a bijection $\mathbb{C} \to S \setminus \{n\}$, which is immediately extendable to a bijection $\mathbb{C} \cup \{\infty\} \to S$ by mapping ∞ on n.

In establishing this correspondence between the elements of $\mathbb{C} \cup \{\infty\}$ and the points of the unit sphere S in $\mathbb{R}^3$, we have not merely found a convenient geometrical object on which to represent $\mathbb{C} \cup \{\infty\}$. The sphere S has a natural topology, as a subspace of $\mathbb{R}^3$, and this can be transferred to $\mathbb{C} \cup \{\infty\}$. With this topology, $\mathbb{C} \cup \{\infty\}$ is homeomorphic to S. Now, $\mathbb{C}$ has a natural metric topology, with metric given by $d(z_1, z_2) = |z_1 - z_2|$, and if $z_1 = \xi_1 + i\eta_1$, $z_2 = \xi_2 + i\eta_2$,

$$|z_1 - z_2| = +\{(\xi_1 - \xi_2)^2 + (\eta_1 - \eta_2)^2\}^{1/2},$$

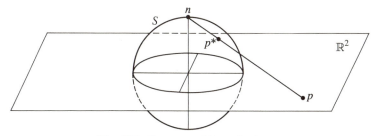

Fig. 17.1. Stereographic projection.

agreeing with the natural metric in $\mathbb{R}^2$, and showing that the bijection $\mathbb{C} \to \mathbb{R}^2$ is a homeomorphism. Moreover, the bijection $\mathbb{R}^2 \to S\backslash\{n\}$ is a homeomorphism, so the bijection $\mathbb{C} \to S\backslash\{n\}$ is a homeomorphism, in terms of the natural topologies of the spaces concerned. The homeomorphism $\mathbb{C} \cup \{\infty\} \to S$ thus extends the natural homeomorphism $\mathbb{C} \to S\backslash\{n\}$. The sphere S (and hence $\mathbb{C} \cup \{\infty\}$ and $\mathbb{R}^2 \cup \{\omega\}$ with the topologies induced from S) is a compact space, being a closed bounded subspace of $\mathbb{R}^3$. We have therefore made the (non-compact) space $\mathbb{C}$ (or $\mathbb{R}^2$) compact by adding a single point ∞ (or ω) and suitably extending the family of open sets. This is an instance of an entirely general procedure of 'one-point compactification' which can be applied to any topological space. The special feature of the present situation is that the one-point compactification of the punctured sphere is extremely obvious.

The space $\mathbb{R}^2 \cup \{\omega\}$, with the topology induced from S, is called the *Gauss plane*. The sphere S is in this context called the *Riemann sphere*; it is of course a perfectly ordinary Euclidean sphere, but to name it as the Riemann sphere means that it is being used to represent $\mathbb{C} \cup \{\infty\}$ in the way which we have described. The term indicates the representation $\mathbb{C} \cup \{\infty\} \to S$, rather than S itself.

We turn from the topological generalities of the situation to examine the algebra of the bijection $\mathbb{R}^2 \to S\backslash\{n\}$ given by $(\xi, \eta) \mapsto (\lambda, \mu, \nu)$. It is not difficult to derive the relations

$$\lambda = \frac{2\xi}{\xi^2 + \eta^2 + 1}, \quad \mu = \frac{2\eta}{\xi^2 + \eta^2 + 1}, \quad \nu = \frac{\xi^2 + \eta^2 - 1}{\xi^2 + \eta^2 + 1} = 1 - \frac{2}{\xi^2 + \eta^2 + 1},$$

and they may be verified by the observations that

$$\lambda^2 + \mu^2 + \nu^2 = 1 \quad \text{(giving } (\lambda, \mu, \nu) \in S)$$

and that

$$(\lambda, \mu, \nu) = (1 - \kappa)(0, 0, 1) + \kappa(\xi, \eta, 0),$$

where $\kappa = 2/(\xi^2 + \eta^2 + 1)$, showing that (λ, μ, ν), $(0, 0, 1)$ and $(\xi, \eta, 0)$ are collinear. The inverse transformation $(\lambda, \mu, v) \mapsto (\xi, \eta)$ is given by the equations

$$\xi = \lambda/(1-v), \quad \eta = \mu/(1-v).$$

Since $d(n, p^*)/d(n, p) = |\kappa| = 2/(\xi^2 + \eta^2 + 1)$, and $\{d(n, p)\}^2 = \xi^2 + \eta^2 + 1$, we have $d(n, p^*) \cdot d(n, p) = 2$. If $p, q \in \mathbb{R}^2$ correspond to $p^*, q^* \in S$ we obtain the relation

$$d(p^*, q^*) = 2d(p, q)/\{d(n, p) \cdot d(n, q)\}.$$

This is most easily shown by writing $d(n, p) = x$, $d(n, q) = y$, so that $d(n, p^*) = 2/x, d(n, q^*) = 2/y$, and then by elementary trigonometry we obtain

$$\{d(p^*, q^*)\}^2 = (2/x)^2 + (2/y)^2 - 2(2/x)(2/y)\cos\alpha$$

$$= (4/x^2 y^2)(y^2 + x^2 - 2xy\cos\alpha)$$

$$= (2/xy)^2 \{d(p, q)\}^2$$

(see Fig. 17.2). If $q = \omega$, then $q^* = n$, and $d(p^*, q^*) = d(p^*, n) = 2/d(n, p)$.

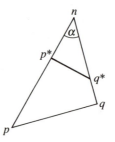

Fig. 17.2. The effect of stereographic projection on distances.

In terms of the complex numbers z_1, z_2 corresponding to p, q, the distance $d(p^*, q^*)$ is $\chi(z_1, z_2)$, say, where χ is given by

$$\chi(z_1, z_2) = 2|z_1 - z_2|/(1 + |z_1|^2)^{1/2}(1 + |z_2|^2)^{1/2},$$

$$\chi(z_1, \infty) = 2/(1 + |z_1|^2)^{1/2},$$

$$\chi(\infty, \infty) = 0.$$

The function χ is often called the chordal distance, as $\chi(z_1, z_2)$ is simply the Euclidean distance between the points of S which correspond to z_1, z_2. The chordal distance is a metric on $\mathbb{C} \cup \{\infty\}$, and the topology which it defines is the same as the topology of $\mathbb{C} \cup \{\infty\}$ induced by the Euclidean topology of the sphere S. The metric topologies defined on $\mathbb{C}$ by χ and the familiar metric d given by $d(z_1, z_2) = |z_1 - z_2|$ are identical.

The function χ is bounded; indeed, $\chi(z_1, z_2) \leqslant 2$, and $\chi(z_1, z_2) = 2$ if and only if the points of S which represent z_1, z_2 are diametrically opposite. The condition $\chi(z_1, z_2) = 2$ gives

$$(1 + z_1\bar{z}_1)(1 + z_2\bar{z}_2) = (z_1 - z_2)(\bar{z}_1 - \bar{z}_2),$$

which on re-arranging gives

$$(1 + z_1\bar{z}_2)(1 + \bar{z}_1 z_2) = 0;$$

hence $\chi(z_1, z_2) = 2$ if and only if $z_2 = -1/\bar{z}_1$.

It is clear that the points of a line l in $\mathbb{R}^2$ correspond under the stereographic projection to points of S which lie in the plane containing the 'north pole' n and the line l. This plane meets S in a circle through n, and the points of this circle correspond to the points of $l \cup \{\omega\}$. It is less obvious that the points of a circle on S which does not contain n correspond to the points of a circle in $\mathbb{R}^2$. The following algebraic argument establishes both of these results.

We consider loci in $\mathbb{R}^2$ of the form

$$\{(x, y) \mid a(x^2 + y^2) + 2gx + 2fy + c = 0\},$$

where $a, g, f, c \in \mathbb{R}$, and a, g, f are not all zero. If $a = 0$, the locus is a line of $\mathbb{R}^2$. If $a \neq 0$, the equation of the locus may be written as

$$(x + (g/a))^2 + (y + (f/a))^2 = (g^2 + f^2 - ac)/a^2.$$

If $g^2 + f^2 - ac > 0$, this is the equation of a circle, with centre at $(-g/a, -f/a)$. If $g^2 + f^2 - ac = 0$, the locus consists of the single point $(-g/a, -f/a)$; and if $g^2 + f^2 - ac < 0$, the locus is empty. If now (X, Y, Z) is the point of S corresponding to the point (x, y) of the locus, we have $x = X/(1 - Z)$, $y = Y/(1 - Z)$, where $Z \neq 1$, and thus obtain

$$a(X^2 + Y^2) + (1 - Z)(2gX + 2fY) + c(1 - Z)^2 = 0.$$

But $X^2 + Y^2 = 1 - Z^2$, so this gives

$$a(1 + Z) + 2gX + 2fY + c(1 - Z) = 0,$$

which we rewrite as

$$2gX + 2fY + (a - c)Z + a + c = 0 \qquad (*)$$

This is the equation of a plane, which must of course have non-empty intersection with S if the original locus in $\mathbb{R}^2$ is non-empty. In fact, using some elementary 3-dimensional co-ordinate geometry, we see that the plane $AX + BY + CZ + D = 0$ meets the unit sphere S if the shortest (perpendicular) distance from the origin to the plane is $\leqslant 1$, and that the condition for this is $A^2 + B^2 + C^2 \geqslant D^2$. For the plane given by $(*)$, this condition is $g^2 + f^2 \geqslant ac$, which is the condition that the locus in $\mathbb{R}^2$ given by the equation

$$a(x^2 + y^2) + 2gx + 2fy + c = 0$$

is non-empty.

This discussion suggests that we should consider together all the loci in the Gauss plane $\mathbb{R}^2 \cup \{\omega\}$ which correspond to circles on the Riemann sphere. We therefore make the following definition. A *circle* in the Gauss plane is either (i) a Euclidean circle in $\mathbb{R}^2$, or (ii) a line in $\mathbb{R}^2$ together with the point ω. If it is desired to distinguish these types of circle, they will be called respectively *central* and *linear* circles. We include in the central circles the *point-circles* consisting of a single point; all other circles will be called *proper* circles.

The equation $a(x^2 + y^2) + 2gx + 2fy + c = 0$ associated with a circle of the Gauss plane may be written in terms of the complex coordinate $z = x + iy$ in the form

$$azz̄ + \bar{\lambda}z + \lambda\bar{z} + c = 0,$$

where $\lambda = g + if \in \mathbb{C}$, $a, c \in \mathbb{R}$. The condition that this equation represents a proper circle is $\lambda\bar{\lambda} > ac$. It is straightforward but tedious to calculate the effect of a general Möbius transformation $z \mapsto z^*$ and to show that z^* satisfies an equation of the same form. We shall avoid heavy algebra by the observation that any Möbius transformation is a product of transformations of the following forms:

(i) $\qquad\qquad\qquad\qquad z \mapsto \alpha z + \beta, \quad \alpha, \beta \in \mathbb{C}, \alpha \neq 0,$

(ii) $\qquad\qquad\qquad\qquad z \mapsto 1/z.$

For a general Möbius transformation $z \mapsto (\alpha z + \beta)/(\gamma z + \delta)$, we see that if $\gamma = 0$, then the transformation is already of type (i). If $\gamma \neq 0$, then the transformation is the product $\mu_1\mu_2\mu_3$ of

$$\mu_1 : z \mapsto \gamma z + \delta,$$

$$\mu_2 : z \mapsto 1/z,$$

$$\mu_3 : z \mapsto \xi z + \eta,$$

where $\xi = (\beta\gamma - \alpha\delta)/\gamma$, $\eta = \alpha/\gamma$. Transformations of type (i) map ∞ to ∞. If $\alpha = \rho e^{i\theta}$ ($\rho \in \mathbb{R}$, $\rho > 0$) then the transformation $z \mapsto \alpha z + \beta$ is the product of

$$z \mapsto \rho z, \quad z \mapsto e^{i\theta}z, \quad z \mapsto z + \beta,$$

and the effects of these on $\mathbb{R}^2$ are respectively a dilatation, a rotation, and a translation. We are now in a position to prove

THEOREM 17.1: *The proper circles of the Gauss plane are an equivalence class of figures under the action of the Möbius group.*

Proof. To show that any Möbius transformation maps a given proper circle to a proper circle, it is sufficient to establish this property for transformations of types (i) and (ii). For those of type (i), the result follows immediately from the fact that dilatations, rotations, and transformations of $\mathbb{R}^2$ map lines to lines, and Euclidean circles to Euclidean circles. For the transformation $z \mapsto 1/z$, we observe that this maps the proper circle with equation

$$a z \bar{z} + \bar{\lambda} z + \lambda \bar{z} + c = 0 \quad (\lambda \bar{\lambda} > ac)$$

on the locus with equation

$$c z \bar{z} + \lambda z + \bar{\lambda} \bar{z} + a = 0$$

and that this is again a proper circle of the Gauss plane.

We need to show further that, given any two proper circles Γ_1, Γ_2, then there is a Möbius transformation μ which maps Γ_1 to Γ_2. Recalling that the Möbius group is (sharply) 3-transitive on the points of the Gauss plane (see Exercise 13.7), we take any three distinct points a_1, b_1, c_1 on Γ_1, and any three distinct points a_2, b_2, c_2 on Γ_2, and consider the Möbius transformation μ such that $(a_1, b_1, c_1)\, \mu = (a_2, b_2, c_2)$. Then μ maps Γ_1 to a locus through a_2, b_2, c_2 and as this locus must be a circle it coincides with Γ_2. This completes the proof.

The proper circles of the Gauss plane occupy a position in the geometry of the complex projective line which is analogous to that of lines in the geometry of the real affine (or projective) plane. Just as in affine geometry we are interested in identifying the collineations (bijections of the plane which map lines to lines), so here we ask about the circle-preserving bijections of the Gauss plane. There are such bijections which are not Möbius transformations, for the *conjugation map* $\kappa : \mathbb{C} \cup \{\infty\} \to \mathbb{C} \cup \{\infty\}$ given by $z \mapsto \bar{z}$, $\infty \mapsto \infty$, is a Euclidean reflexion of $\mathbb{R}^2$ in the x-axis, and in the Gauss plane maps central circles to central circles and linear circles to linear circles. Its fixed points are the points of the x-axis, together with ω; these are the points with complex coordinates in the set $\mathbb{R} \cup \{\infty\}$. We shall call this set the *real axis* of the Gauss plane. We shall also extend the functions $\mathbb{C} \to \mathbb{R}$ denoted by re and im to functions $\mathbb{C} \cup \{\infty\} \to \mathbb{R} \cup \{\infty\}$, by defining $\mathrm{im}\,\infty = 0$, $\mathrm{re}\,\infty = \infty$. The real axis is then

$$\{z \in \mathbb{C} \cup \{\infty\} \mid \mathrm{im}\, z = 0\}.$$

The map κ is involutory (that is, $\kappa^2 = 1$), and if $\mu \in M$, then $\kappa \mu \kappa = \bar{\mu} \in M$, it being easily verified that if $\mu : z \mapsto (\alpha z + \beta)/(\gamma z + \delta)$, then $\bar{\mu} : z \mapsto (\bar{\alpha} z + \bar{\beta})/(\bar{\gamma} z + \bar{\delta})$, and that $\alpha \delta \neq \beta \gamma$ implies $\bar{\alpha} \bar{\delta} \neq \bar{\beta} \bar{\gamma}$. Hence $\mu \kappa = \kappa \bar{\mu}$, and for every element θ in the group $P = \langle \kappa, M \rangle$ generated by κ and M, we have either $\theta = \mu$ or $\theta = \kappa \mu$ for some $\mu \in M$. Hence $P = M \cup \kappa M$, and M is a normal subgroup of P, of index 2. Every element of P is a circle-preserving transformation of the Gauss plane. We can in fact prove that these are the *only* circle-preserving transformations:

THEOREM 17.2: *Every circle-preserving bijection of the Gauss plane is a member of the inversive group P.*

Proof. Let θ be a circle-preserving bijection of $\mathbb{R}^2 \cup \{\omega\}$, and suppose that θ maps the points $(0, 0)$, $(1, 0)$, ω (with complex coordinates 0, 1, ∞) on a, b, c respectively. Let μ be the Möbius transformation which maps a, b, c on $(0, 0)$, $(1, 0)$, ω respectively and let $\phi = \theta\mu^{-1}$. Then ϕ is a circle-preserving bijection of $\mathbb{R}^2 \cup \{\omega\}$, and has fixed points $(0, 0)$, $(1, 0)$, ω. It therefore maps circles through ω to circles through ω. These are the linear circles, and hence $\phi^* = \phi | \mathbb{R}^2$ is a collineation of $\mathbb{R}^2$. From Theorem 12.8, we infer that ϕ^* is an affine transformation having $(0, 0)$ as a fixed point, and is thus of the form

$$(x\ y) \mapsto (\alpha x + \beta y,\ \gamma x + \delta y), \quad \alpha,\ \beta,\ \gamma,\ \delta \in \mathbb{R},\ \alpha\delta \neq \beta\gamma.$$

Further, $(1, 0)$ is a fixed point, giving $\alpha = 1$, $\gamma = 0$, so that the tranformation is

$$(x, y) \mapsto (x + \beta y,\ \delta y).$$

Finally, ϕ^* maps central circles to central circles, and in particular maps the unit circle $x^2 + y^2 = 1$ to the locus

$$(x + \beta y)^2 + (\delta y)^2 = 1.$$

For this to be a circle, we must have $\beta = 0$, $\delta = \pm 1$. Hence ϕ^* is either the identity or Euclidean reflexion in the x-axis, and $\phi = 1$ or $\phi = \kappa$. Hence $\theta = \mu$ or $\theta = \kappa\mu$; in either case, $\theta \in P$.

In the course of the proof of Theorem 17.2, we have established the useful

LEMMA 17.3:   *The only element of P which has the points of the real axis as its set of fixed points is the conjugation map κ.*

This leads us to

THEOREM 17.4:   *The fixed points of $\theta \in P$ are the points of a proper circle if and only if θ is a conjugate (in P) of κ.*

Proof.   First let $\theta = \sigma^{-1}\kappa\sigma$ $(\sigma \in P)$, and let the proper circle Γ be the image of the real axis under σ. Then every point of Γ is fixed by θ, and if p is fixed by θ then $p\sigma^{-1}$ is fixed by κ, and is therefore on the real axis, giving $p \in \Gamma$. Hence Γ is the set of fixed points of θ.

Now suppose that the set of fixed points of $\theta \in P$ is a proper circle Γ, and let μ be a Möbius transformation which maps the real axis to Γ. Then the fixed points of $\mu\theta\mu^{-1}$ are the points of the real axis, and by Lemma 17.1 we have $\mu\theta\mu^{-1} = \kappa$, and $\theta = \mu^{-1}\kappa\mu$. Hence θ is conjugate to κ.

It follows from the second part of the proof of Theorem 17.4 that there is just one element of P which has a given circle Γ as its set of fixed points. This transformation of the Gauss plane will be called *reflexion* in Γ, and will be denoted by ρ_Γ. If Γ is linear, then $\rho_\Gamma = \mu^{-1}\kappa\mu$, where μ may be taken as a Euclidean transformation $\mu: z \mapsto \alpha z + \beta$, $|\alpha| = 1$. Hence $\rho_\Gamma | \mathbb{R}^2$ is simply the

Euclidean reflexion in the line $\Gamma\setminus\{\omega\}$ of $\mathbb{R}^2$. If Γ is a central circle, with equation

$$(z-c)(\bar{z}-\bar{c}) = r^2, \quad r\in\mathbb{R},$$

then the transformation $\theta: z \mapsto z^*$ where

$$z^* - c = r^2/(\bar{z}-\bar{c})$$

has as fixed points precisely the points of Γ. Also, $\theta = \kappa v$, where v is the Möbius transformation

$$v: z \mapsto (r^2/(z-\bar{c})+c);$$

hence $\theta \in P$, and $\theta = \rho_\Gamma$. From the relation $z^* - c = r^2/(\bar{z}-\bar{c})$, we obtain

$$\arg(z^*-c) = \arg(z-c), \quad |z^*-c| = r^2/|z-c|.$$

The first of these tells us that c, z, z^* are collinear, with z, z^* on the same side of c. The second says that $d(z^*, c)\, d(z, c) = r^2$, where r is the radius of Γ (see Fig. 17.3). The importance of the reflexions is shown by the following theorem.

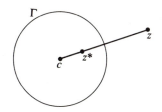

Fig. 17.3. Reflexion in a circle in the Gauss plane.

THEOREM 17.5: *Every Möbius transformation is a product of an even number of reflexions.*

Proof. It is sufficient to prove the result for a dilatation, a rotation, a translation, and the transformation $z \mapsto 1/z$. A dilatation is a product of reflexions in two concentric circles. A rotation is a product of reflexions in two intersecting lines. A translation is a product of reflexions in two parallel lines. The transformation $z \mapsto 1/z$ is the product of reflexion in the unit circle ($z \mapsto 1/\bar{z}$) and reflexion in the real axis ($z \mapsto \bar{z}$).

Since the Möbius group and κ generate P, we obtain

COROLLARY 17.6: *The reflexions generate the inversive group P.*

The transformation of reflexion in a central circle appears in elementary Euclidean geometry under the name of *inversion* (being considered as a transformation of $\mathbb{R}^2\backslash\{c\}$, where c is the centre of the circle). We have avoided the term 'inversion', which some writers take to include reflexion in a line, while others restrict its meaning to reflexion (as we have called it) in a central circle. However, the terms 'inversive group' and 'inversive geometry' do appear to be standard usage, and we have adopted them. It is also convenient to describe the image v of u under the reflexion ρ_Γ as the *inverse* of u with respect to Γ. Since ρ_Γ is involutory, u is the inverse of v, and we say that u, v are mutually inverse with respect to Γ. The following easy result is sufficiently important to be enunciated as a theorem.

THEOREM 17.7: *If u, v are mutually inverse with respect to Γ and if $\sigma \in P$, then $u\sigma$, $v\sigma$ are mutually inverse with respect to $\Gamma\sigma$.*

Proof.   The transformation $\sigma^{-1}\rho_\Gamma\sigma$ is in P, and it is easily checked that $\Gamma\sigma$ is its set of fixed points. Hence $\sigma^{-1}\rho_\Gamma\sigma = \rho_{\Gamma\sigma}$. Now $u\rho_\Gamma = v$, hence

$$(u\sigma)\rho_{\Gamma\sigma} = (u\sigma)(\sigma^{-1}\rho_\Gamma\sigma) = v\sigma.$$

In one-dimensional projective geometry (over any field F) the cross-ratio (a, b, c, d) defined by

$$(a, b, c, d) := \frac{(a-b)(c-d)}{(a-d)(c-b)}, \quad a, b, c, d \in F \cup \{\infty\},$$

is an important invariant of the projective group (see Exercise 5.12), and we expect it to appear in any discussion of the Gauss plane under the action of the Möbius group. If σ is the Möbius transformation which maps a, b, c to 0, 1, ∞ respectively, and $z\sigma = w$, then

$$(z, a, b, c) = (w, 0, 1, \infty) = w,$$

so that (z, a, b, c) is the image of z under σ. If Γ is the circle through a, b, c, then σ maps Γ to the real axis, and z lies on Γ if and only if w lies on the real axis. The circle Γ may therefore be described as

$$\{z \in \mathbb{C} \cup \{\infty\} \mid \mathrm{im}(z, a, b, c) = 0\},$$

We now consider *directed* circles in the Gauss plane. Intuitively, a direction on a circle is prescribed by an arrow indicating a sense of description. Any circle divides the Gauss plane (or the Riemann sphere) into two regions, which it is natural to distinguish as the *left-hand* and *right-hand* regions (with respect to the given sense of description), and which we denote by L, R respectively (see Fig. 17.4).

If now $\sigma \in P$ maps Γ_1 to Γ_2, and a direction is prescribed on Γ_1, the mapping prescribes a corresponding direction on Γ_2. If (L_1, R_1) and (L_2, R_2) are the left-hand and right-hand regions of Γ_1 and Γ_2, then *either* σ maps L_1 to L_2 and R_1 to R_2, *or* σ maps L_1 to R_2 and R_1 to L_2. In the first case we say that σ preserves the orientation of the directed circle Γ_1, in the second case we say that σ reverses the orientation. We shall find that σ either preserves, or

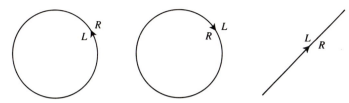

Fig. 17.4. Orientation and circles in the Gauss plane.

reverses, the orientations of all directed circles, and that the orientation-preserving transformations in P are precisely those which belong to the Möbius group M.

Our first task is to give less intuitive definitions of a direction on a circle Γ, and of the corresponding left-hand and right-hand regions of the Gauss plane. A direction on Γ will be specified by an ordered set (a, b, c) of three points on Γ. For the ordered set $(0, 1, \infty)$ on the real axis, it is natural to define the left-hand region as the upper half-plane $\{z \mid \text{im } z > 0\}$. We therefore make the following definition.

If Γ is a directed circle, with direction specified by the ordered triad $(a, b\ c)$ of points on Γ, then the *left-hand* and *right-hand regions* of Γ are $L := \{z \mid \text{im}(z, a, b, c) > 0\}$, $R := \{z \mid \text{im}(z, a, b, c) < 0\}$, respectively. We can now establish the results previously announced.

THEOREM 17.8:   *The conjugation map κ (reflexion in the real axis) reverses the orientation of any directed circle.*

Proof.   Let Γ_1 be a circle with direction prescribed by (a, b, c), and let L_1, R_1 be the corresponding regions. If $\Gamma_2 = \Gamma_1 \kappa$, the corresponding direction on Γ_2 is prescribed by $(\bar{a}, \bar{b}, \bar{c})$, and the left-hand and right-hand regions of Γ_2 are

$$L_2 = \{z \mid \text{im}(z, \bar{a}, \bar{b}, \bar{c}) > 0\}$$
$$R_2 = \{z \mid \text{im}(z, \bar{a}, \bar{b}, \bar{c}) < 0\}.$$

But $(\bar{z}, \bar{a}, \bar{b}, \bar{c}) = \overline{(z, a, b, c)}$, giving $\text{im}(\bar{z}, \bar{a}, \bar{b}, \bar{c}) = -\text{im}(z, a, b, c)$; hence if $z \in L_1$, then $\bar{z} \in R_2$, and if $z \in R_1$, then $\bar{z} \in L_2$.

THEOREM 17.9:   *A Möbius transformation preserves the orientation of any directed circle.*

Proof.   Let Γ_1 be a circle with direction prescribed by (a, b, c), and let L_1, R_1 be the corresponding regions. Let $\Gamma_2 = \Gamma_1 \mu$, where μ is any Möbius transformation, and let μ map a, b, c on a^*, b^*, c^* respectively. Then the direction on Γ_2 prescribed by (a^*, b^*, c^*) corresponds (under the transformation μ) to the given direction on Γ_1, and the left-hand and right-hand regions of Γ_2 are

$$L_2 = \{z \mid \text{im}(z, a^*, b^*, c^*) > 0\},$$
$$R_2 = \{z \mid \text{im}(z, a^*, b^*, c^*) < 0\}.$$

But if $z\mu = z^*$, it follows from the invariance of the cross-ratio under μ that $(z^*, a^*, b^*, c^*) = (z, a, b, c)$. Hence if $z \in L_1$, then $z^* \in L_2$, and if $z \in R_1$, then $z^* \in R_2$.

COROLLARY 17.10: *The orientation-preserving elements of P are those in the subgroup M; the orientation-reversing elements are those in the coset κM.*

Since all reflexions belong to the coset κM, we also have

COROLLARY 17.11: *Every reflexion is orientation-reversing.*

It is easy to see intuitively that a rotation of the Riemann sphere S about an axis through its centre is not only circle-preserving, but also preserves the orientation of any directed circle on the sphere. We therefore expect a rotation of S to correspond to a Möbius transformation of the Gauss plane. This is confirmed by the following calculation of the transformation of the Gauss plane with fixed points c, $-1/\bar{c}$, which represents a rotation of the sphere through an angle $\theta \in \mathbb{R}/2\pi\mathbb{Z}$ about the directed axis $\hat{u}$, where $\hat{u}$ is the unit vector in $\mathbb{R}^3$ determined by the point of S which corresponds to c.

We consider first the Möbius transformation

$$\sigma : z \mapsto \frac{z - c}{\bar{c}z + 1}$$

which maps c, $-1/\bar{c}$ to 0, ∞ respectively. We can show that this corresponds to a Euclidean transformation of the sphere by proving that it preserves the χ-metric; the verification that $\chi(z\sigma, w\sigma) = \chi(z, w)$ is left to the reader. The transformation has just two fixed points, given by $z^2 = -c/\bar{c}$, hence it is not a reflexion or rotatory reflexion on S, and therefore must be a rotation.

Now it is clear that a rotation of the sphere through an angle θ about the polar axis directed towards the *south* pole (represented by $z = 0$) corresponds to the rotation of the Gauss plane about the origin given by $\rho : z \mapsto e^{-i\theta}z$. Hence the Möbius transformation $\tau = \sigma\rho\sigma^{-1}$ represents a rotation of the sphere S with the required angle and directed axis.

The transformation τ can now be calculated, and we find that it can be written in the form

$$\tau : z \mapsto \frac{\alpha z + \beta}{-\bar{\beta}z + \bar{\alpha}},$$

where

$$\alpha = (c\bar{c}e^{\frac{1}{2}i\theta} + e^{-\frac{1}{2}i\theta})/(1 + c\bar{c}),$$
$$\beta = c(e^{\frac{1}{2}i\theta} - e^{-\frac{1}{2}i\theta})/(1 + c\bar{c}).$$

We see that $\alpha\bar{\alpha} + \beta\bar{\beta} = 1$, and hence that the matrix

$$T = \begin{bmatrix} \alpha & \beta \\ -\bar{\beta} & \bar{\alpha} \end{bmatrix}$$

which represents τ satisfies $T\bar{T}' = I$, det $T = 1$, and is therefore a special unitary matrix. Since $\theta \in \mathbb{R}/2\pi\mathbb{Z}$ gives two values of $e^{\frac{1}{2}i\theta}$ (differing only in sign), we obtain two special unitary matrices $\pm T$ which represent a given rotation in SO(3). We have thus by another route arrived at the unitary representation of SO(3) which was obtained in Chapter 16 by way of the quaternion algebra and which is expressed by the isomorphism

$$SO(3) \cong SU(2)/\{I, -I\}.$$

We return now to the study of general Möbius transformations, and we show that these transformations of the Gauss plane have the further important property of preserving the angle of intersection of two circles. We must first define clearly what we mean by the angle of intersection. If we consider two intersecting lines in a physical plane, we see that the notion of angle of intersection is ambiguous, but if we give directions to the lines and take them in a specific order, then the angle through which the first line must be turned in order to make it coincide in position and direction with the second line, is uniquely determined modulo 2π. These observations motivate the formal definition for lines in $\mathbb{R}^2$, using complex coordinates. If l_1, l_2 are lines meeting at $a \in \mathbb{C}$ and directions on the lines are established by naming points on l_1, l_2

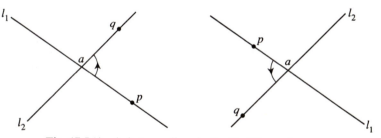

Fig. 17.5. Angle between lines in the Euclidean plane.

with complex coordinates p, q (See Fig. 17.5), then we define

$$\angle(l_1, l_2) := \arg\left(\frac{q-a}{p-a}\right) \in \mathbb{R}/2\pi\mathbb{Z}.$$

Note that if we reverse the directions on *both* lines, we do not alter the value of $\angle(l_1, l_2)$. Note also that the definition implies

$$\angle(l_1, l_2) = -\angle(l_2, l_1).$$

Now, in the Gauss plane two linear circles Λ_1, Λ_2 which meet at a meet again at ∞, and

$$\frac{q-a}{p-a} = (a, q, \infty, p).$$

We therefore define

$$\angle_a(\Lambda_1, \Lambda_2) := \arg(a, q, \infty, p),$$

the directions of Λ_1, Λ_2 being specified by the triads $(a, p, \infty), (a, q, \infty)$. In this definition, our notation mentions the point a, as there are two points of intersection. Note that, for two given linear circles, $\arg(a, q, \infty, p)$ depends only on the directions specified by (a, p, ∞) and (a, q, ∞) and otherwise does not depend on the particular positions of p, q on Λ_1, Λ_2. If (a, p', ∞), (a, q', ∞) specify directions opposite to those determined by (a, p, ∞), (a, q, ∞), we have

$$\arg(a, q, \infty, p') = \arg(a, q', \infty, p) = \arg(a, q, \infty, p) + \pi,$$

and

$$\arg(a, q', \infty, p') = \arg(a, q, \infty, p).$$

If now we have any two circles Γ_1, Γ_2 in the Gauss plane which meet at distinct points a, b, and directions on Γ_1, Γ_2 are specified by triads (a, p, b), $(a, q, b)\,(p \in \Gamma_1, q \in \Gamma_2)$ (Fig. 17.6), we define the angle of intersection of Γ_1, Γ_2

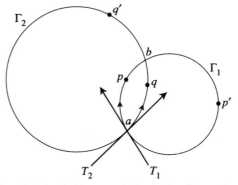

Fig. 17.6. Angle between circles in the Gauss plane.

at a by

$$\angle_a(\Gamma_1, \Gamma_2) := \arg(a, q, b, p).$$

The points a, b separate each of Γ_1, Γ_2 into two arcs, and the value of $\arg(a, q, b, p)$ depends on the arcs of Γ_1, Γ_2 on which p, q lie. If p, p' lie on different arcs of Γ_1 and q, q' lie on different arcs of Γ_2, then $(b, p', a), (b, q', a)$ specify the same directions on Γ_1, Γ_2 as $(a, p, b), (a, q, b)$, and we have

$$\angle_b(\Gamma_1, \Gamma_2) = \arg(b, q', a, p') = \arg(b, q, a, p) = -\angle_a(\Gamma_1, \Gamma_2).$$

The invariance of the cross-ratio ensures that an angle of intersection of two directed circles is preserved by all Möbius transformations. On the other hand, the transformation κ (and hence any transformation in the coset κM of the Möbius group) changes the sign of the angle, because

$$\arg(\bar{a}, \bar{q}, \bar{b}, \bar{p}) = \arg\overline{(a, q, b, p)} = -\arg(a, q, b, p).$$

Although the definition of $\angle_a(\Lambda_1, \Lambda_2)$ clearly expresses what we normally mean by the angle between two linear circles (or lines in $\mathbb{R}^2$), it is less obvious that the definition of $\angle_a(\Gamma_1, \Gamma_2)$ does the same for central circles. To satisfy ourselves on this point, we take the tangents T_1, T_2 to Γ_1, Γ_2 at a, and give them directions agreeing with those of Γ_1, Γ_2. (The tangent T to a circle Γ at a is the unique linear circle which meets Γ at a and at no other point. The direction of T agrees with that of Γ if *either* the left-hand region of Γ lies in the left-hand region of T *or* the right-hand region of Γ lies in the right-hand region of T.) We proceed to show that $\angle_a(\Gamma_1, \Gamma_2) = \angle_a(T_1, T_2)$.

We apply a Möbius transformation which maps a to ∞ and ∞, b, T_1, T_2, Γ_1, Γ_2 to ∞^*, b^*, T_1^*, T_2^*, Γ_1^*, Γ_2^* respectively. Then T_1^*, T_2^*, Γ_1^*, Γ_2^* are all linear circles, and T_1^*, Γ_1^* meet only at ∞, and hence are parallel; similarly T_2^*, Γ_2^* are parallel (see Fig. 17.7). Now, T_1^*, T_2^* meet at ∞^*, and Γ_1^*, Γ_2^*

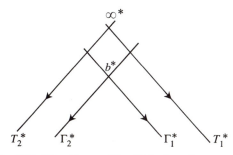

Fig. 17.7. The effect on Fig. 17.6 of sending a to ∞.

meet at b^*, so we have

$$\angle_{b^*}(\Gamma_1^*, \Gamma_2^*) = \angle_{\infty^*}(T_1^*, T_2^*),$$

hence

$$\angle_b(\Gamma_1, \Gamma_2) = \angle_\infty(T_1, T_2),$$

and

$$\angle_a(\Gamma_1, \Gamma_2) = \angle_a(T_1, T_2).$$

A transformation which preserves the angle of intersection of directed curves is said to be *conformal*, so that the outcome of this discussion is

THEOREM 17.12: *The Möbius transformations are conformal on circles.*

The property of conformality may be formulated in much more general terms. Given two directed differentiable arcs in $\mathbb{R}^2$, we can define the angle between the arcs at a point of intersection as the angles between the tangents (with directions agreeing with those of the arcs). We can then discuss conformality at the point of intersection. In terms of the complex coordinatisation of $\mathbb{R}^2$, we are looking for transformations $z \mapsto f(z)$ which preserve the angle of intersection of any two directed differentiable arcs meeting at $z = c$ (say), and it is a major and well-known result of the complex differential calculus that all functions $f(z)$ having a non-zero derivative at $z = c$ have the required property. The Möbius transformation given by $f(z) = (\alpha z + \beta)/(\gamma z + \delta)$ has a non-zero derivative, namely $f'(z) = (\alpha\delta - \beta\gamma)/(\gamma z + \delta)^2$, for all $z \in \mathbb{C}$, $z \neq -\delta/\gamma$, and is thus conformal in the wider sense at all such points.

Complex analysis is usually developed in terms of functions $f:\mathbb{C} \to \mathbb{C}$, but there are considerable advantages in extending the discussion to functions $f:\mathbb{C} \cup \{\infty\} \to \mathbb{C} \cup \{\infty\}$. The space $\mathbb{C} \cup \{\infty\}$ is topologised by the χ-metric, but this agrees on $\mathbb{C}$ with the topology defined by the ordinary d-metric ($d(z_1, z_2) = |z_1 - z_2|$). A relation $\chi(z, \infty) < \delta$ is equivalent to a relation $|z| > \Delta$, hence the latter defines a neighbourhood of ∞, and we can handle the topology of $\mathbb{C} \cup \{\infty\}$ in terms of the modulus. There is no difficulty in saying what we mean by a continuous function $f:\mathbb{C} \cup \{\infty\} \to \mathbb{C} \cup \{\infty\}$. In discussing differentiability at a point $c \in \mathbb{C}$ we now admit ∞ as a value of the derivative: $f'(c)$ exists and is equal to ∞ provided that, given $\Delta \in \mathbb{R}$, $\Delta > 0$, there exists $\delta \in \mathbb{R}$, $\delta > 0$, such that

$$\left| \frac{f(z) - f(c)}{z - c} \right| > \Delta \text{ whenever } |z - c| < \delta.$$

Differentiability at $z = \infty$ is satisfactorily defined by considering $g(z) := f(1/z)$, and saying that $f(z)$ is differentiable at ∞ if and only if $g(z)$ is differentiable at $z = 0$.

If U is a connected open set of $\mathbb{C} \cup \{\infty\}$, and $f:U \to \mathbb{C} \cup \{\infty\}$ is differentiable at all points of U, we say that f is *analytic* on U. It should be noted that a treatment not admitting the value ∞ would describe such a function as *meromorphic*. Points c such that $f(c) = \infty$ would be singular points at which $f(c)$ is not defined; such points would be the type of singularity called a *pole*, and a function f is meromorphic if all its singularities are poles.

It is proved in standard texts (for example, Ahlfors, *Complex Analysis*, pp. 42–6) that the only analytic functions $f:\mathbb{C} \cup \{\infty\} \to \mathbb{C} \cup \{\infty\}$ are the rational functions given by $f(z) = p(z)/q(z)$, where p, q are polynomial functions. Such a function is bijective if and only if p, q are linear polynomials, and $p(z)/q(z)$ is not a constant. Hence the only analytic bijections $f:\mathbb{C} \cup \{\infty\} \to \mathbb{C} \cup \{\infty\}$ are the Möbius transformations. An analytic bijection has an inverse which is also analytic, and is therefore called *bi-analytic*. The fact that the Möbius transformations are the bi-analytic transformations of the Gauss plane gives them the important position that they occupy in the general theory of complex analytic functions.

Exercises

17.1. Show that if a, b are mutually inverse with respect to a circle Γ, and $c \in \Gamma$, then the (unique) Möbius transformation which maps a, b, c to 0, ∞, 1 respectively maps Γ on the unit circle.

Hence show that, given two circles Γ, Γ^* with $p \in \Gamma$, $q \notin \Gamma$ and $p^* \in \Gamma^*$, $q^* \notin \Gamma^*$, there is a unique Möbius transformation which maps Γ to Γ^*, p to p^* and q to q^*.

17.2. Find all Möbius transformations which map the upper half plane (im $z > 0$) on the interior of the unit circle ($|z| < 1$), and map the imaginary axis (re $z = 0$) on the real axis (im $z = 0$).

17.3. Find a Möbius transformation which maps the region

$$\{(x, y) \mid x^2 + y^2 < 1, x^2 - 4x + y^2 + 1 < 0, y > 0\}$$

on the region

$$\{(x, y) \mid x^2 + y^2 > 1, x > 0, y > 0\}.$$

17.4. We define a χ-circle with χ-centre $a \in \mathbb{C} \cup \{\infty\}$ and χ-radius $r \in \mathbb{R}$, $r > 0$, as the locus

$$\{z \in \mathbb{C} \cup \{\infty\} \mid \chi(z, a) = r\},$$

where χ is the chordal metric defined on p. 212. Show that every proper circle (central or linear) of the Gauss plane is a χ-circle, and has two χ-centres a_1, a_2 and corresponding χ-radii r_1, r_2, such that

$$a_1 \bar{a}_2 = -1, \quad r_1^2 + r_2^2 = 4.$$

Find the χ-centres and χ-radii of the following circles:

(i) im $z = 0$; (ii) re $z = 1$; (iii) $|z - 1| = 1$.

17.5. Show that a Möbius transformation $\neq 1$ has either one or two fixed points. Show that the following are a complete set of canonical representatives of the conjugacy classes of the Möbius group:

(i) $z \mapsto z + 1$;

(ii) $z \mapsto az, |a| = 1, \arg a \in (0, \pi]$;

(iii) $z \mapsto az, |a| > 1$;

(iv) $z \mapsto z$.

(Transformations in these conjugacy classes are called (respectively)
(i) *parabolic*, (ii) *elliptic*, (iii) *hyperbolic* if a is real, otherwise
loxodromic. A class of hyperbolic transformations with canonical
representative $z \mapsto \lambda z$ is said to be *proper* if $\lambda > 1$ and *improper* if
$\lambda < -1$.)

17.6.   For a Möbius transformation $\theta : z \mapsto (az+b)/(cz+d)$ we define $\sigma(\theta) \in \mathbb{C}$
by

$$\sigma(\theta) := ((a-d)^2 + 4bc)/(ad-bc).$$

Show that $\sigma(\theta) = 0$ if and only if θ is either parabolic or the identity,
and that $\sigma(\theta) = \sigma(\phi) \neq 0$ if and only if θ and ϕ are conjugate in the
Möbius group.

17.7.   With the notations of Exercise 17.6, suppose that $\sigma(\theta)$ is real and
$\sigma(\theta) \neq 0$. Show that

(i)   if $\sigma(\theta) > 0$, then θ is proper hyperbolic;
(ii)   if $-4 \leqslant \sigma(\theta) < 0$, then θ is elliptic;
(iii) if $\sigma(\theta) < -4$, then θ is improper hyperbolic.

Show also that if $\sigma(\theta)$ is not real, then θ is loxodromic.

17.8.   Show that, given four distinct elements $a, b, c, d \in \mathbb{C} \cup \{\infty\}$, the 24
ordered sets which can be formed from the four elements give only six
(at most) distinct cross-ratios, and that these are of the form

$$\lambda, \quad 1/\lambda, \quad 1-\lambda, \quad 1-(1/\lambda), \quad 1/(1-\lambda), \quad \lambda/(\lambda-1).$$

Show that these six complex numbers are distinct except when they
take the three values $-1, 2, \tfrac{1}{2}$ or the two values $-\omega, -\omega^2$, where
$\omega = \exp(2\pi i/3)$. (In these cases, the four elements are said to form a
harmonic or an *equianharmonic* set, respectively.)

17.9.   With the notations of Exercise 17.8, suppose that $\{a, b, c, d\}$ is neither
harmonic nor equianharmonic. The group $G = S_4$ of all permutations
of $\{a, b, c, d\}$ acts in a natural way on the set Ω of six cross-ratios. Show
that G acts transitively, and find the stabiliser of the element
$\lambda = (a, b, c, d)$ in Ω.

Show that the action of G is not primitive on Ω, and find all sets of
imprimitivity systems (that is, ρ-classes of congruences ρ on Ω, as
described in Chapter 7).

Show that if $\{a, b, c, d\}$ is a harmonic set, then G acts transitively and
primitively on the set Ω^* of three cross-ratios. Find the stabiliser of a
point of Ω^* and verify that it is a maximal subgroup of G (see Corollary
7.6).

17.10. If a, b, c, d are points on a circle in the Gauss plane, we know that the cross-ratio $(a, b, c, d) \in \mathbb{R}$. If $(a, b, c, d) < 0$, we say that the pair $\{a, c\}$ separates the pair $\{b, d\}$. Show that this defines a property of a partition of $\{a, b, c, d\}$ into two (unordered) pairs, that there are three such partitions, and one in which one pair separates the other pair.

17.11. Let $a, b, c, d \in \mathbb{C}$ represent points on a circle in the Gauss plane, and let $(a, b, c, d) < 0$. Show that $(a, c, b, d) = 1 - (a, b, c, d) > 0$, and hence that

$$|a-d|.|b-c| + |a-b|.|c-d| = |a-c|.|b-d|.$$

[For a central circle, this is Ptolemy's theorem, a classical result of Euclidean geometry about cyclic quadrilaterals.]

17.12. Let (a, b, c), (a', b', c') be ordered triads of distinct points on a circle Γ in the Gauss plane. We define homeomorphisms $\phi: \Gamma \to \mathbb{R} \cup \{\infty\}$, $\phi': \Gamma \to \mathbb{R} \cup \{\infty\}$ by $z \mapsto (a, b, c, z)$, $z \mapsto (a', b', c', z)$, and let $\chi = \phi^{-1}\phi'$. Then $\chi: \mathbb{R} \cup \{\infty\} \to \mathbb{R} \cup \{\infty\}$ is a homeomorphism. Now let I be an open interval of $\mathbb{R}$, with $\infty \notin I\chi$, so that $J = I\chi$ is an open interval of $\mathbb{R}$. Then $\chi|I = \chi^*: I \to J$ is a homeomorphism, and is monotonic. Show that χ^* is either increasing or decreasing for every such interval I.

We may thus describe χ itself as either increasing or decreasing. We say that (a, b, c), (a', b', c') have the same *direction* on Γ if χ is increasing, and opposite directions if χ is decreasing. Show (from this definition of direction) that if (a', b', c') is a permutation of (a, b, c), then their directions are the same or opposite according as that permutation is even or odd.

17.13. Show that the elements of $P \setminus M$ $(= \kappa M)$ which are of order 2 are those which can be represented in the form $z \mapsto (a\bar{z}+b)/(c\bar{z}+d)$ with $b, c \in \mathbb{R}$, and $d = -\bar{a}$.

Show that if $a\bar{a} + bc > 0$, then the transformation is a reflexion. Describe in geometrical terms the transformations for which $a\bar{a} + bc < 0$.

17.14. Show that if a rotation in SO(3) of angle $\theta \in \mathbb{R}/2\pi\mathbb{Z}$ is represented by a unitary matrix A, then $\text{tr} A = \pm 2 \cos \frac{1}{2}\theta$. Let

$$A = \begin{bmatrix} e^{\frac{1}{2}i\theta} & 0 \\ 0 & e^{-\frac{1}{2}i\theta} \end{bmatrix}, \quad B = \begin{bmatrix} \alpha & \beta \\ -\bar{\beta} & \bar{\alpha} \end{bmatrix},$$

where $\text{tr} B = \alpha + \bar{\alpha} = 2 \cos(\frac{1}{2}\theta)$, $\det B = \alpha\bar{\alpha} + \beta\bar{\beta} = 1$, and let $C = ABA^{-1}B^{-1}$. Show that $\text{tr} C = 2(\alpha\bar{\alpha} + \beta\bar{\beta} \cos \theta)$. Hence show that SO(3) is simple (cf. Exercise 16.11).

17.15. Two distinct lines l, l' of $\mathbb{R}^2$ meet at p, and $(p, a, b), (p, a', b')$ are triads of distinct points on l, l' respectively. The lines $(a, a'), (b, b')$ meet at $q \in \mathbb{R}^2$. The unique Möbius transformation of the Gauss plane $\mathbb{R}^2 \cup \{\omega\}$ which maps p to p, a to a', and b to b' is denoted by μ. Show that

(i) q is a second fixed point of μ;

(ii)   if $x \in l$, $x' \in m$ and q, x, x' are collinear, then $x\mu = x'$.

17.16. Show that every element $\neq 1$ of the Möbius group which is of finite order has two distinct fixed points.

Show that the analysis of poles and their orbits for finite subgroups of SO(3) which was presented in Chapter 15 (p. 177 et seq.) may be applied to the study of finite subgroups of the Möbius group.

17.17. By considering positive-definite Hermitian inner products in $V_n(\mathbb{C})$, it may be shown that every finite subgroup of $GL(n, \mathbb{C})$ is conjugate to a subgroup of the unitary group $U(n)$ (cf. Lemma 15.2 and the discussion which follows it). Assuming this result, and using the representation of elements of the Möbius group by unimodular 2×2 matrices, show that any finite subgroup of the Möbius group is conjugate to a group of transformations of the Gauss plane corresponding to a group of rotations of the Riemann sphere, and is therefore isomorphic to a finite subgroup of SO(3).

Show that any finite subgroup of the inversive group P is isomorphic to a subgroup of O(3).

Show also that for every finite subgroup of SO(3) there is an isomorphic subgroup of M, and that for every finite subgroup of O(3) there is an isomorphic subgroup of P.

Chapter 18

Topological considerations

The aim of this chapter is to give a brief indication of the way in which topological notions may be brought into play in the study of G-spaces. These ideas are of great importance for further developments in the theory of manifolds, Lie groups and global analysis. Here we can only try to give a preliminary glimpse of the results which flow from the fruitful union of algebra and topology.

Familiarity with the elements of general topology will be assumed. As we are attempting a description rather than a formal exposition, many of our statements are not supported by proof. In some cases, the proof is a straightforward verification from the definitions, which the reader should be able to supply. Some less obvious results are included (with guidance) in the Exercises at the end of this chapter. For the harder results, the reader is referred to the literature and a short bibliography appears at the end of the chapter.

In Chapters 10–17 we made several references to the topology of the space Ω in the geometry (Ω, G) under discussion. The spaces Ω all exhibited a natural topology, being the topology of $\mathbb{R}^n$, or (as in the case of the Riemann sphere) induced by that topology on a subspace. The topology was thus in all cases metric, and hence Hausdorff. In earlier chapters the preoccupation was with finite spaces Ω. For such spaces, the only Hausdorff topology is the discrete topology, which adds nothing to the set-theoretic structure of the space, and is consequently uninteresting. For the geometrical G-spaces, the topology is relevant, because the elements of G are continuous transformations (indeed, homeomorphisms) of Ω.

It is less obvious that the group G which acts on Ω may also be given a relevant topology. It is however quite natural (in relation to the Euclidean group, for instance) to talk of a *small* transformation θ, in the sense that θ maps any point $\omega \in \Omega$ to a point $\omega\theta$ which is *near* to ω (a notion which can be expressed in terms of the topology of Ω). This suggests that we have in mind some measure of the separation of $\theta \in G$ from the identity $1 \in G$, and thus leads us on to think of groups endowed with a topology which is meaningfully related to the group operation. Intuitively, if $\theta, \phi \in G$ are *near*, this must mean that $\theta\phi^{-1}$ and $\theta^{-1}\phi$ are small; that if θ is small, then its inverse θ^{-1} is small; and that if θ is small and $\phi \in G$, then $\theta\phi$ and $\phi\theta$ are near to ϕ. Considerations of this sort lead to the following formal definition.

Definition.   A *topological group* G is a group, together with a topology on the set of elements of G, such that
   (i) the map $G \to G$ given by $\theta \mapsto \theta^{-1}$ is continuous,
   (ii) the map $G \times G \to G$ given by $(\theta, \phi) \mapsto \theta\phi$ is continuous.

Continuity must of course be interpreted in terms of the topology on G and the corresponding product topology on $G \times G$. Any subgroup H has an induced topology, and is a topological group with respect to this topology.
   It follows from the definition that the left and right translations

$$\lambda_\theta : \phi \mapsto \theta\phi, \quad \rho_\theta : \phi \mapsto \phi\theta$$

determined by any $\theta \in G$ are continuous, and indeed homeomorphisms of G to itself (since they have continuous inverses $\lambda_{\theta^{-1}}$ and $\rho_{\theta^{-1}}$). A topological space T such that for any x, $y \in T$ there is a homeomorphism $T \to T$ which maps x to y, is said to be *homogeneous*. Any topological group has this property, and we may therefore prove topological results for an arbitrary point by considering some particular point, the most convenient point to take being the identity element $1 \in G$. One such result, and an important one, is that the topology of any topological group is *regular*. This means that, given a closed set C and a point $p \notin C$, then there are open sets U, V such that $p \in U$, $C \subseteq V$ and $U \cap V = \varnothing$. This is equivalent to the property that, if X is open and $p \in X$, then there is an open set Y, with $p \in Y$ and $\bar{Y} \subseteq X$, where $\bar{Y}$ denotes the closure of Y (see Exercise 18.4). Regularity does not ensure the Hausdorff property, but if a space is T_1 (which means that every singleton subset is closed) and also regular, then it is Hausdorff. It follows from homogeneity that a topological group is Hausdorff if and only if the set $\{1\}$ is closed.
   The algebraic theory of G-spaces pointed out the importance of the right coset spaces $\cos(G : H)$, on which G acts in a natural way. If G is a topological group, then $\cos(G : H)$ inherits a topology from G, in the following manner. If T is any topological space, and $\sim$ an equivalence relation on T, we have a natural surjection $T \to T/\sim$, where $T/\sim$ is the set of equivalence classes, which maps each element of T on the equivalence class which contains it. We topologise $T/\sim$ by giving it the largest topology which makes the surjection continuous. In other words, the open sets of $T/\sim$ are precisely those for which the inverse image is open in T. The inverse image of a set $\{X_\lambda\}$ of equivalence classes is of course the set $\bigcup X_\lambda$ of their elements. Since closed sets are complements of open sets, $\{X_\lambda\}$ is closed in $T/\sim$ if and only if $\bigcup X_\lambda$ is closed in T.
   Clearly $\cos(G : H)$ is $G/\sim$, where $x \sim y$ if and only if $xy^{-1} \in H$, and we topologise $\cos(G : H)$ accordingly. This makes $\cos(G : H)$ a homogeneous space, the appropriate homeomorphisms being provided by right multiplication of cosets by elements of G. Moreover it can be shown that $\cos(G : H)$ is regular (see Exercise 18.5), and hence that it is Hausdorff if and only if its singleton sets are closed. For this (by homogeneity) it is sufficient that $\{H\}$

shall be closed in $\cos(G:H)$, which is so if and only if H is closed in G. The *closed* subgroups of a topological group therefore occupy a very important position; indeed, some writers require as part of the definition of a topological subgroup H of G that H shall be closed in the topology of G.

If we have a topological group G acting on a topological space Ω, we certainly wish the topologies to conform, in the sense that if ω_1, ω_2 are near (in Ω) and g_1, g_2 are near (in G), then $\omega_1^{g_1}, \omega_2^{g_2}$ are near (in Ω). Formally expressed, this is the condition that the action $\mu:\Omega \times G \to \Omega$ shall be continuous (where $\Omega \times G$ has the product topology derived from the topologies of Ω and G). We therefore make the following definition.

Definition.   A *topological G-space* is a triple (Ω, G, μ) where Ω is a topological space, G is a topological group, and the action $\mu:\Omega \times G \to \Omega$ is continuous.

It is a consequence of the definition that, for a given $g \in G$, the maps $\omega \mapsto \omega^g$, $\omega \mapsto \omega^{g^{-1}}$ are mutually inverse bijective continuous maps $\Omega \to \Omega$, and hence homeomorphisms. If G acts faithfully on Ω, we may therefore identify G with a group of homeomorphisms of Ω. If G acts transitively on Ω, then Ω is a homogeneous space.

We recall from the algebraic theory of G-spaces that if G acts transitively on Ω, and G_α is the stabiliser of any point $\alpha \in \Omega$, then there is a bijection

$$\theta : \cos(G:G_\alpha) \to \Omega$$

which is an (algebraic) G-isomorphism. Now it follows from the continuity of the action μ that the surjective map $\phi:G \to \Omega$ given by $\phi:g \mapsto (\alpha, g)\mu = \alpha^g$ is continuous. The stabiliser G_α is the inverse image of the set $\{\alpha\} \subset \Omega$; if the topology of Ω is Hausdorff, $\{\alpha\}$ is closed and hence G_α is closed in G.

Now the natural (continuous) projection $\pi:G \to \cos(G:G_\alpha)$ can be shown to be an open map, that is, it maps open sets to open sets (see Exercise 18.2). Using this fact, and the commutativity of the following diagram, we see

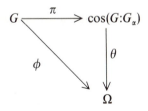

without difficulty that θ is continuous. It is not in general the case that θ is a homeomorphism; an obvious necessary and sufficient condition for this to hold is that ϕ is an open map, and we can easily construct counter-examples in which G has a strong topology (for example, the discrete topology). There are however some very general conditions on the topologies of G and Ω which do ensure that θ is a homeomorphism. The simplest of these is derived from the well-known general result that if X is a compact space and Y is a Hausdorff space, then a continuous bijection $X \to Y$ is necessarily a homeomorphism. If G is compact, then its continuous image $\cos(G : G_\alpha)$ is also compact. Hence if G is compact and Ω is Hausdorff (in which case the existence of the continuous surjection ϕ tells us that both G and Ω are compact and Hausdorff), then θ is a homeomorphism.

Many of the groups which occur in geometry are not compact, but do have the property of *local compactness*, and this, together with a *countability* condition which requires that the space has a covering by a countable family of compact sets, is sufficient to ensure the desired homeomorphism property. We quote from the classical treatise of Pontryagin (1946, Theorem 20) the following theorem.

THEOREM 18.1:   *If Ω is a transitive G-space such that (i) Ω and G are both locally compact Hausdorff spaces, and (ii) G satisfies the countability condition, then $\theta : \cos(G : G_\alpha) \to \Omega$ is a homeomorphism.*

The groups which particularly concern us in geometry are derived from $GL(n, \mathbb{R})$ by taking subgroups and homomorphic images. (It should be noted that $GL(n, \mathbb{C})$ may be regarded as a subgroup of $GL(2n, \mathbb{R})$.) Such groups have acquired the family name of *classical groups*, and they all carry a topology inherited from the natural topology of $GL(n, \mathbb{R})$ which we shall now describe. There is an obvious bijection $\phi : M_n(\mathbb{R}) \to \mathbb{R}^{n^2}$ from the set $M_n(\mathbb{R})$ of $n \times n$ real matrices to the space $\mathbb{R}^{n^2}$ which carries a natural (Euclidean) topology. We topologise $M_n(\mathbb{R})$ by requiring that this bijection shall be a homeomorphism. The topology thus received by $M_n(\mathbb{R})$ is a metric topology with metric d given by

$$d([a_{ij}], [b_{ij}]) = \max_{i, j} |a_{ij} - b_{ij}|.$$

It is not difficult to see that, with this topology, the group $GL(n, \mathbb{R})$ of invertible $n \times n$ real matrices is a topological group. It is only necessary to observe that if $[a_{ij}]^{-1} = [a_{ij}^*]$ and $[a_{ij}][b_{ij}] = [c_{ij}]$, then a_{ij}^* is a rational function of the elements a_{ij} with non-vanishing denominator, and c_{ij} is a polynomial function of the elements a_{ij} and b_{ij}, and that such functions are continuous.

Now, $A = [a_{ij}] \in M_n(\mathbb{R})$ is in $GL(n, \mathbb{R})$ if and only if $\det A \neq 0$, and the function $\det : M_n(\mathbb{R}) \to \mathbb{R}$ is continuous (since $\det A$ is a polynomial function of the elements a_{ij}). Hence, since $\mathbb{R} \setminus \{0\}$ is an open subset of $\mathbb{R}$, $GL(n, \mathbb{R})$ is an open subset of $M_n(\mathbb{R})$. As such it is *locally Euclidean*, that is to say: every point of $GL(n, \mathbb{R})$ is contained in an open subset of $GL(n, \mathbb{R})$ which is homeomorphic to an open subset of $\mathbb{R}^{n^2}$. Spaces of this kind are called *manifolds*; a topological manifold of dimension m is simply a topological space which has a covering by open sets each of which is homeomorphic to an open set of $\mathbb{R}^m$. For each open set U of the covering, the associated homeomorphism ϕ_U maps a point $p \in U$ to a point $p\phi_U = (\alpha_1, \ldots, \alpha_m) \in \mathbb{R}^m$, and thus sets up a *local coordinate system* in U. For points in the intersection $U \cap V$ of two sets of the covering we have two local coordinate systems, and if $p\phi_U = (\alpha_1, \ldots, \alpha_m)$, $p\phi_V = (\beta_1, \ldots, \beta_m)$, then the coordinate map $\chi_{U,V} : (\alpha_1, \ldots, \alpha_m) \mapsto (\beta_1, \ldots, \beta_m)$ is a homeomorphism between two open sets of $\mathbb{R}^m$. The coordinates α_i are continuous functions of the coordinates β_j, and conversely. A particularly important type of manifold occurs when these functions are not merely continuous but are infinitely differentiable. Such a manifold is called *smooth*. It is easily seen that $GL(n, \mathbb{R})$ is a smooth manifold of dimension n^2, because each homeomorphism ϕ_U can be taken as the restriction of ϕ (which, recall, was an identification of $M_n(\mathbb{R})$ with $\mathbb{R}^{n^2}$) to U, and the coordinate maps $\chi_{U,V}$ are then identity maps. If A, B are manifolds, then so is $A \times B$, and

$$\dim(A \times B) = \dim A + \dim B.$$

It is also straightforward to show that if A, B are smooth manifolds, then so is $A \times B$. For two smooth manifolds X, Y we can define a *smooth map* $X \to Y$ by the requirement that, when expressed as a map $(x_1, \ldots, x_m) \mapsto (y_1, \ldots, y_n)$ in terms of local coordinate systems, the coordinates y_i shall be given in terms of the coordinates x_j by smooth (that is, infinitely differentiable) functions $y_i = f_i(x_1, \ldots, x_m)$. A very important class of topological groups is that for which G is a smooth manifold, and the maps $\theta \mapsto \theta^{-1}, (\theta, \phi) \mapsto \theta\phi$ are smooth maps $G \to G$ and $G \times G \to G$ respectively. Such groups are called *Lie groups*, in honour of the Norwegian mathematician Sophus Lie. It will be seen immediately (by looking at the analytical expressions of the relevant maps) that $GL(n, \mathbb{R})$ is a Lie group.

If we have a transitive G-space Ω, with Ω homeomorphic to $\cos(G:G_\alpha)$, it is natural to ask how the topology of G is related to the topologies of Ω and of G_α. As a *set*, G is the cartesian product of Ω and G_α, but it is not in general homeomorphic to the *topological* product of Ω and G_α. Under very wide conditions, and always when G is a Lie group, G is what is known as a *fibre bundle* with base Ω and fibre G_α. We shall not attempt to describe what a fibre bundle is, and for this we must refer the reader to the literature (see Steenrod 1951). We do however note that this structure provides the means of applying the techniques of algebraic topology to the study of the *global* topology of G, and that it has some easily comprehensible consequences for the *local* topology of G. In fact, a fibre bundle is locally a product space, i.e., there is an open covering $\{U_\lambda\}$ of the base space Ω, and a corresponding open covering $\{V_\lambda\}$ of G such that each V_λ is homeomorphic to $U_\lambda \times G_\alpha$. The open set V_λ is simply the inverse image $\phi^{-1}(U_\lambda)$ of U_λ under the continuous map $\phi : G \to \Omega$ (see p. 231), and the restriction $\phi | V_\lambda : V_\lambda \to U_\lambda$ is continuous. In order to establish that V_λ is homeomorphic to $U_\lambda \times G_\alpha$, we need to exhibit a second continuous surjection $\psi : V_\lambda \to G_\alpha$, such that, for any $\omega \in U_\lambda$, and $g \in G_\alpha$, $\phi^{-1}(\omega)$ and $\psi^{-1}(g)$ intersect in a single point of V_λ. Each 'fibre' $\phi^{-1}(\omega)$ is a coset $G_\alpha \times$

of G_α $(x \in G)$, and right multiplication by any element of $x^{-1}G_\alpha$ maps $G_\alpha x$ homeomorphically on G_α. Our hope is to construct a continuous map $\psi \colon V_\lambda \to G_\alpha$ such that $\psi \,|\, \phi^{-1}(\omega) \colon G_\alpha x \to G_\alpha$ is one of these homeomorphisms, for each $\omega \in U_\lambda$. This problem appears in the literature as that of constructing a *local cross-section* of the fibre bundle.

There are many theorems to the effect that, if spaces X and Y have a certain topological property, then so has their topological product $X \times Y$. Such theorems concerning local properties may be expected to hold for a fibre bundle B with base X and fibre Y. One such result is that, if X, Y are manifolds, then B is a manifold and dim $B =$ dim $X +$ dim Y. If G is a Lie group, then any closed subgroup H of G is also a Lie group, and the coset space $\cos(G\colon H)$ is a smooth manifold, and we have dim $G =$ dim $\cos(G\colon H) +$ dim H.

It should be observed that many results asserting the inheritance by G of topological properties enjoyed by a subgroup H and the space $\cos(G\colon H)$ can be proved for topological groups in general, and do not depend on the existence of a fibre bundle structure. It is also the case that for particular classical groups we can frequently find ad hoc methods of proof which avoid any appeal to the powerful general results of the theory of Lie groups.

The rest of this chapter is devoted to a more detailed topological description of some of the spaces and groups which we have encountered in earlier chapters. We shall describe these spaces in terms of subspaces of $\mathbb{R}^n$, and the following 'standard spaces' will be used.

(i)   The open disc (or ball), $D^n = \{(x_1, \ldots, x_n) \,|\, x_1^2 + \cdots + x_n^2 < 1\}$;

(ii)   the closed disc (or ball), $\bar{D}^n = \{(x_1, \ldots, x_n) \,|\, x_1^2 + \cdots + x_n^2 \leqslant 1\}$;

(iii)   the sphere, $S^{n-1} = \{(x_1, \ldots, x_n) \,|\, x_1^2 + \cdots + x_n^2 = 1\}$.

These are all bounded sets of $\mathbb{R}^n$, hence the closed sets $\bar{D}^n$ and S^{n-1} are compact subspaces of $\mathbb{R}^n$. Using $\approx$ to denote homeomorphism, we have $\mathbb{R}^n \approx D^n$. Also, if $\mathbb{R}^+ := \{x \in \mathbb{R} \,|\, x > 0\}$, we have $\mathbb{R}^+ \approx \mathbb{R} \approx D^1$.

Real projective space P^n is yet another useful 'standard space'. The points of n-dimensional real projective space are in bijective correspondence with the 1-dimensional subspaces $\{(\lambda\alpha_0, \ldots, \lambda\alpha_n) \,|\, \lambda \in \mathbb{R}\}$ of $\mathbb{R}^{n+1}$, and there are just two (diametrically opposite) points of the sphere

$$S^n = \{(x_0, \ldots, x_n) \,|\, x_0^2 + \cdots + x_n^2 = 1\}$$

which lie in such a 1-dimensional subspace. The points of projective space P^n thus correspond to pairs of diametrically opposite points of S^n. If now $\sim$ denotes the equivalence relation on S^n given by $a \sim b$ if and only if $a = b$ or $a = -b$, then there is a bijection $P^n \to S^n/\sim$, and we give P^n the topology which $S^n/\sim$ inherits from S^n. Since $S^n/\sim$ is the image of the compact space S^n under the continuous identification map $S^n \to S^n/\sim$, the projective space P^n is compact.

The topology of P^n may equally be derived by representing the points of P^n

on the 'upper closed hemisphere' of S^n given by $x_0 \geqslant 0$, which is homeomorphic to $\bar{D}^n$. The correspondence is now one–one, except for points on the boundary of the hemisphere, which is

$$\{(x_0, \ldots, x_n) \mid x_0 = 0, x_1^2 + \cdots + x_n^2 = 1\} \approx S^{n-1}.$$

Two diametrically opposite points of the boundary correspond to the same point of P^n, and we obtain $P^n \approx \bar{D}^n/\sim$, where $\sim$ is now the equivalence relation given by $a \sim b$ if and only if either $a \in D^n$ and $a = b$ or $a \in \bar{D}^n \setminus D^n$ and $a = \pm b$.

We have already seen that $\mathrm{GL}(n, \mathbb{R})$ is the open subset of $M_n(\mathbb{R})$ defined by

$$\mathrm{GL}(n, \mathbb{R}) := \{A \mid A \in M_n(\mathbb{R}),\ \det A \neq 0\}.$$

It contains as a subgroup the set

$$\mathrm{GL}^+(n, \mathbb{R}) := \{A \mid A \in M_n(\mathbb{R}),\ \det A > 0\}.$$

This subgroup is the inverse image under the continuous surjection $\det: \mathrm{GL}(n, \mathbb{R}) \to \mathbb{R} \setminus \{0\}$ of the set $\mathbb{R}^+$, which is both open and closed in $\mathbb{R} \setminus \{0\}$. Hence $\mathrm{GL}^+(n, \mathbb{R})$ is a closed subgroup of $\mathrm{GL}(n, \mathbb{R})$, and its complement in $\mathrm{GL}(n, \mathbb{R})$ is also closed; therefore $\mathrm{GL}(n, \mathbb{R})$ is disconnected. The complement of $\mathrm{GL}^+(n, \mathbb{R})$ is its coset $\{XA \mid X \in \mathrm{GL}^+(n, \mathbb{R})\}$, where A is any element of $\mathrm{GL}(n, \mathbb{R})$ with $\det A < 0$. The map $X \mapsto XA$ is a homeomorphism of $\mathrm{GL}^+(n, \mathbb{R})$ onto its complement. We shall see later that $\mathrm{GL}^+(n, \mathbb{R})$ is connected. It is clearly locally Euclidean of dimension n^2, since it is an open subset of $M_n(\mathbb{R})$.

The special linear group $\mathrm{SL}(n, \mathbb{R})$ is a topological group, with the topology inherited from $\mathrm{GL}(n, \mathbb{R})$. It is a closed subgroup of $\mathrm{GL}^+(n, \mathbb{R})$, because it is the inverse image of the closed set $\{1\}$ under the continuous surjection $\det: \mathrm{GL}^+(n, \mathbb{R}) \to \mathbb{R}^+$. It is not compact, since it is not bounded in the metric topology of $M_n(\mathbb{R})$. We can show that $\mathrm{SL}(n, \mathbb{R})$ is connected, and locally Euclidean of dimension $(n^2 - 1)$ (see Exercises 18.14 and 18.15).

The orthogonal group $\mathrm{O}(n)$ is a closed subgroup of $\mathrm{GL}(n, \mathbb{R})$, being the inverse image of the closed set $\{I\}$ under the continuous map $\mathrm{GL}(n, \mathbb{R}) \to \mathrm{GL}(n, \mathbb{R})$ given by $A \mapsto AA'$. The group $\mathrm{O}(n)$ is compact, because the conditions on the elements a_{ij} of a real orthogonal matrix require that $|a_{ij}| \leqslant 1$, and this means that $\mathrm{O}(n)$ is bounded in the metric topology of $M_n(\mathbb{R})$. The group is disconnected, as we have a continuous surjection $\det: \mathrm{O}(n) \to \{1, -1\}$ onto the discrete space $\{1, -1\}$.

Much useful information about $\mathrm{O}(n)$ can be obtained by considering the action of $\mathrm{O}(n)$ on the sphere S^{n-1}. The action is faithful and transitive, and the stabiliser of a point of S^{n-1} is a subgroup of $\mathrm{O}(n)$ which is isomorphic to $\mathrm{O}(n-1)$. Since $\mathrm{O}(n)$ is compact and S^{n-1} is Hausdorff, we obtain

$$\cos(\mathrm{O}(n){:}\mathrm{O}(n-1)) \approx S^{n-1}.$$

Using the facts that $\mathrm{O}(1) \approx \{1, -1\}$ and is locally Euclidean of dimension 0, and that S^r is locally Euclidean of dimension r, we obtain by induction (relying on fibre bundle properties) that $O(n)$ is locally Euclidean of dimension $\frac{1}{2}n(n-1)$. (An alternative line of proof is indicated in Exercise 16.)

The special orthogonal group $SO(n)$ is $SL(n, \mathbb{R}) \cap O(n)$, and is therefore a closed subgroup of $GL(n, \mathbb{R})$, and is compact (since $O(n)$ is compact). By considering the action of $SO(n)$ on S^{n-1}, we obtain

$$\cos(SO(n):SO(n-1)) \approx S^{n-1}.$$

Now, if $\cos(G:H)$ and H are connected, then so is G (see Exercise 18.6). Using the facts that $SO(1)$ is the identity (and hence connected), and that S^r is connected, we can show by induction that $SO(n)$ is connected. We know that $O(n)$ is the union of $SO(n)$ and its coset, and that the map $X \mapsto XA$ (where $A \in O(n) \backslash SO(n)$) maps $SO(n)$ homeomorphically on its coset, which is thus also connected. Hence $SO(n)$ and its coset are the components of the disconnected space $O(n)$.

The topology of $GL(n, \mathbb{R})$ can be related to the topology of $O(n)$ by using the algebraic result (which emerges from the Gram–Schmidt orthogonalisation process) that any real invertible matrix A can be expressed uniquely as $A = ST$, where S is a real orthogonal matrix and T is a real upper triangular matrix with positive elements on the diagonal; that is, $T = [\tau_{ij}]$, where $\tau_{ij} = 0$ for $i > j$ and $\tau_{ii} > 0$ (see Exercise 18.11). Such triangular matrices form a subgroup Δ of $GL^+(n, \mathbb{R})$, and the representation identifies $GL(n, \mathbb{R})$ *as a set* with the cartesian product $O(n) \times \Delta$, which can be given the product topology of the natural topologies of $O(n)$ and Δ as subspaces of $M_n(\mathbb{R})$. In order to show that this agrees with the topology of $GL(n, \mathbb{R})$ as a subspace of $M_n(\mathbb{R})$, we need to verify that the maps

$$A = ST \mapsto (S, T), \quad (S, T) \mapsto ST = A$$

are continuous (see Exercise 18.12). We then have

$$GL(n, \mathbb{R}) \approx O(n) \times \Delta.$$

Now, for a matrix in the group Δ, $\frac{1}{2}n(n-1)$ of the elements take values in $\mathbb{R}$, and n of the elements take values in $\mathbb{R}^+$ (the remaining elements being zero), so that

$$\Delta \approx \mathbb{R}^{n(n-1)/2} \times (\mathbb{R}^+)^n,$$

and since $\mathbb{R}^+ \approx \mathbb{R}$, we have $\Delta \approx \mathbb{R}^{n(n+1)/2}$ and thus obtain

$$GL(n, \mathbb{R}) \approx O(n) \times \mathbb{R}^{n(n+1)/2},$$

which is consistent with our earlier observation that $O(n)$ is locally Euclidean of dimension $\frac{1}{2}n(n-1)$. Since $O(n)$ has two homeomorphic components, and $\mathbb{R}^{n(n+1)/2}$ is connected, it follows that $GL(n, \mathbb{R})$ has two homeomorphic components which must be $GL^+(n, \mathbb{R})$ and its complement. We have in fact

$$GL^+(n, \mathbb{R}) \approx SO(n) \times \mathbb{R}^{n(n+1)/2},$$

and have shown that $GL^+(n, \mathbb{R})$ is connected.

It will be clear from our discussion that a central question in the study of the classical groups is the determination of the topological structure of $SO(n)$. To conclude this chapter, we examine briefly the topologies of $SO(n)$, for $n = 2, 3$, and 4.

We know that $SO(2)$ is the multiplicative group of matrices R_α ($\alpha \in \mathbb{R}/2\pi\mathbb{R}$), which is isomorphic to the additive group $\mathbb{R}/2\pi\mathbb{Z}$, which in turn is isomorphic to the circle group S^1 (see Chapter 2, Exercise 2.14). It is not difficult to see that the bijection $SO(2) \to S^1$ given by the isomorphism of groups is a homeomorphism of topological spaces. The result also comes immediately from the fact that $SO(1)$ is trivial, so that the homeomorphism $\cos(SO(2):SO(1)) \approx S^1$ reduces to $SO(2) \approx S^1$.

We recall from Chapter 16 that $SO(3) \cong U/\{E, -E\}$, where U is the group of unit quaternion matrices. The matrices in U are of the form $\lambda E + \alpha I + \beta J + \gamma K$, where $\lambda^2 + \alpha^2 + \beta^2 + \gamma^2 = 1$. There is thus a natural bijection from U to S^3, and it is easily seen that this is a homeomorphism between U (with the topology inherited from $M_4(\mathbb{R})$) and S^3 (with the natural topology inherited from $\mathbb{R}^4$). Now $SO(3)$, as a factor group of U (and thus a coset space), has the topology derived from that of U by identifying pairs $\{Q, -Q\}$ of elements of U. Such pairs are represented by pairs of diametrically opposite points of S^3. Hence $SO(3)$ is homeomorphic to the space obtained from S^3 by identification of diametrically opposite points, and this is projective space P^3. We thus have $SO(3) \approx P^3$.

An element of $SO(4)$ is associated with an ordered pair (Q_1, Q_2) of unit quaternions, and $(Q_1, Q_2), (-Q_1, -Q_2)$ represent the same element of $SO(4)$. With the help of this representation we can show that $SO(4)$ is homeomorphic to the space obtained from $S^3 \times S^3$ by identification of pairs $\{(a, b), (-a, -b)\}$ of 'opposite' points.

Bibliography

Higgins, P. J. (1974). *An introduction to topological groups*, London Mathematical Society Lecture Notes 15. Cambridge University Press.

Pontryagin, L. S. (1946). *Topological groups*, 2nd edn. Princeton University Press.

Steenrod, N. (1951). *The topology of fibre bundles*. Princeton University Press.

Chevalley, C. (1946). *Lie groups*. Princeton University Press.

For initial reading, Chapters 1 and 2 of Higgins (1974) and §§1–7 of Part I of Steenrod (1951) are suggested.

Exercises

18.1. If A, B are subsets of a topological group G, and $AB := \{ab \mid a \in A,\ b \in B\}$, $A^{-1} := \{a^{-1} \mid a \in A\}$, show
(i) that if A is open, then AB and BA are open;
(ii) that if A is closed and B is finite, then AB and BA are closed;
(iii) that A^{-1} is open if and only if A is open.

18.2. Show that the projection map $\pi: G \to \cos(G:H)$ is open. [Take an open set $S \subset G$, and consider the inverse image of $S\pi$. Show that this is SH, and is open, and deduce that $S\pi$ is open.]

18.3. Show that, if a, b, $c \in G$, and $ab = c$, then given any open neighbourhood W of c there are open neighbourhoods U, V of a, b respectively, such that $UV \subset W$.

Deduce that if U is an open neighbourhood of 1 then there is an open neighbourhood V of 1 such that $VV^{-1} \subset U$.

18.4. Let G be a topological group, and let $A \subset G$ be a closed set such that $1 \notin A$. Show that there is an open neighbourhood V of 1 such that $VV^{-1} \cap A = \emptyset$.

Show that AV is open, and deduce that the topology of G is regular.

18.5. Let G be a topological group, and H a subgroup. Let A be a closed set of $\cos(G:H)$, with $H \notin A$. Consider the inverse image $A^* = \bigcup A$ of A under the projection map $\pi: G \to \cos(G:H)$. Show that A^* is closed and $1 \notin A^*$.

Let V be an open neighbourhood of 1 such that $VV^{-1} \cap A^* = \emptyset$. By considering $V\pi$ and $A^*V\pi$, show that $\cos(G:H)$ is a regular space.

18.6. Let H be a connected subgroup of the topological group G. Suppose that G is disconnected, so that $G = U \cup V$, where U and V are non-empty open subsets of G, and $U \cap V = \emptyset$. Show that any coset of H is contained either in U or in V, and deduce that $\cos(G:H)$ is disconnected.

Hence show that if H and $\cos(G:H)$ are connected, then so is G.

18.7. Prove the statements (p. 234) that $\mathbb{R}^+ \approx \mathbb{R}$ and $\mathbb{R}^n \approx D^n$, by constructing suitable maps explicitly in terms of the coordinates.

18.8. Let T be any topological space, and let $T^* = T \cup \{\omega\}$ be the space obtained by adding one further point ω. Show that the prescription that U is an open subset of T^* if (and only if) *either* U is an open subset of T *or* $U = X \cup \{\omega\}$ where $X \subset T$ and $T \setminus X$ is a closed compact subset of T, defines a compact topology on T^*. (We call T^*, with this topology, the one-point compactification of T).

Show that the one-point compactification of $\mathbb{R}^n$ is homeomorphic to S^n. [Hint: use stereographic projection.]

18.9.   (i) Show that $P^1 \approx S^1$.

(ii) Try to draw a picture of a surface (in 3-dimensional space) which is homeomorphic to P^2. [Such a surface must cross itself; in other words, in a 3-dimensional picture of P^2 there have to be some points which represent two distinct points of P^2.]

18.10. Recall that the set of lines of a real projective plane is in natural bijective correspondence with the set of points of the dual plane (Exercise 13.10). Using this bijection, show that the set Ω of lines of $\mathbb{R}^2$ can be topologised so that Ω is homeomorphic to the punctured projective plane (that is, the space P^2 with one point removed).

Show that, if $G = \mathrm{AGL}(2, \mathbb{R})$ and μ is the natural action of G on Ω, then (Ω, G, μ) is a topological G-space.

Show that the punctured projective plane is homeomorphic to the open Möbius band, which is obtained from the set

$$\{(x, y) \in \mathbb{R}^2 \mid 0 \leqslant x \leqslant 1, 0 < y < 1\}$$

by identifying $(0, \lambda)$ with $(1, 1 - \lambda)$ for all $\lambda \in (0, 1)$.

18.11. Prove the statement made on p. 236 that an invertible matrix may be expressed in the form ST there described.

Show that $\mathrm{O}(n) \cap \Delta = \{I\}$, and hence that the expression of a matrix in this form is unique.

18.12. Show (using the notations of Exercise 18.11) that the maps $(S, T) \mapsto ST$ and $ST \mapsto (S, T)$ are continuous. [Hint: consider the elements of $A = ST$ as functions of the elements of S and of T, and conversely.]

18.13. Show that $\mathrm{AGL}(n, \mathbb{R})$ is isomorphic to the group of matrices

$$\begin{bmatrix} A & 0 \\ b & 1 \end{bmatrix},$$

where $A \in \mathrm{GL}(n, \mathbb{R})$ and $b \in \mathbb{R}^n$. Show that, if $\mathrm{AGL}(n, \mathbb{R})$ is given the topology inherited from $M_{n+1}(\mathbb{R})$, then

$$\mathrm{AGL}(n, \mathbb{R}) \approx \mathrm{GL}(n, \mathbb{R}) \times \mathbb{R}^n \approx \mathrm{O}(n) \times \mathbb{R}^{n(n+3)/2}.$$

18.14. Show that any matrix $A \in \mathrm{GL}^+(n, \mathbb{R})$ may be written uniquely in the form $A = BC$, where $B = \mathrm{diag}(\lambda, 1, \ldots, 1)$ with $\lambda > 0$, and $C \in \mathrm{SL}(n, \mathbb{R})$.

Show that the group of all such matrices B is a topological group homeomorphic to $\mathbb{R}^+$.

Deduce that $\mathrm{GL}^+(n, \mathbb{R}) \approx \mathbb{R} \times \mathrm{SL}(n, \mathbb{R})$, and hence show that $\mathrm{SL}(n, \mathbb{R})$ is connected.

18.15. Let S be the subspace of $\mathbb{R}^k$ given by

$$S := \{(x_1, \ldots, x_k) \mid f(x_1, \ldots x_k) = 0\},$$

where $f(0, \ldots, 0) = 0$ and the partial derivatives $D_i f (= \partial f / \partial x_i)$ exist and are continuous in some open neighbourhood in $\mathbb{R}^k$ of $(0, \ldots, 0)$. Given that $D_i f(0, \ldots, 0) = 0$ for $i = 1, \ldots, k-1$, and $D_k f(0, \ldots, 0) \neq 0$, show that there is an open neighbourhood X of $(0, \ldots, 0)$ in S such that the map

$$(x_1, \ldots, x_{k-1}, x_k) \mapsto (x_1, \ldots, x_{k-1})$$

is a homeomorphism of X onto its image Y in $\mathbb{R}^{k-1}$. [Hint: use the implicit function theorem of real analysis.]

Hence show that, if $D_i f(0, \ldots, 0) \neq 0$ for some $i = 1, \ldots, n$ then there is an open neighbourhood of $(0, \ldots, 0)$ in S which is homeomorphic to an open set of $\mathbb{R}^{k-1}$. [Hint: apply an appropriate orthogonal transformation to $\mathbb{R}^k$.]

Show that $SL(n, \mathbb{R})$ is locally Euclidean of dimension $n^2 - 1$. [Hint: consider the subspace $\{A \mid \det A = 1\}$ of $M_n(\mathbb{R})$ in a neighbourhood of the unit matrix.]

18.16. Let S be a skew-symmetric matrix in $M_n(\mathbb{R})$. Show

(i) that $I + S$ and $I - S$ are invertible;
(ii) that the matrices $I + S$, $I - S$, $(I + S)^{-1}$, $(I - S)^{-1}$ commute; and
(iii) that $U := (I - S)(I + S)^{-1}$ is orthogonal.

Show that the map $S \mapsto (I - S)(I + S)^{-1}$ from skew-symmetric to orthogonal matrices is one–one, and that its image consists of the orthogonal matrices which do not have -1 as an eigenvalue.

Deduce that $O(n)$ is locally Euclidean of dimension $\frac{1}{2}n(n-1)$.

18.17. Show that for any $A \in GL(n, \mathbb{C})$, a matrix $\exp A \in GL(n, \mathbb{C})$ is defined by

$$\exp A = \operatorname*{Lt}_{n \to \infty} (I + A + \frac{1}{2!} A^2 + \cdots + \frac{1}{n!} A^n).$$

[The argument follows that for real matrices given in Chapter 14.] Show that $\exp(T^{-1}AT) = T^{-1}(\exp A)T$, and by choosing for the given matrix A the matrix T so that $T^{-1}AT$ is triangular, show that

$$\det(\exp A) = \exp(\operatorname{tr} A).$$

18.18. Show that, for $A \in M_n(\mathbb{R})$, the map $A \mapsto \exp A$, restricted to a sufficiently small open neighbourhood X of the zero matrix, maps X bijectively onto an open neighbourhood of the unit matrix in $GL^+(n, \mathbb{R})$ and is a homeomorphism of X to Y. [Hint: use the Inverse Function Theorem of real analysis.]

Show that the subspace $\{A \mid \operatorname{tr} A = 0\}$ of $M_n(\mathbb{R})$ is Euclidean of dimension $n^2 - 1$, and hence that $\mathrm{SL}(n, \mathbb{R})$ is locally Euclidean of dimension $n^2 - 1$.

Show that the subspace $\{A \mid A + A' = 0\}$ of $M_n(\mathbb{R})$ is Euclidean of dimension $\frac{1}{2}n(n-1)$, and hence that $\mathrm{SO}(n)$ is locally Euclidean of dimension $\frac{1}{2}n(n-1)$.

18.19. An element of $\mathrm{SO}(3)$ is determined by a directed axis $\hat{u}$ and angle of rotation $\theta \in \mathbb{R}/2\pi\mathbb{Z}$. By taking a representative $\bar{\theta}$ of θ, with $\bar{\theta} \in [-\pi, \pi]$, and considering the points $(\bar{\theta}/\pi) . \hat{u} \in \mathbb{R}^3$, obtain a surjective map $\bar{D}^3 \to \mathrm{SO}(3)$, and a bijective map $P^3 \to \mathrm{SO}(3)$. Show that this bijective map is a homeomorphism.

18.20. The group $\mathrm{GL}(V_n(\mathbb{R}))$ is topologised by taking a basis for V_n and using the associated bijection $\mathrm{GL}(V_n(\mathbb{R})) \to M_n(\mathbb{R})$. Show that the transformation $M_n(\mathbb{R}) \to M_n(\mathbb{R})$ given by $A \mapsto T^{-1}AT$ is a homeomorphism, and hence that the topology of $\mathrm{GL}(V_n(\mathbb{R}))$ is independent of the basis selected.

18.21. The (additive) topological group $\mathbb{R}^2$ has the subgroup $\mathbb{Z}^2$, consisting of pairs (m, n) with $m, n \in \mathbb{Z}$. Show that $\mathbb{Z}^2$ is closed in $\mathbb{R}^2$, and that $\mathbb{R}^2/\mathbb{Z}^2$ is the topological group $S^1 \times S^1$. (As a topological space, this is a torus, and the group is known as the torus group.)

18.22. Show that the stabiliser in the Möbius group $M = \mathrm{PGL}(2, \mathbb{C})$ of a proper circle in the Gauss plane is isomorphic to the real Möbius group $\mathrm{PGL}(2, \mathbb{R})$.

18.23. Show that the set Ω of directed proper circles in the Gauss plane, acted on by the Möbius group M, is a topological G-space in which $\Omega \approx S^2 \times \mathbb{R}$.

18.24. Recall that the elements of the real Möbius group are uniquely representable by real 2×2 matrices of determinant ± 1. Hence show that the group is disconnected.

Recall that a real Möbius transformation is uniquely determined by specifying the image of a point not on the real axis, and the image of a point on the real axis. Hence show that the real Möbius group has two components, each homeomorphic to $S^1 \times \mathbb{R}^2$.

Obtain the same result by considering the representation of real 2×2 matrices with determinant ± 1 in the form ST of Exercises 18.11 and 18.12.

Chapter 19

The group theory of Rubik's magic cube

The cube with the *m*ystifying *a*bility to *g*enerate *i*nstant *c*onfusion provides an excellent example of groups acting on sets. You have probably seen the puzzle—indeed, we hope you own one. But in case not, here is a brief description. It is a cube divided into 27 small cubes, arranged $3 \times 3 \times 3$ as they must be. Inside there is an ingenious arrangement which holds the small cubes together in such a way that any one of the 'faces' of the magic cube may be rotated about its centre, (Fig. 19.1).

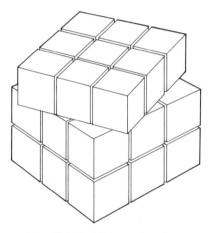

Fig. 19.1. Rubik's magic cube.

The visible faces of the little cubes are coloured. In its pristine form each face of the large cube, made up of nine little squares, is monochrome, and the six faces have different colours. If one face is rotated, then another, and another, after four or five such moves the colours have become totally confused. Two questions invariably come to mind. The first is, how does it work? We leave it to the reader to find that out for himself. He will need the sparkling imagination of the cube's inventor, Professor Ernö Rubik of Budapest, if he is to find an answer by pure thought. The second is, how can one solve it? That is, what workable, easily memorable rules are there for returning the cube to its original state from any configuration? Although by now there are many different algorithms available, we do not wish to spoil your fun by giving details. These few pages are intended merely to indicate the role that group theory can play in a study of the magic cube, and to use the cube as an excellent illustration of the theory of groups acting on sets.

The cube that we shall analyse has its faces originally coloured red, sapphire, turquoise, umber, violet and white. Now a *basic move*, that is, the rotation of one outer layer of the cube through 90° about its centre, does not move the central little cube of that layer relative to the big cube. Whatever happens, whichever way round the cube is held in the hand, we can recognise by a glance at the central squares which face should be which colour. Thus the six little squares in the centres of the faces provide us with a satisfactory set of axes fixed in the cube relative to which we can describe what happens. We shall use R, S, T, U, V, W as names for the six basic moves: in whatever configuration the cube happens to be, R is a clockwise rotation through 90° of the face whose central little square is red; and S, T, U, V, W are defined analogously. Since we are interested in what happens when we make a succession of basic moves we are led immediately to the group M that is generated by the six moves R, S, T, U, V, W. This group M, whose elements we shall call *compound moves* or, simply, *moves*, is known as the *group of the cube*.

The effect of a basic move is to permute amongst themselves the 27 little cubes (and also to rotate them, but for the moment we ignore that). But of course the central cube that we cannot see, and each of the six cubes in the middle of the faces, remain where they are—these seven little cubes provide us, after all, with our 'fixed' axes. So let us write Γ for the set consisting of the other 20 little cubes. A basic move, and therefore any move, acts to permute Γ, and thus we have an action of M on Γ. Now the cubes in Γ can be classified naturally as corner cubes, of which there are 8, one at each corner of the main cube, and edge cubes, of which there are 12, one in the middle of each edge. Obviously, a move takes corner cubes to corner cubes and edge cubes to edge cubes. Furthermore, it is easy to check that by a succession of basic moves we can take any given corner cube to any corner position, or any given edge cube to any edge position. Thus we have

LEMMA 19.1: *The group M is intransitive on Γ. It has two orbits, the set Γ_c of 8 corner cubes and the set Γ_e of 12 edge cubes.*

To understand M more fully we need to be able to identify first, exactly which permutations of Γ are produced by elements of M; and secondly, what the kernel of the action of M on Γ is. We shall indicate what the answers to these questions are in a series of observations, exercises and hints.

The group of all permutations of Γ that permute the corner cubes amongst themselves and the edge cubes amongst themselves is the direct product $\mathrm{Sym}(\Gamma_c) \times \mathrm{Sym}(\Gamma_e)$, which we shall identify with $S_8 \times S_{12}$ for notational simplicity. So if P denotes the group of permutations of Γ induced by M then $P \leqslant S_8 \times S_{12}$.

OBSERVATION 19.2: *If X, Y are the basic moves associated with a pair of adjacent faces of the cube then the commutator move $X^{-1}Y^{-1}XY$ acts as a 3-cycle on Γ_e, permuting three edge cubes cyclically and leaving the remaining 9 edge cubes fixed. It acts as a double transposition on the corner cubes, transposing two pairs of them and leaving the remaining 4 corner cubes fixed.*

It follows that $(X^{-1}Y^{-1}XY)^2$ acts as a 3-cycle on Γ_e while returning all the corner cubes to where they were before (but beware—it does not leave them with their original orientations); and $(X^{-1}Y^{-1}XY)^3$ acts as a double-transposition on Γ_c while fixing every edge cube.

EXERCISE 19.3: *Show that the 3-cycles obtained in this way from commutator moves generate the group A_{12} of all even permutations of Γ_e. Show also that the double-transpositions obtained in this way from commutator moves generate the group A_8 of all even permutations of Γ_c.*

Hint. Consider the group generated by two 3-cycles that have just one entry in common. You should be able to show that this is A_5. Then consider the group generated by this together with one further 3-cycle that has just one entry in common with these 5; and build up knowledge of the whole group generated by the available 3-cycles in steps like these.

We are now in a position to identify the group P (of all permutations of Γ induced by M). For, by what you have just shown, it certainly contains all of $A_8 \times A_{12}$. Furthermore, a basic move such as R acts as a 4-cycle on Γ_c and this is an odd permutation. Therefore every permutation of Γ_c can be produced by some element of M. Similarly any permutation of Γ_e can be achieved by a suitable move. On the other hand, each basic move acts as a 4-cycle on both Γ_e and Γ_c, hence as an even permutation of Γ. It follows that every move gives an even permutation of Γ, that is, that

$$A_8 \times A_{12} \leqslant P \leqslant (S_8 \times S_{12}) \cap A_{20}.$$

Now $(S_8 \times S_{12}) \cap A_{20}$ consists of those permutations that are even on each of Γ_e and Γ_c or odd on each of Γ_e and Γ_c; it contains $A_8 \times A_{12}$ as a subgroup of index 2. These observations suffice to prove

LEMMA 19.4: *The group P is $(S_8 \times S_{12}) \cap A_{20}$; it consists of all permutations of Γ that have the same parity on Γ_e and Γ_c; its order is $\frac{1}{2}.8!.12!$.*

We turn our attention now to the kernel of the action of M on Γ. This is the group N consisting of all moves that leave each individual small cube in its original place, but perhaps turned round relative to its original position. Clearly, N is normal in M and $P \cong M/N$.

To study N, and to discuss M in greater detail than we have done up to now, we consider the effect of moves on the visible faces of the small cubes. Since each face of the magic cube is divided into 9 little squares, there are 54 of these.

But we are ignoring the central squares of each face (or rather, we are using the six central squares to provide a frame of reference). So let Ω be the set consisting of the remaining 48 squares. Obviously M acts on Ω to permute the little squares, and it acts faithfully by definition: thus we think of M as being a group of permutations of Ω.

Members of Ω are again of two kinds. Some of them are faces of corner cubes and some are faces of edge cubes. Thus we may write Ω as a disjoint union $\Omega_c \cup \Omega_e$. Since each corner cube has 3 visible faces, and each edge cube has 2 visible faces, we have

$$|\Omega_c| = |\Omega_e| = 24.$$

Obviously, any move takes corner squares to corner squares and edge squares to edge squares. It is not hard to check that any corner square can be moved to any other corner position and that any edge square can be moved to any other edge position. Thus we have

OBSERVATION 19.5:   *The group M has two orbits in Ω, namely Ω_c and Ω_e.*

The actions of M on Γ_c and on Γ_e are excellent examples of transitive but imprimitive actions in the sense discussed in Chapter 7. For, obviously the triples of squares that surround a corner are permuted bodily amongst themselves, and the pairs of squares that meet at an edge are permuted amongst themselves. This corresponds to the fact that the map $f: \Omega \to \Gamma$ which assigns a square to the little cube of which it is a face is an M-morphism. It can be seen as constituted of two M-morphisms of transitive M-spaces, $f_c: \Omega_c \to \Gamma_c$, $f_e: \Omega_e \to \Gamma_e$. The former is a 3-to-1 mapping, the latter a 2-to-1 mapping.

Before we return to our study of M and N let us consider a larger group Q. Try the thought-experiment in which you imagine that you can dismantle the magic cube (leaving just the axes consisting of the central cube and the six little cubes in the middle of the faces) and then re-assemble it with the corner cubes and the edge cubes back in any position you wish. The collection of all such operations forms a group that we call Q (because in practice such steps are only to be taken on the quiet). Obviously $M \leqslant Q$ and we leave you to do

EXERCISE 19.6:   *The group Q has order $8! \cdot 3^8 \cdot 12! \cdot 2^{12}$. It is a direct product $Q_c \times Q_e$, where $|Q_c| = 8! \cdot 3^8$ and $|Q_e| = 12! \cdot 2^{12}$.*

Notice that Q_c acts on Γ_c as the full symmetric group, and the kernel of the action is a direct product of 8 copies of the cyclic group Z_3 because each corner cube can be imagined to be rotated in situ through an angle 0, $2\pi/3$, or $4\pi/3$. We call this group K_c. It is an abelian group; indeed, an elementary abelian group in which every non-identity element has order 3. Similarly, Q_e acts on Γ_e as the full symmetric group S_{12}, and the kernel K_e of the action is an elementary abelian 2-group, a direct product of 12 copies of Z_2 (because each

edge cube could be replaced in its correct position in either of two ways round).
In fact one can easily show that Q_c and Q_e are so-called 'wreath products' (as
defined in Exercise 8.3), and so the structure of Q can be exhibited:

$$Q \cong (Z_3 \text{ wr } S_8) \times (Z_2 \text{ wr } S_{12}).$$

To compute the group N consisting of all compound moves that leave all
the small cubes in their original position (though perhaps re-oriented), we
write N_c, N_e for the groups of permutations induced by N on Ω_c and Ω_e
respectively. Notice that $N_c \leqslant K_c$ and $N_e \leqslant K_e$; therefore N_c is an elementary
abelian 3-group and N_e is an elementary abelian 2-group. Now an easy
exercise:

EXERCISE 19.7:   *Prove that $N = N_c \times N_e$.*

Suppose, for the moment, that the visible faces of the corner cubes were not
coloured, but that each was numbered 0, 1, or 2. This should have been done in
such a way that each corner cube had its three faces numbered 0, 1, 2 in cyclic
order, going clockwise as one looks straight at that corner. Then we could
assign to each move, indeed, to each operation in Q, a number modulo 3 in the
following way. For each corner of the cube in its new configuration we ask how
much the numbering of the little corner cube has been rotated relative to the
numbering of the corner cube that was originally there. This gives us an
angle 0 or $2\pi/3$ or $4\pi/3$. We add all the 8 angles computed in this way and get
$2k\pi/3$, where k is 0, 1, or 2 modulo 3, and k depends, of course, on the move (or
the Q-transformation) in question.

EXERCISE 19.8:   *Show that for any move k is 0.*

Hint.   Show that the map we have just described from Q to the additive group
of integers modulo 3 is a homomorphism. Show also that for any of the six
basic moves k is 0. It then follows that k is 0 for any sequence of basic moves,
that is, for any compound move.

LEMMA 19.9:   *The group N_c consists of those elements of K_c which, as
rotations of the 8 corner cubes, have total angle sum 0 (modulo 2π). Thus N_c has
index 3 in K_c, so $|N_c| = 3^7$.*

Proof.   Certainly each element of N_c acts simply to rotate each individual
corner cube through an angle 0, $2\pi/3$, or $4\pi/3$ about the axis which is the
diagonal of the magic cube through the relevant corner; and from
Exercise 19.8 it follows that the sum of these 8 angles must be 0 modulo 2π.

Now suppose that Red, Sapphire, Turquoise are three faces meeting at one
of the corners. Observe that the compound move

$$(RS^{-1}R^{-1}S)^3(ST^{-1}S^{-1}T)^3(TR^{-1}T^{-1}R)^3$$

(chosen not for its simplicity but because its derivation is not far to seek) is in
the group N_c. It leaves 5 of the corner cubes unchanged, and it rotates each of

the other 3 corner cubes through an angle $2\pi/3$. It is now an exercise, similar to Exercise 19.3, to check that by a succession of such moves one can achieve any configuration where the corner cubes have simply been individually rotated, and the total angle sum of these 8 rotations is 0 modulo 2π.

A similar, though rather easier, analysis proves

LEMMA 19.10:   *The group N_e consists of those members of K_e which rotate an even number of the edge cubes through an angle π.*

The proof is helped by the observation that a move such as

$$((RS^{-1}R^{-1}S)^2(ST^{-1}S^{-1}T)^2(TR^{-1}T^{-1}R)^2)^3$$

lies in N_e and turns just 2 of the edge cubes round.

To summarise: we have found that N_c is elementary abelian of order 3^7, that N_e is elementary abelian of order 2^{11}, and that $N = N_c \times N_e$, so that $|N| = 3^7 . 2^{11}$. Since we now know both N and M/N (which is P) we know a great deal about M. For example,

$$|M| = \tfrac{1}{2} . 8! . 3^7 . 12! . 2^{11}.$$

Perhaps the most useful way to understand M is to think of it as a subgroup of Q, that is, of $(Z_3 \text{ wr } S_8) \times (Z_2 \text{ wr } S_{12})$. Comparing the orders of Q and M you will see that the index $|Q:M|$ is 12. This means that, if you re-assemble the cube randomly if it falls apart, then your chances of being able to solve it thereafter are just 1 in 12.

We hope that, if you do not already know how to solve Rubik's cube, the hints you can pick up from our analysis will be enough to start you on the development of an algorithm of your own. We hope also that you now have enough understanding of group actions that you can analyse the group theory of the $4 \times 4 \times 4$, $5 \times 5 \times 5$, ... cubes yourself.

Index